ACPL ITEM
DISCARDED

621
TER
OP AMPS

DO NOT REMOVE
CARDS FROM POCKET

ALLEN COUNTY PUBLIC LIBRARY
FORT WAYNE, INDIANA 46802

You may return this book to any agency, branch,
or bookmobile of the Allen County Public Library.

DEMCO

OP AMPS
DESIGN, APPLICATION, & TROUBLESHOOTING

David L. Terrell

Prentice Hall, Englewood Cliffs, New Jersey 07632

Library of Congress Cataloging-in-Publication Data

TERRELL, DAVID L.
   OP AMPS : design, application, and troubleshooting / David L. Terrell.
      p, cm.
   Includes index.
   ISBN 0-13-638685-7
    1. Operational amplifiers.   I. Title.
   TK7871.58.O6T47 1992
   621.39′5--dc20   91-17294
                                                                                              CIP

Editorial/production supervision
  and interior design: **BARBARA MARTTINE**
Cover designer: **BEN SANTORA**
Prepress buyer: **MARY ELIZABETH MCCARTNEY**
Manufacturing buyer: **SUSAN BRUNKE**
Acquisitions editor: **GEORGE KUREDJIAN**

© 1992 by Prentice-Hall, Inc.
A Simon & Schuster Company
Englewood Cliffs, New Jersey 07632

The publisher offers discounts on this book when ordered
in bulk quantities. For more information, write:

Special Sales/Professional Marketing
Prentice-Hall, Inc.
Professional and Technical Reference Division
Englewood Cliffs, New Jersey 07632

All rights reserved. No part of this book may be
reproduced, in any form or by any means.
without permission in writing from the publisher.

Printed in the United States of America

10 9 8 7 6 5 4 3 2 1

ISBN 0-13-638685-7

PRENTICE-HALL INTERNATIONAL (UK) LIMITED, *London*
PRENTICE-HALL OF AUSTRALIA PTY. LIMITED, *Sydney*
PRENTICE-HALL CANADA INC., *Toronto*
PRENTICE-HALL HISPANOAMERICANA, S.A., *Mexico*
PRENTICE-HALL OF INDIA PRIVATE LIMITED, *New Delhi*
PRENTICE-HALL OF JAPAN, INC., *Tokyo*
SIMON & SCHUSTER ASIA PTE. LTD., *Singapore*
EDITORA PRENTICE-HALL DO BRASIL, LTDA., *Rio de Janeiro*

This work is dedicated to my greatest source of motivation—Linda. Her charm, beauty, wit, intelligence, and stunning personality are without equal. I want to thank her for the continuous support and encouragement necessary to complete this book, and I want to thank her for writing this dedication.

CONTENTS

PREFACE   xi

1   BASIC CONCEPTS OF THE INTEGRATED OPERATIONAL AMPLIFIER   1

     1.1   Overview of Operational Amplifiers   1

         1.1.1   Brief history, 1   1.1.2   Review of differential voltage amplifiers, 2   1.1.3   A quick look inside the OP AMP, 3   1.1.4   A survey of op amp applications, 4

     1.2   Review of Important Basic Concepts   5

         1.2.1   Ohm's law, 5   1.2.2   Kirchoff's current law, 7   1.2.3   Kirchoff's voltage law, 8   1.2.4   Thevenin's theorem, 10   1.2.5   Norton's theorem, 12   1.2.6   Superposition theorem, 12

     1.3   Basic Characteristics of Ideal Op Amps   14

         1.3.1   Differential voltage gain, 14   1.3.2   Common-mode voltage gain, 15   1.3.3   Bandwidth, 16   1.3.4   Slew rate, 16   1.3.5   Input impedance, 16   1.3.6   Output impedance, 17   1.3.7   Temperature effects, 18   1.3.8   Noise generation, 18   1.3.9   Troubleshooting tips, 18

     1.4   Introduction to Practical Op Amps   19

         1.4.1   Differential voltage gain, 20   1.4.2   Common-mode voltage gain, 20   1.4.3   Bandwidth, 22   1.4.4   Slew rate 22   1.4.5   Input impedance 23   1.4.6   Output impedance 24   1.4.7   Temperature effects 24   1.4.8   Noise generation 24   1.4.9   Power supply requirements 25   1.4.10   Troubleshooting tips 26

1.5 Circuit Construction Requirements 26
1.5.1 Prototyping methods, 27 1.5.2 Component placement, 27 1.5.3 Routing of leads, 27 1.5.4 Power supply distribution, 28 1.5.5 Power supply decoupling, 28 1.5.6 Grounding considerations, 31 1.5.7 Actual circuit performance, 32

1.6 Electrostatic Discharge 35

2 AMPLIFIERS 37

2.1 Amplifier Fundamentals 37
2.1.1 Gain, 37 2.1.2 Frequency response, 38 2.1.3 Feedback, 39

2.2 Inverting Amplifier 40
2.2.1 Operation, 40 2.2.2 Numerical analysis, 42 2.2.3 Practical design techniques, 54

2.3 Noninverting Amplifier 59
2.3.1 Operation, 59 2.3.2 Numerical analysis, 60 2.3.3 Practical design techniques, 69

2.4 Voltage Follower 74
2.4.1 Operation, 74 2.4.2 Numerical analysis, 74 2.4.3 Practical design techniques, 78

2.5 Inverting Summing Amplifier 80
2.5.1 Operation, 80 2.5.2 Numerical analysis, 83 2.5.3 Practical design techniques, 90

2.6 Noninverting Summing Amplifier 95
2.6.1 Operation, 95 2.6.2 Numerical analysis, 95 2.6.3 Practical design techniques, 97

2.7 AC Coupled Amplifier 97
2.7.1 Operation, 97 2.7.2 Numerical analysis, 99 2.7.3 Practical design techniques, 109

2.8 Current Amplifier 113
2.8.1 Operation, 113 2.8.2 Numerical analysis, 114 2.8.3 Practical design techniques, 118

2.9 Hi-Current Amplifier 120
2.9.1 Operation, 120 2.9.2 Numerical analysis, 121 2.9.3 Practical design techniques, 125

2.10 Troubleshooting Tips for Amplifier Circuits 132
2.10.1 Basic troubleshooting concepts, 132 2.10.2 Specific techniques for op amps, 133

3 VOLTAGE COMPARATORS 136

3.1 Voltage Comparator Fundamentals 136
3.2 Zero-Crossing Detector 137

Contents vii

		3.2.1 Operation, 137 3.2.2 Numerical analysis, 138
		3.2.3 Practical design techniques, 140
3.3	Zero-Crossing Detector with Hysteresis	144

        3.3.1 Operation, 144 3.3.2 Numerical analysis, 146
        3.3.3 Practical design techniques, 149

  3.4   Voltage Comparator with Hysteresis   152

        3.4.1 Operation, 152 3.4.2 Numerical analysis, 153
        3.4.3 Practical design techniques, 156

  3.5   Window Voltage Comparator   160

        3.5.1 Operation, 160 3.5.2 Numerical analysis, 160
        3.5.3 Practical design techniques, 162

  3.6   Voltage Comparator with Output Limiting   165

        3.6.1 Operation, 165 3.6.2 Numerical analysis, 166
        3.6.3 Practical design techniques, 169

  3.7   Troubleshooting Tips for Voltage Comparators   174
  3.8   Nonideal Considerations   175

4 OSCILLATORS   177

  4.1   Oscillator Fundamentals   177
  4.2   Wein-bridge   178

        4.2.1 Operation, 178 4.2.2 Numerical analysis, 179
        4.2.3 Practical design techniques, 181

  4.3   Voltage-Controlled Oscillator   185

        4.3.1 Operation, 185 4.3.2 Numerical analysis, 186
        4.3.3 Practical design techniques, 191

  4.4   Variable Duty Cycle   198

        4.4.1 Operation, 198 4.4.2 Numerical analysis, 199
        4.4.3 Practical design techniques, 202

  4.5   Triangle Wave Oscillator   209

        4.5.1 Operation, 209 4.5.2 Numerical analysis, 209
        4.5.3 Practical design techniques, 210

  4.6   Troubleshooting Tips for Oscillator Circuits   214
  4.7   Nonideal Considerations   215

5 ACTIVE FILTERS   217

  5.1   Filter Fundamentals   217
  5.2   Low-Pass Filter   219

        5.2.1 Operation, 220 5.2.2 Numerical analysis, 221
        5.2.3 Practical design techniques, 222

  5.3   High-Pass Filter   226

        5.3.1 Operation, 226 5.3.2 Numerical analysis, 228
        5.3.3 Practical design techniques, 229

  5.4   Band-Pass Filter   233

5.4.1 *Operation*, 233   5.4.2 *Numerical analysis*, 235
5.4.3 *Practical design techniques*, 236

5.5   Band-Reject Filter   241
5.5.1 *Operation*, 241   5.5.2 *Numerical analysis*, 243
5.5.3 *Practical design techniques*, 244

5.6   Troubleshooting Tips for Active Filters   251

6   POWER SUPPLY CIRCUITS   253

6.1   Voltage Regulation Fundamentals   253
6.1.1 *Series regulation*, 254
6.1.2 *Shunt regulation*, 254   6.1.3 *Switching regulation*, 255   6.1.4 *Line and load regulation*, 255
6.1.5 *Voltage references*, 256

6.2   Series Voltage Regulators,   261
6.2.1 *Operation*, 261   6.2.2 *Numerical analysis*, 263
6.2.3 *Practical design techniques*, 264

6.3   Shunt Voltage Regulation   271
6.3.1 *Operation*, 271   6.3.2 *Numerical analysis*, 272
6.3.3 *Practical design techniques*, 274

6.4   Switching Voltage Regulators   280
6.4.1 *Principles of operation*, 280   6.4.2 *Switching versus linear-voltage regulators*, 283   6.4.3 *Classes of switching regulators*, 284

6.5   Over-Current Protection   285
6.5.1 *Load interruption*, 285   6.5.2 *Constant-current limiting*, 285   6.5.3 *Foldback current limiting*, 287

6.6   Over-Voltage Protection   289
6.7   Power-Fail Sensing   289
6.8   Troubleshooting Tips for Power Supply Circuits   291

7   SIGNAL PROCESSING CIRCUITS   295

7.1   Concept of the Ideal Diode   295
7.2   Ideal Rectifier Circuits   297
7.2.1 *Operation*, 297   7.2.2 *Numerical analysis*, 298
7.2.3 *Practical design techniques*, 300

7.3   Ideal, Biased Clipper   306
7.3.1 *Operation*, 306   7.3.2 *Numerical analysis*, 308
7.3.3 *Practical design techniques*, 309

7.4   Ideal Clamper   314
7.4.1 *Operation*, 315   7.4.2 *Numerical analysis*, 316
7.4.3 *Practical design techniques*, 321

7.5   Peak Detectors   323
7.5.1 *Operation*, 324   7.5.2 *Numerical analysis*, 325

Contents ix

        7.5.3   *Practical design techniques, 326*
- 7.6   Integrator   331
  - 7.6.1   *Operation, 332*   7.6.2   *Numerical analysis, 333*
  - 7.6.3   *Practical design techniques, 334*
- 7.7   Differentiator   338
  - 7.7.1   *Operation, 338*   7.7.2   *Numerical analysis, 340*
  - 7.7.3   *Practical design techniques, 340*
- 7.8   Troubleshooting Tips for Signal Processing Circuits   344

8   DIGITAL-TO-ANALOG AND ANALOG-TO-DIGITAL CONVERSION   346

- 8.1   D/A and A/D Conversion Fundamentals   347
  - 8.1.1   *Analog-to-digital converters, 347*
  - 8.1.2   *Preamplifiers, 349*   8.1.3   *Sample-and-hold circuits, 349*   8.1.4   *Multiplexers, 350*
  - 8.1.5   *Digital-to-analog converters, 352*
- 8.2   Weighted D/A   354
- 8.3   R2R Ladder D/A   356
- 8.4   Parallel A/D   358
- 8.5   Tracking A/D   360
- 8.6   Dual-Slope A/D Conversion   362
- 8.7   Successive Approximation A/D   367

9   ARITHMETIC FUNCTION CIRCUITS   370

- 9.1   Adder   370
  - 9.1.1   *Operation, 370*   9.1.2   *Numerical analysis, 371*
  - 9.1.3   *Practical design techniques, 373*
- 9.2   Subtractor   375
  - 9.2.1   *Operation, 376*   9.2.2   *Numerical analysis, 376*
  - 9.2.3   *Practical design techniques, 378*
- 9.3   Averaging Amplifier   380
- 9.4   Absolute Value Circuit   381
  - 9.4.1   *Operation, 382*   9.4.2   *Numerical analysis, 382*
  - 9.4.3   *Practical design techniques, 385*
- 9.5   Sign-Changing Circuit   388
  - 9.5.1   *Operation, 388*   9.5.2   *Numerical analysis, 388*
  - 9.5.3   *Practical design techniques, 389*
- 9.6   Troubleshooting Tips for Arithmetic Circuits   391

10   NONIDEAL OP AMP CHARACTERISTICS   394

- 10.1   Nonideal DC Characteristics   394
  - 10.1.1   *Input bias current, 394*   10.1.2   *Input offset*

		current, 398 10.1.3 Input offset voltage, 399 *10.1.4 Drift, 400 10.1.5 Input resistance, 401* *10.1.6 Output resistance, 401*
	10.2	Nonideal AC Characteristics 402 *10.2.1 Frequency response, 402 10.2.2 Slew rate, 403* *10.2.3 Noise, 404 10.2.4 Frequency compensation, 405* *10.2.5 Input resistance, 412 10.2.6 Output resistance, 412*
	10.3	Summary and Recommendations 413 *10.3.1 AC-coupled amplifiers, 413 10.3.2 DC-coupled amplifiers, 413 10.3.3 Relative magnitude rule, 414* *10.3.4 Safety margins on frequency compensation, 414*

11 SPECIALIZED DEVICES   416

- 11.1 Programmable Op Amps   416
- 11.2 Instrumentation Amplifiers   417
- 11.3 Logarithmic Amplifiers   421
- 11.4 Antilogarithmic Amplifiers   424
- 11.5 Multipliers/Dividers   425
- 11.6 Single Supply Amplifiers   428
- 11.7 Multiple Op Amp Packages   434
- 11.8 Hybrid Operational Amplifiers   434

APPENDICES (1-10)   436

INDEX   489

PREFACE

What is the value of pi (π)? Is it 3? Is it 3.1? How about 3.14? Or perhaps you think 3.14159526535897932384626433832790 is more appropriate. Each of these answers is correct just as each of these answers is incorrect; they vary in their degree of resolution and accuracy. The degree of accuracy is often proportional to the complexity or difficulty of computation. So it is with operational amplifier circuits or all electronic circuits for that matter. The goal of this text is to provide workable tools for analysis and design of operational amplifier circuits that are free from the shrouds of complex mathematics and yet produce results which have a *satisfactory degree of accuracy*.

This book offers a subject coverage that is fairly typical for texts aimed at the postsecondary school market. The organization of each circuits chapter, however, is very consistent and provides the following information on each circuit presented:

1. *Theory of operation*. A discussion that explains what the circuit does, why it behaves the way it does, and identifies the purpose of each component. This section contains no mathematics, promotes an intuitive understanding of circuit operation, and is based on an application of basic electronics principles such as series and parallel circuits, Ohm's Law, Kirchoff's Laws, and so on.
2. *Numerical Analysis*. Techniques are presented that allow calculation of most key circuit parameters for an existing op amp circuit design. The mathematics is strictly limited to basic algebra and does not require (although it permits) the use of complex numbers.
3. *Practical Design*. A sequential design procedure is described that is based on the preceding numerical analysis and application of basic electronics principles. The goals of each design are contrasted with the actual circuit performance measured in laboratory tests.

In addition to these areas that are presented for each individual type of circuit, each circuits chapter has a discussion on troubleshooting techniques as they apply to the type of circuits discussed in that particular chapter.

The majority of this text treats the op amp as a quasi-ideal device. That is, only those nonideal parameters that have a significant impact on a particular design are considered. In Chap. 10, a more complete discussion of nonideal behavior is offered and includes both AC and DC considerations.

The analytical and design methods provided in the text are not limited to a particular op amp. The standard 741 and its higher performance companion the MC1741SC are frequently used as example devices since these devices are still used in major electronics schools. However, the equations and methodologies directly extend to newer, more advanced devices. In fact, since newer devices typically have performance that is closer to the ideal op amp, the equations and methods frequently work even better for the newer op amps. To provide the reader with a perspective regarding the range of op amp performance that is available, Chap. 11 includes a comparison between a general-purpose op amp and a hybrid op amp having, for example, a 5500 volts per microsecond slew rate as compared to the 0.5 volt per microsecond slew rate often found in general-purpose devices.

Literally, every circuit in every circuits chapter has been constructed and tested in the laboratory. In the case of circuit design examples, the actual performance of the circuit was captured in the form of oscilloscope plots. The following test equipment was used to measure circuit performance:

1. Hewlett Packard Model 8116A Pulse/Function Generator
2. Hewlett Packard Model 54501A Digitizing Oscilloscope
3. Hewlett Packard Think Jet Plotter
4. Heath Model 2718 Triple Output Power Supply

Items 1 to 3 were provided courtesy of Hewlett Packard. This equipment delivered exceptional ease of use, accuracy of measurement and produced a camera-ready plotter output of the scope displays. The oscilloscope plots that are presented in the text are unedited and represent the actual circuit performance. This alleviates the confusion that is frequently encountered when the ideal waveform drawings typically presented in many textbooks are contrasted with the actual results in the laboratory. Any deviations from the ideal that would have been masked by an artist's ideal drawings are there for your examination in the actual oscilloscope plots presented throughout this book.

Although the text is appropriate for use in a resident electronics school, the consistent and independent nature of the discussions for each circuit make it equally appropriate as a reference manual or handbook for working engineers and technicians.

So what is considered to be a *satisfactory degree of accuracy* in this text? Based on over 20 years of experience as a technician, an engineer, and a classroom instructor, it is apparent to the author that most practical designs require tweaking in the laboratory before a final design is evolved. That is, the engineer can design a circuit using the most appropriate models and the most extensive analysis, but the exact performance is rarely witnessed the first time the circuit is constructed. Rather, the paper design generally puts us close to the desired performance. Actual measurements on the circuit in a laboratory environment will then allow optimization of component values. The methods presented in this text, then, will produce designs that can deliver performance that is close to the original design goals. If tighter performance is required, then tweaking can be done in the laboratory ... a step that would generally be required even if more elaborate methods were employed.

My sincere thanks goes to Rick Lane of Hewlett Packard whose assistance on this project was instrumental. I also want to thank the thousands of students who continue to ask, "Why didn't they just say that in the book?" and who reward their instructors by saying, "That's not so hard. It's easy when you do it that way!"

CHAPTER 1

BASIC CONCEPTS OF THE INTEGRATED OPERATIONAL AMPLIFIER

1.1 OVERVIEW OF OPERATIONAL AMPLIFIERS

1.1.1 Brief History

Operational amplifiers began in the days of vacuum tubes and analog computers. They consisted of relatively complex differential amplifiers with feedback. The circuit was constructed in a way that the characteristics of the overall amplifier were largely determined by the type and amount of feedback. Thus, the complex differential amplifier itself had become a building block that could function in different "operations" by altering the feedback. Some of the operations that were used included adding, multiplying, and logarithmic operations.

The operational amplifier continued to evolve through the transistor era and continued to decrease in size and increase in performance. The evolution continued through molded or modular devices and finally in the mid 1960s a complete operational amplifier was integrated into a single integrated circuit (IC) package.

Since that time, the performance has continued to improve dramatically and the price has generally decreased as the benefits of high-volume production are evidenced. The performance increases include such items as higher operating voltages, lower current requirements, higher current capabilities, more tolerance to abuse, lower noise, greater stability, greater power output, higher input impedances, and higher frequencies of operation.

In spite of all the improvements however, the high performance, integrated operational amplifier of today is still based on the fundamental differential amplifier. Although the individual components in the amplifier are not accessible to you, it will enhance your understanding of the op amp if you have some appreciation for the internal circuitry.

1.1.2 Review of Differential Voltage Amplifiers

You will recall from your basic electronics studies that a differential amplifier has two inputs and either one or two outputs. The amplifier circuit is not directly affected by the voltage on **either** of its inputs alone, but it is affected by the **difference** in voltage between the two inputs. This voltage difference is amplified by the amplifier and appears at the output in its amplified form. The amplifier may have a single output which is referenced to common or ground. If so, it is called a single-ended amplifier. On the other hand, the output of the amplifier may be taken between two points neither of which is common or ground. In this case, the amplifier is called a double-ended or differential output amplifier.

Figure 1.1 shows a simple transistor differential voltage amplifier. Specifically, it is a single-ended differential amplifier. Transistors Q_1 and Q_2 have a shared emitter bias so the combined collector current is largely determined by the -20-volt source and the 10 kilohm emitter resistor. The current through this resistor then divides (Kirchoff's Current Law) and becomes the emitter current for the two transistors. Within limits, the total emitter current remains fairly constant and simply diverts from one transistor to the other as the signal or changing voltage is applied to the bases. In a practical differential amplifier, the emitter network generally contains a constant current source.

Now consider the relative effect on the output if the input signal is increased in the polarity shown. This will decrease the bias on Q_2 while increasing the bias on Q_1. Thus, a larger portion of the total emitter current is diverted through Q_1 and less through Q_2. This decreased current flow through the collector resistor for Q_2 produces less voltage drop and allows the output to become more positive.

If the polarity of the input were reversed, then Q_2 would have more current flow and the output voltage would decrease (i.e., become less positive).

Suppose now that both inputs are increased or decreased in the same direction. Can you see that this will affect the bias on both of the transistors in the same way? Since the total current is held constant and the relative values for each transistor did not change, then both collector currents remain constant. Thus the output does not reflect a change

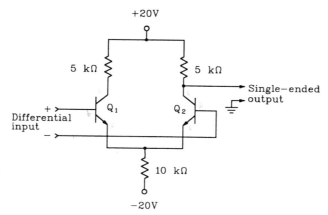

Figure 1.1 A simple differential voltage amplifier based on transistors.

when both inputs are altered in the same way. This latter effect gives rise to the name differential amplifier. It only amplifies the difference between the two inputs, and is relatively unaffected by the absolute values applied to each input. This latter effect is more pronounced when the circuit uses a current source in the emitter circuit.

In certain applications, one of the differential inputs is connected to ground and the signal to be amplified is applied directly to the remaining input. In this case the amplifier still responds to the difference between the two inputs, but the output will be in or out of phase with the input signal depending on which input is grounded. If the signal is applied to the (+) input, with the (−) input grounded, as labeled in Fig. 1.1, then the output signal is essentially in phase with the input signal. If, on the other hand, the (+) input is grounded and the input signal is applied to the (−) input, then the output is essentially 180 degrees out of phase with the input signal. Because of the behavior described, the (−) and (+) inputs are called the inverting and noninverting inputs, respectively.

1.1.3 A Quick Look Inside the Op Amp

Figure 1.2 shows the schematic diagram of the internal circuitry for a common integrated circuit op amp. This is the 741 op amp which is common in the industry. It is not particularly important for you to understand the details of the internal operation. Nor is it worth your while to trace current flow through the internal components. The internal diagram is shown here for the following reasons:

1. to emphasize the fact that the op amp is essentially an encapsulated circuit composed of familiar components,
2. to show the differential inputs on the op amp,
3. to gain an understanding of the type of circuit driving the output of the op amp.

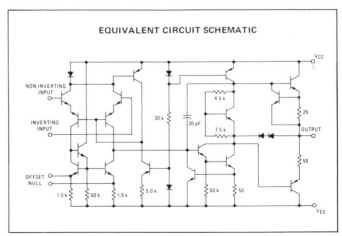

Figure 1.2 The internal schematic for an MC1741 op amp. (Copyright of Motorola, Inc. Used by permission)

You can see that the entire circuit is composed of transistors, resistors and a single 30 picofarad capacitor. A closer examination shows that the inverting and noninverting inputs go directly to the bases of two transistors connected as a differential amplifier. The emitter circuit of this differential pair is supplied by a constant current source.

If you examine the output circuitry, you can see that it resembles that of a complementary-symmetry amplifier or perhaps a totem-pole output as used in digital circuits. In any case, the output is pulled closer to the positive supply whenever the upper output transistor conducts harder. Similarly, if the lower output transistor were to turn on harder, then the output would be pulled in a negative direction. Also note, the low values of resistances in the output circuit. The inputs labeled offset null are provided to allow compensation for imperfect circuitry. Use of these inputs is discussed at a later point.

1.1.4 A Survey of Op Amp Applications

So now you know where they came from, what they are made of, and a few of their characteristics. But what uses are there for an op amp in the electronics industry? Although the following is certainly not an exhaustive list, it does serve to illustrate the range of op amp applications.

Amplifiers. Op amps are used to amplify signals that range from DC through the higher RF frequencies. The amplifier can be made to be frequency selective (i.e., act as a filter) much like the tone control on your favorite stereo system. It may be used to maintain a constant output in spite of changing input levels. The output can produce a compressed version of the input to reduce the range needed to represent a certain signal. The amplifier may respond to microvolt signals originating in a transducer which is used to measure temperature, pressure, density, acceleration, and so on. The gain of the amplifier can be controlled by a digital computer thus extending the power of the computer into the analog world.

Oscillators. The basic op amp can be connected to operate as an oscillator. The output of the oscillator may be sinusoidal, square, triangular, rectangular, sawtooth, exponential, or other shape. The frequency of oscillation may be stabilized by a crystal or controlled by a voltage or current from another circuit.

Regulators. Op amps can be used to improve the regulation in power supplies. The actual output voltage is compared with a reference voltage and the difference is amplified by an op amp and used to correct the power supply output voltage. Op amps can also be connected to regulate and/or limit the current in a power supply.

Rectification. Suppose you want to build a half-wave rectifier with a peak input signal of 150 millivolts. This is not enough to forward bias a standard silicon diode. On the other hand, an op amp can be configured to provide the characteristics of an ideal diode with zero forward voltage drop. Thus it can rectify very small signals.

Computer Interfaces. The op amp is an integral part of many circuits used to convert analog signals representing real world quantities (e.g., temperature, RPM, pressure, etc.) into corresponding digital signals that can be manipulated by a computer. Similarly, the op amp is also frequently used to convert the digital output of a computer into an equivalent analog form for use by industrial devices (e.g., motors, lights, heating elements).

Fields of Application. Op amps find use in such diverse fields as medical electronics, industrial electronics, agriculture, test equipment, consumer products, and automotive products. It has become a basic building block for analog systems and for the analog portion of digital systems.

1.2 REVIEW OF IMPORTANT BASIC CONCEPTS

Throughout my career in the electronics field, I have known certain individuals whose observable skills in circuit analysis far exceeded most others with similar levels of education and experience. These people all have one definite thing in common; they have an unusually strong mastery of basic—really basic—electronics. They have the ability to look at a complex, unfamiliar circuit and see a combination of simple circuits that can be analyzed with such tools as Ohm's and Kirchoff's Laws.

This portion of the text provides a condensed review of several important laws or theorems that are used to analyze electronic circuits. A mastery of these basic ideas will greatly assist you in understanding and analyzing the operation of the circuits presented in the remainder of the text and encountered in industry.

1.2.1 Ohm's Law

The basic form of Ohm's Law is probably known to everyone who is even slightly trained in electronics. The three forms are

$$E = IR \tag{1.1}$$

$$I = \frac{E}{R} \tag{1.2}$$

$$R = \frac{E}{I} \tag{1.3}$$

where E (or V) represents the applied voltage (volts), R represents the resistance of the circuit (ohms), and I represents the current flow (amperes).

Your concept of Ohm's Law, however, should extend beyond the arithmetic operations required to solve a problem. You need to develop an intuitive feel for the circuit behavior. For example, without the use of mathematics, you should know that if the applied voltage to a particular circuit is increased while the resistance remains the same, then the value of current in the circuit will also increase proportionately. Similarly, without mathematics, it needs to be obvious to you that an increased current flow through a fixed resistance will produce a corresponding increase in the voltage drop across the resistance.

Many equations presented in this text appear to be new and unique expressions to describe the operation of op amp circuits. When viewed more closely, however, they are nothing more than applications of Ohm's Law. For example, consider the following expression:

$$i_x = \frac{V_I - V_D}{R_I}$$

Once the subtraction has been completed in the numerator, which is like computing the value of two batteries in series, the problem becomes a simple Ohm's Law problem Eq. (1.2).

For a test of your intuitive understanding of Ohm's Law as applied to series-parallel circuits, try to evaluate the problem shown in Fig. 1.3 without resorting to the direct use of mathematics. In the figure, no numeric values are given for the various components. The value of R_3 is said to have increased (i.e., has more resistance). What will be the relative effects on the three current meters (increase, decrease, or remain the same)? Try it on your own before reading the next paragraph.

Your reasoning might go something like this. If R_3 increases in value, then the current (I_3) through it will surely decrease. Since R_3 increased in value, the parallel combination of R_2 and R_3 will also increase in effective resistance. This increase in parallel resistance will drop a greater percentage of the applied voltage. This increased voltage across R_2 will cause I_2 to increase. Since the parallel combination of R_2 and R_3 have increased in resistance, the total circuit resistance is greater which means that total current will decrease. Since the total current flows through R_1, the value of I_1 will show a decrease.

This example illustrates an intuitive, nonmathematical method of circuit analysis. Time spent in gaining mastery in this area will pay rewards to you in the form of increased analytical skills for unfamiliar circuits.

Ohm's Law also applies to AC circuits with or without reactive components. In the case of AC circuits with reactive devices, however, all voltages, currents, and imped-

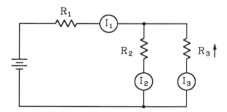

Figure 1.3 How does an increase in the resistance of R_3 affect the currents I_1, I_2, and I_3?

Sec. 1.2 Review of Important Basic Concepts

ances must be expressed in their complex form (e.g., $2 - j5$ would represent a series combination of a 2-ohm resistor and a 5-ohm capacitive reactance).

1.2.2 Kirchoff's Current Law

Kirchoff's Current Law tells us that all of the current entering a particular point in a circuit must also leave that point. Figure 1.4 illustrates this concept with several examples. In each case, the current entering and leaving a given point is the same. This law is generally stated mathematically in the form of

$$I_T = I_1 + I_2 + I_3 + \ldots I_N \tag{1.4}$$

where I_T is the total current leaving a point (for instance) and I_1, I_2, and so on, are the various currents entering the point. In the case of Fig. 1.4(c), we can apply Eq. (1.4) as

$$I_4 = I_1 + I_2 + I_3$$

Here again, though, it is important for you to strive to develop an intuitive, nonmathematic appreciation for what the law is telling you.

Consider the examples shown in Fig. 1.5. Without using your calculator, can you estimate the effect on the voltage drop across R_1 when resistor (R_3) opens? Try it before reading the next paragraph.

In the first case, Fig. 1.5(a), your reasoning might be like this. Since the open resistor (R_3) was initially very small compared to parallel resistor R_2, it will have a dramatic effect on total current when it opens. That is, Kirchoff's Current Law would tell us that the total current (I_1) is composed of I_2 and I_3. Since R_3 was initially much

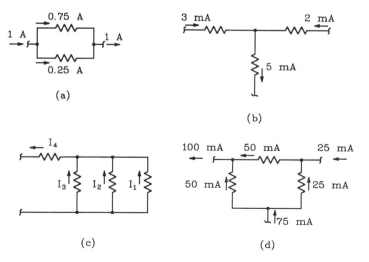

Figure 1.4 Examples of Kirchoff's Current Law which illustrate that the current entering a point must equal the current leaving that same point.

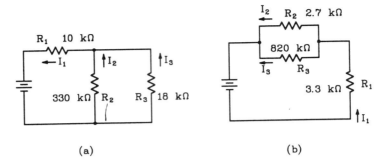

Figure 1.5 Estimate the effect on circuit operation if R_3 were to become open.

smaller than R_2, its current will be much greater (Ohm's Law). Therefore, when R_3 opens, the major component of current I_1 will drop to zero. This reduced value of current through R_1 will greatly reduce the voltage drop across R_1.

In the second case, R_3 is much larger than the parallel resistor R_2, and therefore, contributes very little to the total current I_1. When R_3 opens, there will be no significant change in the voltage across R_1.

Again be reminded of the value of a solid intuitive view of electronic circuits.

Kirchoff's Current Law can also be used to analyze AC circuits with reactive components provided the circuit values are expressed in complex form.

1.2.3 Kirchoff's Voltage Law

Kirchoff's Voltage Law basically says that all of the voltage sources in a closed loop must be equal to the voltage drops. That is, the net voltage (sources + drops) is equal to zero. Figure 1.6 shows some examples. This law is most often stated mathematically in a form such as

$$\boxed{V_1 + V_2 + V_3 - V_{APP} = 0} \quad (1.5)$$

In the case of Fig. 1.6(c), we apply Eq. (1.5) as

$$+ V_1 - V_{R_1} - V_{R_2} - V_2 - V_{R_3} + V_3 = 0$$

Another concept that is closely related to Kirchoff's Voltage Law is the determination of voltages at certain points in the circuit **with respect to** voltages at other points. Consider the circuit shown in Fig. 1.7. It is common to express circuit voltage with respect to ground. Voltages such as $V_B = 5$ volts, $V_D = -2$ volts, and $V_A = 8$ volts are voltage levels with respect to ground. In our analysis of op amp circuits, it will also be important to determine voltages with respect to points other than ground. The following is an easy two-step method:

Sec. 1.2   Review of Important Basic Concepts

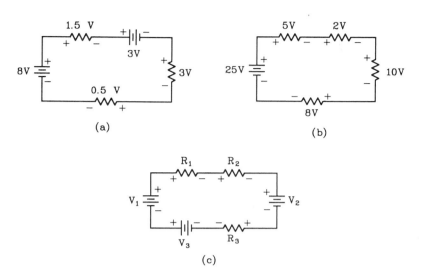

Figure 1.6 Examples of Kirchoff's Voltage Law which illustrate the sum of all voltages in a closed loop must equal zero.

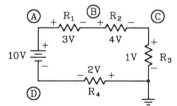

Figure 1.7 A circuit used to illustrate the concept of reference points.

1. Label the polarity of the voltage drops,
2. Start at the reference point and move toward the point in question. As you pass through each component, add (algebraically) the value of the voltage drop using the polarity nearest the end you **exit**.

For example, let us determine the voltage at point A with respect to point C in Fig. 1.7. Step one has already been done. We will begin at point C (reference point) and progress in either direction toward point A combining the voltage drops as we go. Let us choose to go in a counterclockwise direction since that is the shortest path. Upon leaving R_2 we get +4 volts, upon leaving R_1 we get +3 volts which adds to the previous +4 volts to give us a total of +7 volts. Since we are now at point A we have our answer of +7 volts. This is an important concept and one that deserves practice.

Kirchoff's Voltage Law can also be used to analyze AC circuits with reactive components provided the circuit values are expressed in complex form.

1.2.4 Thevenin's Theorem

Thevenin's Theorem is a technique that allows us to convert a circuit (often a complex circuit) into a simple equivalent circuit. The equivalent circuit consists of a constant voltage source and a single series resistor called the Thevenin voltage and Thevenin resistance, respectively. Once the values of the equivalent circuit have been calculated, subsequent analysis of the circuit becomes much easier.

You can obtain the Thevenin equivalent circuit by applying the following sequential steps:

1. Short all voltage sources and open all current sources. (Replace all sources with their internal impedance if it is known.) Also open the circuit at the point of simplification.
2. Calculate the value of Thevenin's resistance as seen from the point of simplification.
3. Replace the voltage and current sources with their original values and open the circuit at the point of simplification.
4. Calculate Thevenin's voltage at the point of simplification.
5. Replace the original circuit component with the Thevenin equivalent for subsequent analysis of the circuit beyond the point of simplification.

Consider, for example, the circuit in Fig. 1.8. Here four different values of R_x are connected to the output of a voltage divider circuit. The value of loaded voltage is to be calculated. Without a simplification theorem such as Thevenin's Theorem, each resistor value would require several computations. Now let us apply Thevenin's Theorem to the circuit.

First we define the point of simplification to be the place where R_x is connected. This is shown in Fig. 1.9(a). The voltage source is shorted and the circuit is opened at the point of simplification in Fig. 1.9(b). We can now calculate the Thevenin resistance (R_{TH}). By inspection, we can see that the 5 kilohm and the 20 kilohm resistors are now in parallel. Thus the Thevenin resistance is found in this case by the parallel resistor equation

$$R_T = \frac{R_1 R_2}{R_1 + R_2} \tag{1.6}$$

In this particular case,

$$R_{TH} = \frac{(5 \times 10^3)(20 \times 10^3)}{(5 \times 10^3) + (20 \times 10^3)}$$

$$= \frac{100 \times 10^6}{25 \times 10^3}$$

$$= 4 \text{ k}\Omega$$

Sec. 1.2 Review of Important Basic Concepts

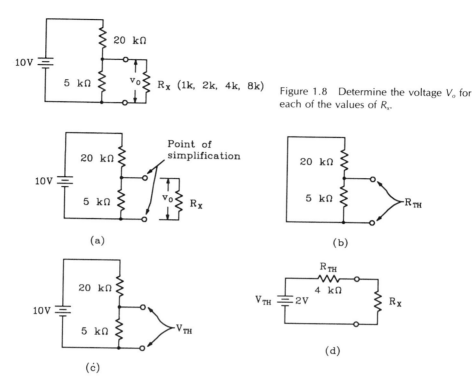

Figure 1.8 Determine the voltage V_o for each of the values of R_x.

Figure 1.9 Thevenizing the circuit of Fig. 1.8.

Next, we determine the Thevenin voltage by replacing the sources (step 3). This is shown in Fig. 1.9(c). The voltage divider equation, Eq. (1.7), is used in this case to give us the value of Thevenin's voltage.

$$V_{R1} = \left(\frac{R_1}{R_1 + R_2}\right) V_{APP} \quad (1.7)$$

That is,

$$V_{TH} = \left(\frac{5 \times 10^3}{5 \times 10^3 + 20 \times 10^3}\right) 10$$
$$= 0.2 \times 10$$
$$= 2 \text{ V}$$

Figure 1.9(d) shows the Thevenin equivalent circuit. Calculations for each of the values of R_x can now quickly be computed by simply applying the voltage divider equation. The value of Thevenin's Theorem would be even more obvious if the original circuit were more complex.

The preceding discussion was centered on resistive DC circuits. The techniques described, however, apply equally well to AC circuits with inductive and/or capacitive components. The voltages and impedances are simply expressed in their complex form (e.g., $5 + j7$ would represent a 5-ohm resistance and a 7-ohm inductance).

1.2.5 Norton's Theorem

Norton's Theorem is similar to Thevenin's Theorem in that it produces an equivalent, simplified circuit. The major difference is that the equivalent circuit is composed of a current source and a parallel resistance rather than a voltage source and a series resistance like the Thevenin equivalent.

The sequential steps for obtaining the Norton equivalent circuit are as follows:

1. Short all voltage sources and open all current sources. (Replace all sources with their internal impedance if it is known.) Also open the circuit at the point of simplification.
2. Calculate the value of Norton's resistance as seen from the point of simplification.
3. Replace the voltage and current sources with their original values and short the circuit at the point of simplification.
4. Calculate Norton's current at the point of simplification.
5. Replace the original circuit component with the Norton's equivalent for subsequent analysis of the circuit beyond the point of simplification.

Figure 1.10 shows these steps as they apply to the circuit given in Fig. 1.8. Once the equivalent circuit has been determined, the various values of R_x can be connected and the resulting voltage calculated. The calculations, though, are simple current divider equations.

Norton's Theorem can also be used to analyze AC circuits with reactive components provided the circuit values are expressed in their complex form.

1.2.6 Superposition Theorem

The Superposition Theorem is most useful in analyzing circuits that have multiple voltage or current sources. Essentially, it states that the net effect of all of the sources can be determined by calculating the effect of each source singly and then combining the individual results. The steps are

1. For each source, compute the values of circuit current and voltage with all of the remaining sources replaced with their internal impedances. (We generally short voltage sources and open current sources.)
2. Combine the individual voltages or currents at the point(s) of interest to determine the net effect of the multiple sources.

Sec. 1.2 Review of Important Basic Concepts

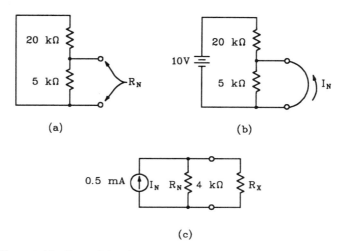

Figure 1.10 Determining the Norton equivalent for the circuit of Fig. 1.8.

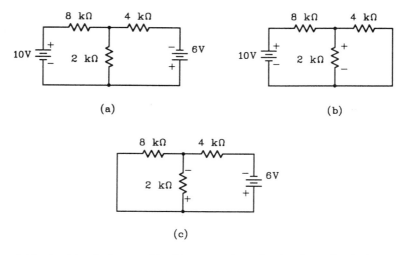

Figure 1.11 Applying the Superposition Theorem to determine the effects of multiple sources.

As an example, let us apply the Superposition Theorem to the circuit in Fig. 1.11(a) for the purpose of determining the voltage across the 2-kilohm resistor. Let us first determine the effect of the 10-volt battery. We short the 6-volt battery and evaluate the resulting circuit, Fig. 1.11(b). Analysis of this series-parallel circuit will show you that the 2-kilohm resistor has approximately 1.43 volts across it with the upper end being positive.

Next, we evaluate the effects of the 6-volt source in Fig. 1.11(c). This is another simple circuit that produces about 1.71 volts across the 2-kilohm resistor with the upper end being negative.

Since the two individual sources produced opposite polarities of voltage across the 2-kilohm resistor, we determine the net effect by subtracting the two individual values. Thus, the combined effect of the 10- and 6-volt sources is $1.43 - 1.71 = -0.28$ volts.

The Superposition Theorem works with any number of sources either AC or DC and can include reactive components as long as circuit values are expressed as complex numbers.

1.3 BASIC CHARACTERISTICS OF IDEAL OP AMPS

Let us now examine some of the basic characteristics of an ideal operational amplifier. By focusing on ideal performance, we are freed from many complexities associated with nonideal performance. For many real applications, the ideal characteristics may be used to analyze and even design op amp circuits. In more demanding cases, however, we must include other operating characteristics which are viewed as deviations from the ideal.

The basic schematic symbol for an op amp is shown in Fig. 1.12. It has the inverting and noninverting inputs labeled $(-)$ and $(+)$, respectively, and has a single output. Although it certainly must have power supply connections, they are not generally included on schematic diagrams.

1.3.1 Differential Voltage Gain

The differential voltage gain is the amount of amplification given to voltage appearing between the input terminals. In the case of the ideal op amp, the differential voltage gain is infinity. You may recall from your studies of transistor amplifiers that the output from an amplifier is limited by the magnitude of the DC supply voltage. If an attempt is made to obtain greater outputs, then the output is clipped or limited at the maximum or minimum levels. Since the ideal op amp has such extreme gain (infinite) this means that with even the smallest input signal the output will be driven to its limits (typically ± 15 volts for ideal op amps).

This is an important concept so be sure to appreciate what is being said. To further clarify the concept, let us compare the ideal op amp with a more familiar amplifier. We will suppose that the familiar amplifier has a differential voltage gain of 5 and has output limits of ± 15 volts. You will recall that the output voltage (v_o) of an amplifier can be determined by multiplying its input voltage times the voltage gain.

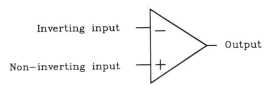

Figure 1.12 The basic operational amplifier symbol.

Sec. 1.3   Basic Characteristics of Ideal Op Amps

$$\boxed{v_o = v_i A_v} \qquad (1.8)$$

Let us compute the output for each of the following input voltages: $-4.0, -2.0, -1.0, -0.5, -0.1, 0.0, 0.1, 0.5, 1.0, 2.0,$ and 4.0

$$
\begin{aligned}
v_o &= -4 \times 5 = -20 \text{ volts [output limited to } -15 \text{ volts]} \\
&= -2 \times 5 = -10 \text{ volts} \\
&= -1 \times 5 = -5 \text{ volts} \\
&= -0.5 \times 5 = -2.5 \text{ volts} \\
&= -0.1 \times 5 = -0.5 \text{ volts} \\
&= 0 \times 5 = 0 \text{ volts} \\
&= 0.1 \times 5 = 0.5 \text{ volts} \\
&= 0.5 \times 5 = 2.5 \text{ volts} \\
&= 1 \times 5 = 5 \text{ volts} \\
&= 2 \times 5 = 10 \text{ volts} \\
&= 4 \times 5 = 20 \text{ volts [output limited to } +15 \text{ volts]}
\end{aligned}
$$

Now let us do similar calculations with the same input voltages applied to an ideal op amp. You can quickly realize that in all cases except 0.0 volts input, the output will try to go beyond the output limit and will be restricted to ±15 volts. For example,

$$v_o = 0.1 \times \infty = \infty \text{ V [output limited to } +15 \text{ V]}$$

At this point you might well be asking, "So what good is it if every voltage we apply causes the output to be driven to its limit?" Well, Sec. 1.1.4 of the text should indicate the usefulness of the op amp in general. In Chap. 2 you will become keenly aware of how the infinite gain can be harnessed into a more usable value. For now, however, it is important for you to remember that **an ideal op amp has an infinite differential voltage gain.**

1.3.2 Common-mode Voltage Gain

Common-mode voltage gain refers to the amplification given to signals that appear on both inputs relative to the common (typically ground). You will recall from a previous discussion that a differential amplifier is designed to amplify the difference between the two voltages applied to its inputs. Thus, if both inputs had +5 volts, for instance, with respect to ground, then the difference would be zero. Similarly, the output would be zero. This defines ideal behavior and is a characteristic of an ideal op amp. In a real op amp, common-mode voltages can receive some amplification and thus depart from the desired behavior. Since we are currently defining ideal characteristics you should remember that **an ideal op amp has a common-mode voltage gain of zero.** This means the output is unaffected by voltages that are common to both inputs (i.e., no difference).

Figure 1.13 further illustrates the measurement of common-mode voltage gains.

1.3.3 Bandwidth

Bandwidth, as you might expect, refers to the range of frequencies that can be amplified by the op amp. Most op amps respond to frequencies down to and including DC. The upper limit is dependent on several factors including the specific op amp being considered. But in the case of an ideal op amp, we shall consider the range of acceptable frequencies to extend from DC through an infinitely high frequency. That is, **the bandwidth of an ideal op amp is infinite.** This is illustrated graphically in Fig. 1.14. The graph shows that all frequencies of input voltage receive equal gains (infinite).

1.3.4 Slew Rate

The output of an ideal op amp can change as quickly as the input voltage changes in order to faithfully reproduce the input waveform. We shall see in a later section that a real op amp has a practical limit to the rate of change of voltage on the output. This limit is called the slew rate of the op amp.

1.3.5 Input Impedance

The input impedance of an op amp can be represented by an internal resistance between the input terminals (refer to Fig. 1.15). As the value of this internal impedance increases, the current supplied to the op amp from the input signal source decreases. That is to say,

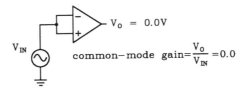

Figure 1.13 The common-mode voltage gain of an ideal op amp is zero.

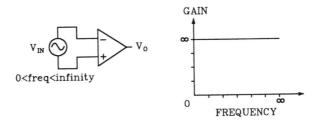

Figure 1.14 The bandwidth of an ideal op amp is infinite.

Sec. 1.3 Basic Characteristics of Ideal Op Amps

higher input impedances produce less loading by the op amp. Ideally, we would want the op amp to present minimum loading effects so we want a high input impedance. It is important to remember that **an ideal op amp has infinite input impedance.** This means that the driving circuit does not have to supply any current to the op amp. Another way to view this characteristic is to say that no current flows in or out of the input terminals of the op amp. They are effectively open circuited.

1.3.6 Output Impedance

Figure 1.16 shows an equivalent circuit that illustrates the effect of output impedance. The output circuit is composed of a voltage source and a series resistance (r_o). The internal voltage source has a value of $A_V v_I$. This says simply that the output has a potential similar to the input but is larger by the amount of voltage gain. Regardless of the absolute value of the internal source, the equivalent circuit shows that this voltage is divided between the external load (R_L) and the internal series resistance r_o. In order to get the most voltage out of the op amp and to minimize the loading effects of external loads, we would want the internal output resistance to be as low as possible. Thus, **the output impedance of an ideal op amp is zero.** Under these ideal conditions, the output voltage will remain constant regardless of the load applied. In other words, the op amp can supply any required amount of current without its output voltage changing. A practical op amp will have limitations, but the output impedance will still be quite low.

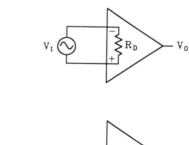

Figure 1.15 The input impedance of an ideal op amp is infinite.

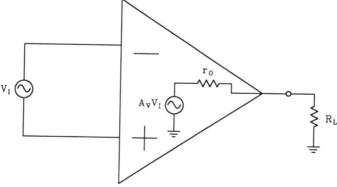

Figure 1.16 The output impedance of an ideal op amp is zero.

1.3.7 Temperature Effects

Since the op amp is constructed from semiconductor material, its behavior is subject to the same temperature effects that plague transistors, diodes, and other semiconductors. Reverse leakage currents, forward voltage drops, and the gain of internal transistors all vary with temperature. For now we shall ignore these effects and conclude that **an ideal op amp is unaffected by temperature changes.** Whether or not we can ignore temperature effects in practice depends on the particular op amp, the application, and the environment.

1.3.8 Noise Generation

Anytime current flows through a semiconductor device, electrical noise is generated. There are several mechanisms that can be responsible for the creation of the noise but in any case it is generally considered as undesirable. In many applications, the noise levels generated are so small as to be insignificant. In other cases, we must take precautions to minimize the effects of the noise generation. For now, however, we shall consider that **an ideal op amp does not generate internal noise.** If we apply a noise-free signal on the input, then we can expect to see a noise-free, high-fidelity signal reproduced at the output.

Table 1.1 summarizes the important characteristics of ideal op amps.

1.3.9 Troubleshooting Tips

Even though we have barely begun to discuss op amps and how they work, we can still begin to extend our troubleshooting skills to include op amps. Figure 1.17 shows a simple op amp circuit. Notice the addition of the power supply connections (+15V and −15V)

TABLE 1.1

IDEAL OP PARAMETER	VALUE OR CHARACTERISTIC
Differential voltage gain	Infinite
Common-mode voltage gain	Zero
Bandwidth	Infinite
Slew rate	Infinite
Input impedance	Infinite
Output impedance	Zero
Temperature effects	Unaffected
Noise generation	No noise generation

Sec. 1.4   Introduction to Practical Op Amps   19

and the pin numbers of the integrated circuit package (741). With reference to the ideal op amp circuit shown in Fig. 1.17, we know the following represent normal operation:

1. There should be positive 15 volts DC on pin 7 with respect to ground.
2. There should be negative 15 volts DC on pin 4 with respect to ground.
3. As long as v_I is greater than zero, the output should be at either of two extreme voltages (approximately ±15 V).

The power supply connections can be referred to as V^+ and V^- for the positive and negative terminals, respectively.

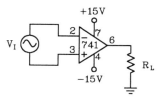

Figure 1.17   Schematic representation of an op amp showing the ±power supply connections.

Items 1 and 2 are essential checks regardless of the circuit being evaluated. Item 3 results from the infinite voltage gain of the ideal op amp. If you applied an AC signal and monitored the output of Fig. 1.17 (under normal conditions) with an oscilloscope, you would see a square wave. The amplitude would be near ±15 volts and the frequency would be identical to the input. Satisfy yourself that this latter statement is true, and you will be well on your way toward understanding op amp operation.

1.4 INTRODUCTION TO PRACTICAL OP AMPS

Now let us consider some of the nonideal effects of an op amp. By understanding the ideal characteristics described in the preceding sections and the nonideal characteristics presented in this section we will be in a position to evaluate and discuss these characteristics as we analyze and design the circuits in the remainder of the text. As the circuits are presented, an ideal approach will be used whenever practical to introduce the concept. We will then identify those nonideal characteristics that should be considered for each application. A more complete discussion of the nonideal performance of op amps is presented in Chap. 10. As each characteristic is discussed in the following paragraphs, we will compare the following items:

1. the ideal value
2. a typical nonideal value
3. the value for a real op amp

Appendix 1 presents the manufacturer's specification sheets for a 741 op amp, one of the most widely used devices. We shall refer to these specifications in the following paragraphs.

1.4.1 Differential Voltage Gain

You will recall that an ideal op amp has an infinite differential voltage gain. That is, any nonzero input signal will cause the output to be driven to its limits. In the case of a real op amp, the voltage gain is affected by several factors including:

1. the particular op amp being considered
2. the frequency of operation
3. the temperature
4. the value of supply voltage

For DC and very low-frequency applications the differential voltage gain will generally be from 50,000 to 1,000,000. Although this is less than the infinite value cited for ideal op amps, it is still a very high gain value. As the frequency increases, the available gain decreases. The point at which this decreasing gain becomes a problem is discussed briefly in a subsequent paragraph and more completely in Chap. 10. For purposes of the present discussion, you should know that **the differential voltage gain of a typical nonideal op amp starts at several hundred thousand and decreases as frequency increases.**

Now let us determine the differential voltage gain for an actual 741 op amp (refer to App. 1). In the specification sheet, the manufacture calls this parameter the Large Signal Voltage Gain. The value is given as ranging from a low of 20,000 to a typical value of 200,000. No maximum value is given. You will also find a number of graphs in App. 1. You should examine those graphs that present *open-loop* voltage gain as a function of another quantity. The terms *open* and *closed* loops are used extensively when discussing op amps. If a portion of the amplifier's output is returned to its input (i.e., feedback), then the amplifier is said to have a closed loop. You can readily see from the graphs in App. 1 that the gain of the op amp is not especially stable. Pay particular attention to the graph showing open-loop voltage gain as a function of frequency. Notice that the gain drops dramatically as the frequency increases.

In Chap. 2 you will learn that the op amp gain can be easily stabilized with a few external components. In fact, the fluctuating gain characteristic can be made insignificant in an actual op amp circuit.

1.4.2 Common-mode Voltage Gain

Although an ideal op amp has no response to voltages that are common to both inputs (i.e., no difference voltage), a practical op amp may have some response to such signals. Figure 1.18 shows how the common-mode voltage gain is measured. In the ideal case, of course, there would be no output and the computed gain would be zero. In the real case, there might be, for example, as much as 2 millivolts generated with a 1 millivolt common-mode input signal. That is, the common-mode voltage gain might be two in a typical case.

Sec. 1.4   Introduction to Practical Op Amps   21

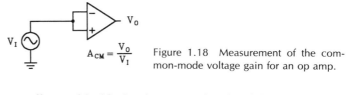

Figure 1.18   Measurement of the common-mode voltage gain for an op amp.

The manufacturers normally provide this data by contrasting the differential voltage gain with the common-mode voltage gain. This parameter, called common-mode rejection ratio (CMRR) is computed as follows:

$$\text{CMRR} = \frac{A_D}{A_{CM}} \tag{1.9}$$

where A_D and A_{CM} are the differential and common-mode gains, respectively. In a specification sheet this is normally written in the decibel form. To convert from the decibel value given in the data sheet to the form shown, the following conversion formula is used.

$$\text{CMRR} = 10^{dB/20} \tag{1.10}$$

where dB is the value of the common-mode rejection ratio expressed in decibels.

Let us refer to App. 1 and determine the common-mode voltage gain for a 741 op amp. The minimum value is listed as 70 dB with 90 dB being cited as typical. Converting the typical value to the standard CMRR ratio form requires application of Eq. (1.10).

$$\begin{aligned}
\text{CMRR} &= 10^{dB/20} \\
&= 10^{90/20} \\
&= 10^{4.5} \\
&= 31{,}622
\end{aligned}$$

To determine the actual common-mode voltage gain, we simply divide the differential voltage gain by the CMRR value [transposed version of Eq. (1.9)].

$$A_{CM} = \frac{A_D}{\text{CMRR}} \tag{1.11}$$

Recall that a typical differential voltage gain for the 741 is 200,000. Thus, a typical common-mode voltage gain can be shown to be

$$A_{CM} = \frac{200{,}000}{31{,}622} = 6.3$$

Whether this value is good or bad, high or low, acceptable or unacceptable is determined by the particular application being considered. For now, you should strive to understand the meaning of common-mode voltage gain and how it differs from the differential voltage gain.

1.4.3 Bandwidth

You may recall from your studies of basic amplifier and/or filter theory that the bandwidth of a circuit is defined as the range of frequencies that can be passed or amplified with less than 3 dB power loss (6 dB voltage loss). In the case of the ideal op amp, we said that the bandwidth was infinite since it could respond equally well to frequencies extending from DC through infinitely high frequencies. As we saw in our discussion of differential voltage gain, however, not all frequencies receive equal gains in a practical op amp.

If you examine the behavior of the op amp itself with no external circuitry, it acts as a basic low-pass filter. That is, the low frequencies (all the way to DC) are passed or amplified maximally. The higher frequencies are attenuated. The bandwidth of practical op amps nearly always begins at DC. The upper edge of the passband, however, may be as low as a few hertz. This would seem to represent a serious op amp limitation. We shall see, however, that this apparently restricted bandwidth can be dramatically increased by the addition of external components.

Now let us determine the bandwidth of a 741 op amp by examining the specification sheets in App. 1. The bandwidth in the data sheet is often given as 1.0 megahertz. This seems like a fairly respectable value but it is misleading when viewed from the basic definition of bandwidth. In the case of op amps, the true open-loop (i.e., no external components) bandwidth is of very little value since it is so very low (a few hertz). The bandwidth generally cited in the data sheet is more appropriately labeled the gain-bandwidth product. Recall that the differential gain decreases as the frequency increases. The gain-bandwidth product indicates the frequency at which the differential gain drops to one (unity). This frequency is also called the unity-gain frequency.

To further illustrate the bandwidth characteristics of the 741, examine the graph showing open-loop voltage gain as a function of frequency. You can see that the amplifier has a gain of about 200,000 at 1 hertz but the gain has dropped dramatically by the time the input frequency reaches 10 hertz. In fact, the actual upper edge of the passband (the half-power or 3 dB frequency) is about 5 hertz. This is the actual bandwidth of the open-loop op amp. Observe that the gain drops steadily until it reaches unity at a frequency of 1.0 megahertz. This is the unity-gain frequency. This same value (1.0 MHz) is obtained by multiplying the DC gain (200,000) by the bandwidth (5 Hz). Thus it is also called the gain-bandwidth product. It is the gain-bandwidth product that is generally labeled "bandwidth" in the data sheets.

1.4.4 Slew Rate

Although the output of an ideal op amp can change levels instantly (as required by changes on the input), a practical op amp is limited to a rate of change specified by the slew rate of the op amp. The slew rate is specified in volts per second and indicates the highest rate of change possible in the output.

Sec. 1.4   Introduction to Practical Op Amps   23

To further appreciate this characteristic, consider a square wave input to an op amp. In the case of an ideal op amp, the output also will be a square wave. In the case of a real op amp, however, the rise and fall times will be limited by the slew rate of the op amp. In the extreme case, a square wave input can produce a triangle output if the slew rate is so low that the output is not given adequate time to fully change states during a given alternation of the input cycle.

The 741 op amp has a slew rate of 0.5 volts per microsecond. There are other op amps that have significantly higher rates.

The slew rate (in conjunction with the output amplitude) limit the highest usable frequency of the op amp. The highest **sinewave** frequency that can be amplified without slew rate distortion is given by Eq. (1-12).

$$\boxed{f_{SRL} = \frac{\text{slew rate}}{\pi \, v_o(\text{max})}} \quad (1.12)$$

where $v_o(\text{max})$ is the maximum peak-to-peak output voltage swing. In the case of a 741 op amp, for example, the 0.5 volts per microsecond slew rate limits the useable frequency range for a ± 10 volt output swing to

$$f_{SRL} = \frac{\text{slew rate}}{\pi \, v_o(\text{max})}$$

$$f_{SRL} = \frac{0.5V / 10^{-6}s}{3.14 \times 20} = 7.96 \text{ kHz}$$

1.4.5 Input Impedance

The input impedance of an op amp is the impedance that is seen by the driving device. The lower the input impedance of the op amp the greater is the amount of current which must be supplied by the signal source. You will recall that we considered an ideal op amp to have an infinite input impedance, and therefore, drew no current from the source.

A real op amp does require a certain amount of input current to operate, but the value is generally quite low compared to the other operating currents in the circuit. You may wish to re-examine Fig. 1.2 and notice that the current for the input terminals is essentially providing base current for the differential amplifier transistors. Since the transistors have a constant current source in the emitter circuit, the input impedance is very high. A typical op amp will have an input impedance in excess of 1 megohm with several megohms being reasonable. If this is still not high enough, then an op amp with a field-effect transistor input may be selected.

Appendix 1 shows the data sheet for a 741 op amp. If you look under the heading of input resistance you will find that these devices have a minimum input resistance rating of 0.3 megohm and a typical value of 2.0 megohms. Further, the input impedance is not constant. It varies with both input frequency and operating temperature. In many applications, we can ignore the nonideal effects of input impedance. As we study the applications presented in the text, we will learn when and how to consider the effects of less than ideal input impedances.

1.4.6 Output Impedance

The output impedance of an ideal op amp is zero. This means that regardless of the amount of current drawn by an external load, the output voltage of the op amp remains unaffected. That is, no loading occurs.

In the case of a practical op amp, there is some amount of output impedance. The ideal output voltage is divided between this internal resistance and any external load resistance. Generally, this is an undesired effect so we prefer the op amp to have a very low-output impedance.

The manufacturer's specification sheet in App. 1 lists the typical value of output impedance for a 741 as 75 ohms. What is not clear from the data sheet is that this value refers to open-loop output resistance. In most practical applications, the op amp is provided with feedback (i.e., closed loop). Under these conditions, the effective output impedance can be dramatically reduced with values as low as 1/100th of the open-loop output impedance being reasonable.

1.4.7 Temperature Effects

Although we want an ideal op amp to be unaffected by temperature, some effects are inevitable since the op amp is constructed from semiconductor material that has temperature-dependent characteristics. In a practical op amp, nearly every parameter is affected to some degree by temperature variations. Whether or not the changes in a particular characteristic are important to us is dependent on the application being considered and the nature of the operating environment. We will examine methods for minimizing the effects of temperature problems as we progress through the remainder of the text.

1.4.8 Noise Generation

Under ideal conditions, an amplifying or signal processing circuit should have no signal voltages at the output that do not have corresponding signal voltages at the input. When the circuit has additional fluctuations in the output we call these changes noise.

There are many sources of electrical noise generation inside of the op amp. A detailed analysis of the contribution of each source to the total circuit noise is a complex subject and well beyond the goal of this text. We will, however, examine techniques that can be used to minimize problems with noise. It is fortunate that noise problems are most prevalent in circuits operating under low-signal conditions. Most other circuits do not require a detailed analysis of the circuit noise and can be adequately controlled by applying some basic guidelines and precautions for minimizing noise.

We can get an appreciation for the noise generated in a 741 by examining the data supplied by the manufacturer and shown in App. 1. Several graphs are provided to describe

Sec. 1.4   Introduction to Practical Op Amps   25

the noise performance of the 741. The op amp noise is effectively added to the desired signal at the input of the op amp. If the input signal is small or even comparable in amplitude to the total op amp noise, then the noise voltages will likely cause erroneous operation. On the other hand, if the desired signal is much larger than the noise signal, then the noise can be ignored for many applications.

1.4.9 Power Supply Requirements

When ideal op amps were discussed, we briefly indicated that all op amps require an external DC power supply. Most op amps are designed for dual supply operation with ± 15 volts being the most common. Other op amps are designed specifically for single supply operation. However, in all cases, we must provide a DC supply for the op amp in order for it to operate.

The power supply connections on the op amp are generally labeled $+V_{CC}$ or V^+ for the positive connection and $-V_{CC}$ or V^- for the negative connection. You will recall from Fig. 1.2 that the DC power source provides the bias and operating voltages for the op amp's internal transistors. The magnitude of the power supply voltage is determined by the application and limited by the specific op amp being considered. A typical op amp will operate with supply voltages as low as 6 volts and as high as 18 volts although neither of these values should be viewed as extremes. Certain devices in specific applications can operate on less than 5 volts. Other high-voltage op amps are designed to operate normally with voltages substantially higher than 18 volts.

The current capability of the power supply is another consideration. The actual op amp draws fairly low currents with 1 to 3 milliamperes being typical. Some low-power op amps require only a few microamperes of supply current to function properly. In most applications, the external circuitry plays a greater role in power supply current requirements than the op amp itself.

Yet, another power supply consideration involves the amount of noise contributed to the circuit by the power supply. There are several forms of power supply noise including the following:

1. power line ripple caused by incomplete filtering,
2. high frequency noise generated within the supply circuitry,
3. switching transients produced by switching regulators,
4. noise coupled to the DC supply line from other circuits in the system,
5. externally generated noise that is coupled onto the DC supply line,
6. noise caused by poor voltage regulation.

Noise that appears on the DC power supply lines can be passed through the internal circuitry of the op amp and appear at the output. Depending on the type and in particular the frequency of the noise voltages, they will undergo varying amounts of attenuation as they pass through the op amp's components. Frequencies below 100 hertz are severely attenuated with losses as great as 10,000 being typical. As the noise frequencies increase, however, the attenuation in the op amp is greatly reduced. Frequencies over 1.0 megahertz may be coupled from the DC supply line to the output of the op amp with no significant reduction in amplitude. The degree to which the output is affected by noise on the DC supply lines is called the **power supply rejection ratio (PSRR).**

Power distribution is a very important consideration in circuit design yet frequently receives only minimal attention. This issue will be addressed in the following section with regard to circuit construction.

1.4.10 Troubleshooting Tips

At this point in our study of op amps, there is little difference between ideal and nonideal devices. You will recall from Sec. 1.3.9 that the following items are necessary for proper operation of the op amp:

1. There should be positive 15 volts DC on the V^+ connection with respect to ground.
2. There should be negative 15 volts DC on the V^- connection with respect to ground.
3. As long as the differential input voltage is greater than zero, the output should be at either of two extreme voltages (approximately ± 15 volts).

Items 1 and 2 are essential checks regardless of the circuit being evaluated and whether it is viewed as ideal or nonideal. In the case of Item 3, we can now refine our expectations of normal operation. We saw from Fig. 1.2 that the output of an op amp has two transistor/resistor pairs between the output and the $\pm V_{CC}$ connections. As you know, when current flows through these components a portion of the supply voltage is dropped. For most bipolar op amps, the internal voltage drop is approximately 2 volts regardless of the polarity of the output voltage. Thus, if the output of an op amp was forced to its positive extreme and was being operated from a ± 15 volt supply, we would expect the output to measure approximately $+13$ volts. This is called the positive saturation voltage ($+V_{SAT}$). Similarly, the negative extreme of the output is called the negative saturation voltage ($-V_{SAT}$) and is about 2 volts above (i.e., less negative than) the negative power supply voltage.

1.5 CIRCUIT CONSTRUCTION REQUIREMENTS

This is one of the most important sections in the text and yet the most likely to be skipped or skimmed. Every technician and engineer believes he or she knows how to build circuits. Perhaps you do know how to build circuits, but you are urged to study the following sections anyway. It is difficult to convince people of the value of many of the techniques discussed. The reason for the lack of acceptance is that many proper construction techniques can be skipped or slighted without any observable deterioration in circuit performance in many cases. But it is equally true that some of the most elusive problems experienced when building and testing circuits are a direct result of poor, or at least inappropriate, circuit construction techniques. So, to repeat, you are urged to apply the

Sec. 1.5 Circuit Construction Requirements

following techniques on a consistent basis whether or not there appears to be an observable change in performance.

1.5.1 Prototyping Methods

There are numerous ways to construct a circuit for purposes of testing prior to committing the design to a printed circuit board. The techniques and precautions described in subsequent paragraphs are universal and are appropriate to all prototyping methods including the following:

1. protoboard
2. wirewrap
3. perforated board
4. copper-clad board

This is not intended as a complete list of prototyping methods or even the best methods. Rather, the list represents some of the most common methods used by technicians and engineers in the industry.

1.5.2 Component Placement

Circuit construction essentially consists of placing the components on some sort of supporting base, then interconnecting the appropriate points. If the circuit being constructed is a noncritical, DC, resistive circuit, then component placement is arbitrary. As the frequency of circuit operation increases, the importance of proper component placement also increases. An op amp is inherently a high-frequency device. Even if it is being used as a DC amplifier, high-frequency noise signals will be present and can adversely affect circuit operation. Therefore, component placement is important when constructing op amp circuits regardless of the application.

The components should be physically placed such that both of the following goals are accomplished:

1. Interconnecting leads can be as short as practical
2. Low-level signals and devices should not be placed adjacent to high-level devices

Although these rules may seem difficult or unnecessarily restrictive at first, they will become second nature if you consistently use these practices.

1.5.3 Routing of Leads

If you consistently achieve the goals cited in Sec. 1.5.2, then the task of properly routing the interconnecting wires is much easier. The wire routing should achieve the following goals:

1. Make all leads as short as practical
2. Avoid routing input leads or low-level signals parallel to output or high-level signals
3. Make straight, direct connections rather than forming cables or bundles of wires

1.5.4 Power Supply Distribution

Power supply distribution refers to the way that the $\pm V_{CC}$ and ground connections are routed throughout the circuit. For consistently good results with prototype operation, you should apply the following power supply distribution techniques:

1. use a wire size that is large enough to minimize impedance,
2. run $\pm V_{CC}$ and ground parallel and as close as possible to each other,
3. avoid longer lead lengths than necessary,
4. do not allow the current for digital or high-current devices to flow through the same ground wires as small signal linear signals (except for the main system ground point),
5. twist the power distribution lines that run between the power supply and the circuit under test.

For many circuits, these rules and practices can be severely abused with no apparent reduction in circuit performance. But why take the chance? If failure to apply these techniques is the cause of poor circuit performance, it may be very difficult to isolate and an otherwise good design may be classed as unpredictable, unreliable, impractical, and so on.

1.5.5 Power Supply Decoupling

Closely associated with power supply distribution is power supply decoupling. Again, this is an area that is difficult to appreciate and frequently gets slighted. To help you understand the mechanisms involved, let us examine the problem of power distribution more closely.

In theory, each device or circuit connected between a DC source and ground receives the same voltage and are unaffected by each other (i.e., they are in parallel). In practice, however, the wires supplying the power contain resistance and inductance. Figure 1.19 shows a simplified representation of the problem.

As the current for circuit 1 flows through the power supply lines, the inductance and resistance cause voltage to be dropped. Thus circuit 1 receives less voltage than expected. Circuit 2 cannot receive more than circuit 1 and, in fact, receives even less due to the voltage drop across the resistances and inductances between circuit 1 and 2.

You may not be alarmed at this point because you know that the resistance in copper wire is very low, and therefore, the resulting voltage drop must surely be very low. You may be right as long as **both** the current and the frequency are low.

Sec. 1.5   Circuit Construction Requirements

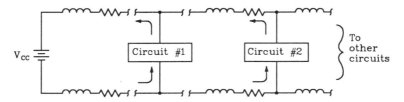

Figure 1.19   Printed circuit traces and wire used for DC power distribution have distributed inductance and resistance which cause voltage drops for high frequency currents.

In the case of an op amp circuit, the frequency is rarely low. Even if it is your intention to build a DC amplifier, there still will be high-frequency noise signals in the circuit. The high frequencies, whether desired or undesired, cause high-frequency fluctuations in the power supply current. These changes in current cause voltage drops across the distributed inductance in the power and ground lines. Since the frequencies are generally in the megahertz range, substantial inductive reactance and therefore voltage drop may result.

The high-frequency voltage drops described cause several problems including the following:

1. The noise and/or high-frequency signal from one circuit affects the supply voltage for another circuit. The circuits are now coupled rather than being independent as theory would suggest.
2. The output of a circuit can be shifted in phase and coupled back to the input. If the circuit has sufficient gain, then we will have all the conditions necessary for sustained oscillation.
3. The overall power distribution circuit tends to behave like a loop antenna and radiates the high-frequency signals into adjacent circuits or systems.

Section 1.5.4 specifies the use of an adequate wire size. This primarily affects the DC resistance of the wire. By running the V_{CC} line and the ground return physically close together as suggested in Sec. 1.5.4, however, you can reduce the actual inductance of the supply lines and thus improve the high-frequency performance. Additionally, by keeping the supply lines close together, you reduce the loop area of an effective loop antenna and dramatically reduce radiations from the power supply loop.

Sections 1.5.2, 1.5.3, and 1.5.4 all recommend the use of short lead lengths. Shorter lead length directly reduces the value of distributed inductance and so reduces the magnitude of the high-frequency voltage drop problem.

You are now in a position to appreciate the value of decoupling components. The intent of decoupling is to further isolate one circuit from another with reference to the DC power distribution lines. We will examine decoupling at two important points in the system:

1. circuit decoupling
2. power-entry decoupling

Circuit decoupling generally consists of a capacitor connected between $\pm V_{CC}$ and ground at a point physically close to the circuit being decoupled. Figure 1.20 illustrates the effect of the decoupling capacitor. Without the decoupling capacitors (refer to Fig. 1.19), surges in current (i.e., high-frequency changes) had to pass through the inductance and resistance of the power supply distribution lines. With the decoupling capacitors in place, short-term demands for increased current (i.e., transients or high-frequency changes) can be supplied by the decoupling capacitor. You may view it as a filter capacitor that disallows sudden changes in voltage across its terminals. You may also consider that the decoupling capacitor has a low reactance to high-frequency signals and bypasses those signals around the circuit being decoupled. In any case, the net result is that the circuits are provided with a more stable, electrically quiet source of DC power and are more effectively isolated from each other.

In most cases, ceramic disc capacitors in the range of 0.01 to 0.1 microfarad are good choices for circuit decoupling capacitors. Electrolytic capacitors are useless for this purpose because of their high internal inductance. It should be clear from Fig. 1.20 that the decoupling capacitor must be connected physically close to the circuit or device being decoupled in order to be effective. Additionally, the leads of the decoupling capacitor should be kept as short as possible. Lengths as small as ¾ inch can nullify the effects of the decoupling capacitor in some cases.

Power-entry decoupling provides a similar function but is applied at the point where the power supply leads attach to the circuit under test. Power-entry decoupling consists of the following:

1. A tantalum electrolytic capacitor connected between each V_{CC} and ground. The ideal value is dependent on the circuit being tested, but generally a value of 25 to 100 microfarad is adequate.
2. A 0.1 microfarad ceramic capacitor connected in parallel with the tantalum decoupling capacitor.
3. An optional, but desirable, ferrite bead or ring slipped over the $\pm V_{CC}$ wires leading to the power supply.
4. Twisted leads between the power supply and the power entry point.

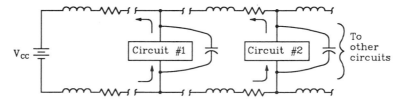

Figure 1.20 Decoupling capacitors placed physically close to the circuit being decoupled helps improve circuit isolation.

Sec. 1.5   Circuit Construction Requirements

1.5.6 Grounding Considerations

Since ground is inherently part of the power distribution system, many of the practices presented in Secs. 1.5.4 and 1.5.5 apply to the ground structure as well. In addition to these practices, though, we must take some additional precautions to insure reliable circuit performance. We shall examine the following techniques:

1. use of a ground plane
2. quiet grounds

The performance of a circuit can nearly always be improved by using a large planar area as the ground connection. In the final product, this ground plane is one entire layer in a multilayer printed circuit board. For prototyping purposes, however, it is not always easy to get a ground plane. Perforated board is probably the most impractical method of prototyping when a ground plane is desired. Wirewrap boards, on the other hand, are available with an integral ground plane.

Copper-clad board prototyping can provide electrical results which exceed that of the other methods when high-frequency operation is required but may take longer to construct. In this case, the entire surface of the copper-clad board is connected to system ground. This minimizes the resistance of the ground path and provides a shield against external interference. Additionally, when the insulated $\pm V_{CC}$ wires are routed against the ground plane, the inductance of the power distribution system is greatly reduced. Radiation into neighboring circuits is also minimized. It should be noted that multilayer printed circuit boards use one or more entire planes for ground. This practice minimizes the ground impedance and dramatically reduces both emissions from the circuit and susceptibility to external radiation.

Protoboard provides one of the fastest methods for prototyping circuits. Additionally, the metal base plate (on some models) provides a ground plane. The ground currents, however, do not flow through this ground plane. The ground and V_{CC} distribution buses are physically close and parallel which lowers the inductance and radiation. One potential problem that you must be aware of with this type of board is that there is a significant amount of capacitance between the various connections on the board. The problems caused by this capacitance increases as the frequency increases and/or the signal amplitudes decrease. Boards of this type are generally best suited for low-frequency circuits. It should be noted, however, that most of the circuits presented in this text were constructed and tested on such a system with excellent results.

The term *quiet ground* is most often used with reference to digital systems or systems that have a combination of analog and digital devices. Digital circuits can generate high levels of transient currents in the ground network. These currents can interfere with the proper operation of low-level analog circuits. The degree of interference is generally worsened for faster rise times and increased current drive in the digital gates. To minimize the ground noise problem (sometimes called ground bounce), the circuit should be constructed such that the analog devices are connected directly to the main ground (power input) connection. That is, the ground currents for the digital devices should not be

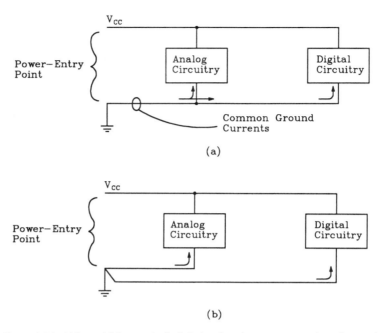

Figure 1.21 Wires which carry both digital and analog currents can introduce noise to the analog circuits.

allowed to flow through the ground wires of the analog devices. Figure 1.21 clarifies this concept. In Fig. 1.21(a), the return, or ground, current for the digital circuitry flows through wires (generally printed circuit traces) which are common to the analog circuitry. This will likely cause interference or noise in the analog circuitry. By contrast, Fig. 1.21(b) shows a similar circuit that utilizes a separate ground path for the digital and analog circuitry.

1.5.7 Actual Circuit Performance

This section is included as a final effort to make a believer of you. Figure 1.22 shows two sets of oscilloscope waveforms. These waveforms are real and were obtained from the plotter output on a digitizing oscilloscope; they are not theoretical drawings. Each set of waveforms illustrates the effect of one of the circuit construction rules presented in preceding sections. The first oscilloscope plot in each set illustrates circuit performance with a construction rule violated. The companion waveform shows the exact circuit with that one rule implemented properly. The results clearly indicate the change in performance. Figure 1.22(a) shows an effect of improper component placement. Figure 1.22(b) illustrates how the circuit performance can deteriorate when excessive lead lengths are used to construct the circuit.

Sec. 1.5 Circuit Construction Requirements

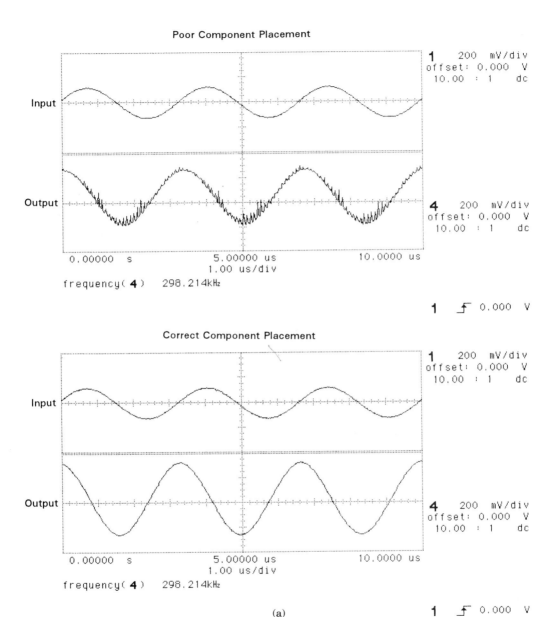

(a)

Figure 1.22 Performance problems can be caused by component placement (a) and excessive lead lengths (b). (Test equipment courtesy of Hewlett-Packard Company) (continued)

34 Basic Concepts of the Integrated Operational Amplifier Chap. 1

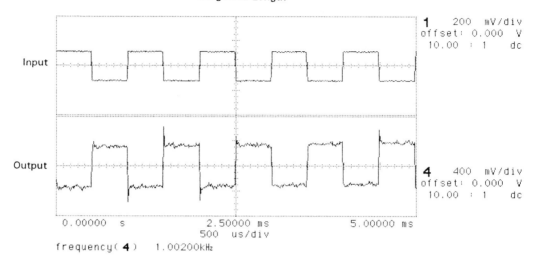

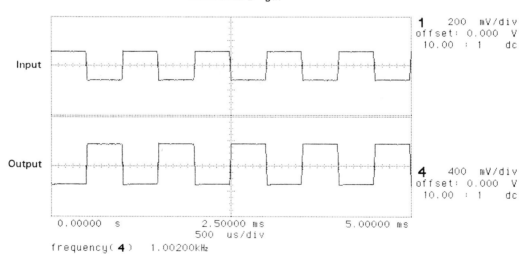

(b)

Figure 1.22b

1.6 ELECTROSTATIC DISCHARGE

Anyone who has been in a cold, dry climate has probably walked across a carpet or slid across a car seat and then witnessed a sizable high-voltage arc jumping between their body and a nearby metal object. This discharge of static electricity is called *electrostatic discharge* or simply ESD, and can pose a serious threat to op amps and other integrated electronic components. The sensitivity of an op amp (or other component) to ESD is largely a function of the technology used to build the device. In general, integrated circuits using MOSFET transistors are more susceptible to damage than circuits employing bipolar transistors. But, even bipolar circuits can be destroyed, or at least weakened, by ESD currents.

When handling integrated circuits—particularly in a cold, dry climate—you should be sure that your body is not allowed to accumulate static charge. This can be accomplished in several ways.

1. Attach a conductive strap to your wrist with the other end connected to a ground potential. Many manufacturers offer straps of this type which provide a discharge path for the static charge, but do not present a safety hazard to the wearer.
2. Always touch ground (e.g., a metal chassis) before contacting the integrated circuit.
3. Condition the environment by either ionizing the air or using an effective humidifier.

REVIEW QUESTIONS

1. How many inputs does a differential amplifier have?
2. If a differential amplifier has a single output pin, it is called a _____ -ended amplifier, and the output is referenced to _____ .
3. Differential amplifiers whose output is taken between two pins (neither of which is ground) is called a _____ -ended amplifier.
4. What is the lowest frequency that can be amplified by a typical op amp?
5. If three 10-megohm resistors were connected in series across a 10-volt power supply and one of the resistors became open, would this open have much effect (e.g., $>5\%$) on the currents in the remaining resistors?
6. What is the effect (significant or negligible) on *total* current flow if the resistors described in Ques. 5 are all connected in parallel with the power source?
7. If two 10-megohm resistors are connected in parallel with a 1.0-kilohm resistor, explain the relative effect on *total* current flow if one of the 10-megohm resistors opens. Repeat this question for the case where the 1.0-kilohm resistor develops an open.
8. If zero volts is applied directly to the inverting input of an ideal op amp at the same time -0.1 volts is applied to the noninverting input, compute or describe the value of output voltage.
9. If a portion of an op amp's output signal is returned to its input (i.e., feedback), we say that the amplifier is operating _____ loop.

10. What is the value of each of the following parameters for an ideal op amp:
    a. bandwidth
    b. input current
    c. open-loop voltage gain
    d. input impedance
    e. lowest operating frequency
    f. highest operating frequency
    g. slew rate
    h. output impedance
    i. common-mode voltage gain
11. Can a real (i.e., practical) op amp be used to amplify DC?

    Refer to App. 1 for Quests. 12 to 14.

    12. What is the open-loop voltage gain of a standard 741 op amp at a frequency of 2 hertz?
    13. What is the minimum input resistance of a 741 op amp?
    14. What is the approximate open-loop voltage gain of a 741 op amp at a frequency of 100 kilohertz?
    15. If the largest output voltage swing required for a particular 741 design is ±5 volts, what is the highest sinewave frequency that can be amplified before slew rate limiting begins to distort the output?

CHAPTER 2

AMPLIFIERS

2.1 AMPLIFIER FUNDAMENTALS

2.1.1 Gain

This chapter focuses on the analysis and design of several basic amplifier circuits. Not only is the amplifier circuit a fundamental building block in linear circuits, but the analytical techniques introduced in this chapter will greatly enhance your ability to analyze the circuits presented in subsequent chapters. Regardless of the specific circuit application (e.g., summing circuit, active filter, voltage regulator, etc.) the op amp itself is simply an amplifier. Therefore, a thorough understanding of op amp behavior in circuits designed specifically as amplifiers will provide us with analytical insight that is applicable to nearly all op amp circuits.

An amplifier generally accepts a small signal at its input and produces a larger, amplified version of the signal at its output. The gain (A) or amplification is expressed mathematically as

$$\text{gain} = A = \frac{\text{output}}{\text{input}} \tag{2.1}$$

We may speak of voltage gain, current gain, or power gain. In each of these cases the equation is valid. If, for example, a particular voltage amplifier produced a 5-volt RMS output when provided with a 2-volt RMS input, we would compute the voltage gain as

$$\text{voltage gain} = A_v = \frac{\text{output voltage}}{\text{input voltage}} = \frac{5}{2} = 2.5$$

There are no units for gain. It is simply a ratio of two numbers.

It is also convenient to express a gain ratio in its equivalent decibel (dB) form. The conversion equations are:

$$\text{voltage gain (dB)} = A_V \text{ (dB)} = 20 \log_{10} \frac{V_{OUT}}{V_{IN}} \qquad (2.2)$$

$$\text{current gain (dB)} = A_I \text{ (dB)} = 20 \log_{10} \frac{I_{OUT}}{I_{IN}} \qquad (2.3)$$

$$\text{power gain (dB)} = A_P \text{ (dB)} = 20 \log_{10} \frac{P_{OUT}}{P_{IN}} \qquad (2.4)$$

The voltage gain of 2.5 on the amplifier as discussed could be expressed in decibels by applying Eq. (2.2).

$$\begin{aligned} A_V(\text{dB}) &= 20 \log_{10} \frac{V_{OUT}}{V_{IN}} \\ &= 20 \log_{10} \frac{5}{2} \\ &= 20 \log_{10} 2.5 \\ &= 20 \times 0.3979 \\ &= 7.96 \text{ dB} \end{aligned}$$

So we see that an amplifier with a voltage gain of 2.5 also has a voltage gain of 7.96 dB. It should be noted that technically, the equations cited previously for calculating voltage and current gains in their decibel form require that the input and output impedances be equal. In practice, this is rarely the case. In spite of this known error, it is common in the industry to calculate and express the gains as described.

You should also be reminded that fractional gains (i.e., losses) are expressed as negative decibel values.

2.1.2 Frequency Response

The frequency response of an amplifier describes how its amplification varies with changes in frequency. We often communicate the frequency response of an amplifier in graphical form. Figure 2.1 shows a typical frequency response curve.

The vertical axis indicates the amplifier's voltage gain expressed in decibels. The horizontal axis shows the input frequency range.

The frequency response curve shown in Fig. 2.1 indicates that the amplifier provides

Sec. 2.1   Amplifier Fundamentals   39

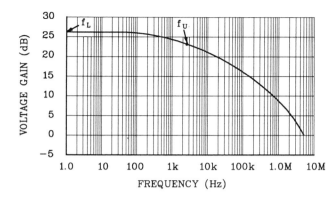

Figure 2.1 The bandwidth of an amplifier is the range of frequencies between the upper (f_U) and lower (f_L) cutoff frequencies.

greater gain for low frequencies. Once the input frequency exceeds a certain value, the amplification begins to reduce significantly. Those frequencies that are amplified to within 3 dB of the maximum output voltage level are considered as having passed the amplifier. Any frequency whose output voltage is lower than the maximum output voltage by more than 3 dB is considered to have been rejected by the amplifier. The frequency which separates the passband frequencies from the stopband frequencies is called the cutoff frequency. And since -3 dB corresponds to a power ratio of 0.5, the cutoff frequency is also called the half-power frequency or simply the half-power point on the frequency response curve.

The bandwidth of an amplifier is measured between the two half-power points. If the frequency response of an amplifier extends to include zero (i.e., DC), then the bandwidth of the amplifier is the same as the upper cutoff frequency. This is the case for the amplifier represented in Fig. 2.1. Here the lower frequency range extends all the way to zero, but in many circuits there will be a lower cutoff frequency which is greater than DC. The bandwidth is expressed as

$$\text{bandwidth} = \text{BW} = f_U - f_L \tag{2.5}$$

where f_U and f_L are the upper and lower cutoff frequencies, respectively.

2.1.3 Feedback

So far in our discussions on op amps, we have only considered the behavior of the op amp itself with no external components. The op amp has been examined only in its open-loop configuration. In most practical applications, a portion of the amplifier's output is returned through external components to the input of the amplifier. The return signal, called feedback, is then mixed with the incoming signal to determine the effective signal applied to the input of the op amp.

The amplitude, frequency, and phase characteristics of the feedback signal can dramatically alter the behavior of the overall circuit. If the feedback signal has a phase relationship that is additive when mixed with the incoming signal, we refer to the return signal as positive feedback. On the other hand, if the feedback signal is out of phase (optimally 180°) with the input signal, then the effective input signal is reduced, and we label the feedback as negative feedback. Both forms of feedback are useful in certain op amp applications, but for the remainder of this chapter we will limit our concerns to negative feedback.

The external components which provide the feedback path may be frequency selective. That is, if some frequencies pass through the feedback circuit with less attenuation than other frequencies, then we have frequency selective feedback. This is a very useful form of feedback, but for the remainder of this chapter we will limit our discussion to nonselective feedback methods.

2.2 INVERTING AMPLIFIER

The first circuit examined in detail is the inverting amplifier and it is one of the most common op amp applications. Figure 2.2 shows the schematic diagram of the basic inverting amplifier.

2.2.1 Operation

Under normal operation, an amplified but inverted (i.e., 180° phase shifted) version of the input signal (v_I) appears at the output (v_O). If the input signal is too large or the amplifier's gain is too high, then the output signal will be clipped at the positive and negative saturation levels ($\pm V_{SAT}$).

Now let us understand how the negative feedback returned through R_F affects the amplifier operation. To begin our discussion, let us momentarily freeze the input signal as it passes through zero volts. At this instant, the op amp has no input voltage (i.e., $v_D = 0$ V). It is this differential input voltage that is amplified by the gain of the op amp to become the output voltage. In this case, the output voltage will be zero.

Now suppose the output voltage tried to drift in a positive direction. Can you see that this positive change would be felt through R_F and would cause the inverting pin ($-$) of the op amp to become slightly positive? Since essentially, no current flows in or out of the op amp input, there is no significant voltage drop across R_B. Therefore the ($+$) input of the op amp is at ground potential. This causes v_D to be greater than zero with the ($-$) terminal being the most positive. When v_D is amplified by the op amp, it appears in the output as a negative voltage (inverting amplifier action). This forces the output,

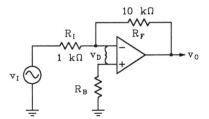

Figure 2.2 A basic inverting amplifier circuit.

Sec. 2.2 Inverting Amplifier

which had initially tried to drift in a positive direction, to return to its zero state. A similar, but opposite, action would occur if the output tried to drift in the negative direction. Thus, as long as the input is held at zero volts, the output is forced to stay at zero volts.

Now suppose we allow the input signal to rise to a +2 volt instantaneous level and freeze it for purposes of the following discussion. With +2 volts applied to R_I and zero at the output of the op amp, the voltage divider made up of R_F and R_I will have 2 volts across it. Since the (−) terminal of the op amp does not draw any significant current, the voltage divider is essentially unloaded. We can see, even without calculating values, that the (−) input will now be positive. Its value will be somewhat less than 2 volts because of the voltage divider action, but it will definitely be positive. The op amp will now amplify this voltage (v_D) to produce a negative-going output. As the output starts increasing in the negative direction, the voltage divider now has a positive voltage (+2 V) on one end and a negative voltage (increasing output) on the other end. Therefore the (−) input may still be positive, but it will be decreasing as the output gets more negative. If the output goes sufficiently negative, then the (−) pin (v_D) will become negative. If, however, this pin ever becomes negative, then the voltage would be amplified and appear at the output as a positive-going signal. So, you see, for a given instantaneous voltage at the input, the output will quickly ramp up or down until the output voltage is large enough to cause v_D to return to its near zero state. All of this action happens nearly instantaneously so that the output appears to be immediately affected by changes at the input.

We can also see from Fig. 2.2 that changes in the output voltage receive greater attenuation than equivalent changes in the input. This is because the output is fed back through a 10-kilohm resistor, but the input is applied to the 1.0 kilohm end of the voltage divider. Thus, if the input makes a 1-volt change, the output will have to make a bigger change in order to compensate and force v_D back to its near zero value. How much the output must change for a given input change is **strictly** determined by the ratio of the voltage divider resistors. Therefore, since the ratio of output change to input change is actually the gain of the amplifier, we can say that the gain of the circuit is determined by the ratio of R_F to R_I.

Recall that the internal gain (open-loop gain) of the op amp is not a constant. It varies with different devices (even with the same part number), it is affected by temperature, and it is different for different input frequencies. Now that we have added feedback to our op amp, the overall circuit gain is determined by external components (R_F and R_I). These can be quite stable and relatively unaffected by temperature, frequency, and so on.

If the input signal is too large or the ratio of R_F to R_I is too great, then the output voltage will not be able to go high enough (positive or negative) to compensate for the input voltage. When this occurs, we say the amplifier has reached saturation, and the output is clipped or limited at the $\pm V_{SAT}$ levels. Under these conditions, the output is unable to rise enough to force v_D back to its near zero level. From this you can safely conclude the following important rules regarding negative feedback amplifiers:

1. If the output is below $+V_{SAT}$ and above $-V_{SAT}$ then v_D will be very near zero volts.
2. If v_D is anything other than near zero volts, then the amplifier will be at one of the two saturation voltages ($\pm V_{SAT}$).

When the feedback and input resistor combination in Fig. 2.2 are viewed as an unloaded voltage divider, it is easy to determine current flow. Since we know that no significant current is allowed to enter or leave the ($-$) pin of the op amp, we can conclude that any current flowing through R_I must also pass through R_F. The polarity of the input and output voltages will determine the direction of this current, but it is important to realize that the value of current through R_I is the same as the current through R_F.

Since very little current flows in or out of the input terminals of the op amp, we saw that there was essentially no voltage drop across R_B which caused the ($+$) input terminal to remain at ground potential. Since v_D is always near zero as long as the amplifier remains unsaturated, this means that the ($-$) input terminal must also remain very near to ground potential. This is an important concept. Although the ($-$) input is not actually grounded, it remains very near ground potential. We commonly refer to this point in the circuit as *virtual ground*.

R_B is included to compensate for errors caused by the fact that some bias current does flow in or out of the op amp terminals. Even though this bias current is small, it can cause a slight voltage drop across R_F and R_I. This voltage drop is then amplified and appears at the output of the op amp as an error voltage. By including R_B in series with the ($+$) terminal and making its value equal to the parallel combination of R_F and R_I, we can generate a voltage which is roughly equal, but the opposite polarity from that caused by the drop across R_F and R_I. The error is generally reduced substantially but will be reduced to zero only if the bias currents in the two input terminals happen to be equal.

2.2.2 Numerical Analysis

Let us now learn to analyze the performance of the inverting amplifier circuit. Calculate all of the following:

1. voltage gain
2. input impedance
3. input current requirement
4. slew-rate limiting frequency
5. maximum output voltage swing
6. maximum input voltage swing
7. output impedance
8. output current capability
9. minimum value of load resistance

Sec. 2.2 Inverting Amplifier

10. bandwidth
11. power supply rejection ratio

For purposes of our first numerical analysis exercise, let us evaluate the performance of the circuit in Fig. 2.3.

Voltage gain. You will recall that the voltage gain for this circuit is determined by the ratio of R_F to R_I or simply

$$\boxed{\text{voltage gain} = A_V = -\frac{R_F}{R_I}} \quad (2.6)$$

The minus sign is used to remind us of the phase inversion since this is an inverting amplifier. Do not interpret the minus as a loss or a reduction in signal strength. For the circuit in Fig. 2.3, the voltage gain will be

$$A_V = -\frac{R_F}{R_I}$$

$$= -\frac{33 \times 10^3}{2.7 \times 10^3}$$

$$= -12.2$$

We can express this as a decibel gain by applying Eq. (2.2).

$$A_V \text{ (dB)} = 20 \log_{10} A_V$$

$$= 20 \log_{10} 12.2$$

$$= 21.7 \text{ dB}$$

In this conversion, be particularly careful not to include the minus sign as part of the voltage gain. First, we will be unable to compute the logarithm. Second, if we tried to put the minus sign in after the calculation the resulting negative decibel answer would be misinterpreted as a loss.

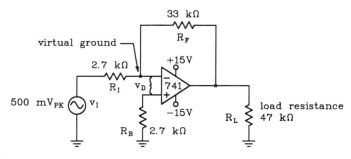

Figure 2.3 An inverting amplifier circuit used for a numerical analysis example.

It is important to note that the voltage gain computed in this section is the ideal closed-loop voltage gain of the circuit. The actual circuit gain will roll off as the input frequency is increased. This effect will be discussed as part of the discussion on bandwidth.

Input impedance. The input impedance of the amplifier shown in Fig. 2.3 is that resistance (or impedance) as seen by the source (v_I). You will recall that the voltage between the (+) and (−) terminals of the op amp (v_D) will always be about zero unless the amplifier is saturated. Since the (+) terminal is connected to ground (via R_B) in the inverting amplifier circuit, it is reasonable to assume that the (−) pin will always be near ground potential even though it is not and cannot be connected directly to ground. But, since the (−) input is essentially at ground potential, we call this point in the circuit a virtual ground.

Considering that the (−) pin is a virtual ground, it becomes apparent that the input impedance seen by the source is simply R_I. That is, as far as current demand is concerned, resistor R_I is effectively connected across the signal source. The equation for input impedance is given by Eq. (2.7).

$$\boxed{\text{input impedance} = Z_I = R_I} \quad (2.7)$$

The case of the inverting amplifier shown in Fig. 2.3, the input impedance is computed as follows:

$$Z_I = R_I$$
$$= 2.7 \text{ k}\Omega$$

In general, as long as the (−) pin remains at a virtual ground potential, the input impedance will be equal to the impedance between this pin and the source. If the impedance is more complex (e.g., resistor capacitor combination), then you must use complex numbers to represent the impedance. The basic method, however, remains the same.

Input current requirement. Ohm's Law can be used to calculate the amount of current that must be supplied by the source. Recall that essentially no current flows into or out of the (−) terminal of the op amp. Therefore, the only current supplied by the source is that drawn by R_I. Since R_I is effectively in parallel with the source caused by the effect of the virtual ground, the input current can be computed as follows:

$$\boxed{\text{input current} = i_I = \frac{v_I}{R_I}} \quad (2.8)$$

$$i_I = \frac{v_I}{R_I}$$
$$= \frac{500 \times 10^{-3}}{2.7 \times 10^3}$$
$$= 185 \text{ }\mu\text{A peak}$$

Sec. 2.2   Inverting Amplifier

Since this is a sinusoidal waveform, we could easily convert this to an RMS value if desired as shown.

$$i_1 \text{ (RMS)} = i_i \text{ (peak)} \times 0.7071 \qquad (2.9)$$

In our present case, the RMS value is found as follows:

$$i_1 \text{ (RMS)} = i_1 \text{ (peak)} \times 0.7071$$
$$= 185 \times 10^{-6} \times 0.7071$$
$$= 130.8 \ \mu A$$

As long as the input source can supply at least this much current without reducing its output, the op amp circuit will not load the source.

Maximum output voltage swing. The output voltage of an op amp is limited by the positive and negative saturation voltages. These can both be approximated as 2 volts less than the DC supply voltage. Since the DC supply in Fig. 2.3 is ± 15 volts, the saturation voltages will be $+13$ volts and -13 volts for the positive and negative limits, respectively. Thus, the maximum output voltage swing is computed as follows:

$$v_O \text{ (max)} = (+V_{SAT}) - (-V_{SAT}) \qquad (2.10)$$

For the circuit in Fig. 2.3, the maximum output voltage swing is found as shown.

$$v_O \text{ (max)} = (+V_{SAT}) - (-V_{SAT})$$
$$= (+13) - (-13)$$
$$= 26 \text{ V}$$

Since both DC supplies are equal, the output can swing equally above and below zero. This is the normal condition.

If you desire to be more accurate in the estimation of output saturation voltage, you may refer to the manufacture's data sheet in App. 1. The manufacturer lists minimum and typical output voltage swings for different values of load resistance.

Slew-rate limiting frequency. It should also be noted that this maximum output is only obtainable for frequencies below the point where slew-rate limiting occurs. This frequency can be estimated with the following equation:

$$f_{SRL} = \frac{\text{slew rate}}{\pi v_O(\text{max})} \qquad (2.11)$$

In the case of the circuit being considered, the highest frequency that can produce a full output swing without distortion caused by slew-rate limiting is computed as

$$f_{SRL} = \frac{\text{slew rate}}{\pi v_O (\text{max})}$$

$$= \frac{0.5 \text{ V}/\mu\text{s}}{26\pi}$$

$$= 6.12 \text{ kHz}$$

If we attempt to amplify frequencies higher than 6.12 kilohertz (and full amplitude) with the circuit shown in Fig. 2.3, then the output will be nonsinusoidal. Once the input frequency goes higher than a certain frequency (about 9 kHz in this case), then the output amplitude begins to drop in addition to the distorted shape.

Maximum input voltage swing. We have computed the voltage gain of the circuit, and we know the maximum output voltage swing. We, therefore, have enough information to compute the largest input signal that can be applied without driving the amplifier into saturation.

$$\boxed{v_I(\text{max}) = \frac{v_O(\text{max})}{A_V}} \quad (2.12)$$

Calculations for the present case are

$$v_I(\text{max}) = \frac{v_O(\text{max})}{A_V}$$

$$= \frac{26}{12.2}$$

$$= 2.13 \text{ volts peak-to-peak}$$

Since we are working with sinusoidal waveforms, we might choose to express this value as peak or RMS as shown.

$$\boxed{v_I(\text{peak}) = \frac{v_I(\text{max})}{2}} \quad (2.13)$$

For our present circuit, we have

$$v_I(\text{peak}) = \frac{v_I(\text{max})}{2}$$

$$= \frac{2.13}{2}$$

$$= 1.07 \text{ volts peak}$$

Sec. 2.2   Inverting Amplifier

Also,

$$\boxed{v_I(\text{RMS}) = v_I(\text{peak}) \times 0.707} \qquad (2.14)$$

In the present case,

$$v_I(\text{RMS}) = v_I(\text{peak}) \times 0.707$$
$$= 1.07 \times 0.707$$
$$= 756 \text{ mV RMS}$$

For the amplifier circuit presented in Fig. 2.3, input signals as great as 756 millivolts RMS can be amplified without saturation clipping. If you attempt to amplify larger signals, then the peaks on the output waveform will be flattened at the output saturation voltage limits.

Output impedance. Recall from Chap. 1 that the output impedance of an op amp is generally quite low. The data sheet in App. 1 lists 75 ohms as a typical output resistance for a 741 op amp. This value, however, is the open-loop output resistance. When negative feedback is added to the amplifier (as in Fig. 2.3), the effective output impedance decreases sharply. The value of effective output impedance can be approximated as shown.

$$\boxed{r_o = \frac{(R_I + R_F) \times \text{output impedance (open loop)}}{A_{OL} R_I}} \qquad (2.15)$$

where A_{OL} is the open-loop gain of the op amp at the specified frequency. This can be read from the manufacturer's graphical data (see App. 1) showing open-loop gain as a function of frequency. Alternately, you may compute it as

$$\boxed{A_{OL} = \frac{\text{unity gain frequency}}{f_{IN}}} \qquad (2.16)$$

where f_{IN} is the specific input frequency being considered.

For the circuit in Fig. 2.3, the closed-loop output impedance can be estimated at 1000 hertz as follows. First we compute the open-loop gain at 1000 hertz by applying Eq. (2.16).

$$A_{OL} = \frac{\text{unity gain frequency}}{f_{IN}}$$
$$= \frac{10^6}{10^3}$$
$$= 1000$$

Next compute the effective output impedance with Eq. (2.15).

$$r_O = \frac{(R_I + R_F) \times \text{output impedance (open loop)}}{A_{OL}R_I}$$

$$= \frac{(2700 + 33000) \times 75}{1000 \times 2700}$$

$$= 0.99 \text{ ohms}$$

This low value approaches our ideal value of zero ohms. Now, recompute the value of output impedance at a higher frequency of 5 kilohertz.

First compute A_{OL} with Eq. (2.16).

$$A_{OL} = \frac{\text{unity gain frequency}}{f_{IN}}$$

$$= \frac{1 \times 10^6}{5 \times 10^3}$$

$$= 200$$

Next compute the effective output impedance with Eq. (2.15).

$$r_O = \frac{(R_I + R_F) \times \text{output impedance (open loop)}}{A_{OL}R_I}$$

$$= \frac{(2.7 \times 10^3 + 33 \times 10^3) \times 75}{200 \times 2.7 \times 10^3}$$

$$= 4.96 \text{ ohms}$$

This value is significantly higher than our first estimate and clearly shows the increase in output resistance as the input frequency is increased.

How does a particular value of output impedance affect the performance of the amplifier circuit? To understand the effects, examine the equivalent circuit shown in Fig. 2.4. Here we see a voltage source labeled v_O driving a series circuit.

The v_O source is that voltage that would be present at the output of the op amp if the output impedance were truly zero ohms. You can see that this ideal voltage (v_O) is divided between the output impedance (r_O), which is internal to the op amp, and R_L which is the op amp load. The voltage reaching the load can be computed with the voltage divider equation.

$$\text{load voltage} = v_{\text{load}} = \frac{v_O R_L}{R_L + r_O} \quad (2.17)$$

Let us compute the actual load voltage in Fig. 2.3 at a frequency of 5 kilohertz. First we compute the ideal output voltage, Eq. (2.1).

Sec. 2.2   Inverting Amplifier

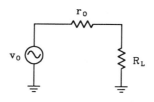

Figure 2.4 The equivalent output circuit of an op amp can be used to judge the effects of output impedance (r_O).

$$v_O = v_I A_V$$
$$= 500 \times 10^{-3} \times 12.2$$
$$= 6.1 \text{ volts peak}$$

We have found the value of r_O at 5 kilohertz to be 4.96 ohms. Using the method shown in Fig. 2.4, we can now determine the actual load voltage with Eq. (2.17).

$$V_{load} = \frac{v_O R_L}{R_L + r_O}$$
$$= \frac{6.1 \times 47 \times 10^3}{47 \times 10^3 + 4.96}$$
$$= 6.099 \text{ volts peak}$$

At a frequency of 5 kilohertz when the output resistance has increased to nearly 5 ohms, the effect of nonideal output resistance is minimal. Problems could be anticipated when the output resistance exceeds 1 percent of the value of load resistance.

Although the preceding calculation illustrates the effects of output resistance, it is valid only if we are below the frequency that causes slew-rate limiting (f_{SRL}). If f_{SRL} is exceeded, we can expect the actual output to be much lower than the value computed with Eq. (2.17), and the output will be nonsinusoidal in shape. Additionally, this method is inappropriate if the output drive capability of the op amp is exceeded.

Output current capability.   The output of the op amp in Fig. 2.3 must supply two currents: the current through the feedback resistor (i_F) and the current to the load resistor (i_L). It is the sum of these currents that flows into or out of the output of the op amp.

The output of many (but not all) op amps is short circuit protected. That is, the output may be shorted directly to ground or to either DC supply voltage without damaging the op amp. For a protected op amp (such as the 741), the output current capability is not determined by the maximum allowable current before damage, but rather depends on the amount of reduced output voltage the application can tolerate.

With no output current being supplied to the load, the output voltage stays at the expected v_O level, and the total output current is equal to i_F. As the load current is increased (load resistance decreased), the actual output voltage begins to drop as shown in the previous section. Finally, if the load resistance is reduced all the way to zero ohms, the output current will be limited to a safe value. This value can be found in the data sheet (App. 1), and is 20 milliamps for the 741 device.

As the load resistance varies from infinity (open) to zero (short), the output current from the op amp varies from i_F to 20 milliamps. The limiting factor is the amount of reduction that can be tolerated on the output voltage.

The amount of current (i_F) flowing through the feedback resistor is easily computed with Ohm's Law as

$$\text{feedback current} = i_F = \frac{v_L}{R_F} \qquad (2.18)$$

On an unprotected op amp, the value of load current plus the value of feedback current must be kept below the stated output current rating. If this value is not supplied in the data sheet, then it can be estimated by using the maximum power dissipation data. Recall that *power = voltage × current.*

Minimum value of load resistance. The minimum value of load resistance is determined by the maximum value of output current (determined in the previous section). The actual computation is essentially Ohm's Law.

$$\text{minimum load resistance} = R_L(\text{min}) = \frac{v_L}{i_L} \qquad (2.19)$$

where i_L is the maximum allowable output current of the op amp minus the current (i_f) flowing through the feedback circuit, and v_L is the minimum acceptable output voltage.

Note that in many, if not most, applications, the value of output current needed for the load is substantially below the limiting value so no significant loading occurs.

Let us assume that the application shown in Fig. 2.3 requires us to have at least 1.19 volts across the load when 100 millivolts is applied to the input terminal. Let us assume that the frequency of interest is 5 kilohertz. From previous calculations we know the voltage gain (A_V) is 12.2 (ignoring the effects of bandwidth described in the next section) and the output resistance at 5 kilohertz is 4.96 ohms.

Figure 2.5 shows the equivalent circuit at this point. The value of i_O can be computed with Ohm's Law.

$$i_O = \frac{v_O - v_L}{r_O} \qquad (2.20)$$

More specifically,

$$i_O = \frac{v_O - v_L}{r_O}$$

Sec. 2.2 Inverting Amplifier

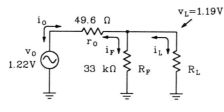

Figure 2.5 An equivalent circuit used to compute the minimum allowable load resistor.

$$= \frac{1.22 - 1.19}{4.96}$$

$$= 6.05 \text{ milliamps}$$

The value of i_F can also be computed using Ohm's Law Eq. (2.18).

$$i_F = \frac{v_L}{R_F}$$

$$= \frac{1.19}{33000}$$

$$= 36.1 \text{ μA}$$

Kirchoff's Current Law can now be used to determine the value of load current (i_L).

$$\boxed{i_L = i_O - i_F} \qquad (2.21)$$

Calculations for the present example are

$$i_L = i_O - i_F$$
$$i_L = 6.05 \text{ mA} - 36.1 \text{ μA}$$
$$= 6.01 \text{ mA}$$

Using Ohm's Law, Eq. (2.19), we can now compute the value of R_L.

$$R_L = \frac{v_L}{i_L}$$

$$= \frac{1.19}{6.01 \times 10^{-3}}$$

$$= 198 \Omega$$

With this small value of load resistance, we would not be able to provide full range voltage swings on the output because of excessive loading.

In the foregoing calculations (as with most calculations presented in this text), it is not important to remember all of the equations. Rather, strive to understand the concept and realize that most of what we are discussing is centered on basic electronics principles that you learned when you studied introductory AC and DC circuits.

Bandwidth. Although the bandwidth of an ideal op amp was considered to be infinite, the bandwidth of real op amps and the associate amplifier circuit are definitely restricted. In the case of the circuit shown in Fig. 2.3, the lower cutoff frequency is essentially zero. That is, since the op amp responds all the way down to DC and since there are no reactive components to reject the lower frequencies, the amplifier circuit will operate with frequencies as low as DC.

The upper cutoff frequency is quite a different story. Figure 2.6 shows the open-loop frequency response (upper curve) for a 741 op amp. This is the same curve presented in the manufacturer's data sheet as open-loop voltage gain as a function of frequency. Also, drawn on the graph in Fig. 2.6 is a line showing a voltage gain of 12.2. This is the ideal closed-loop gain that we calculated for the circuit in Figure 2.3.

Notice that the difference between the open- and closed-loop gain curves is maximum at low frequencies. As the frequency increases, the difference between the two curves becomes less. Near the right side of the graph, the two curves actually intersect. What really happens to the overall circuit gain as the frequency increases?

The derivation of the formula for amplifier voltage gain ($A_V = -R_F/R_I$) was based on the assumption that the op amp had an infinite (or at least a very high) voltage gain. This allowed us to make the assumption that the differential input voltage (v_D) was zero. As you can see from the graph in Fig. 2.6, our assumptions are reasonable for low frequencies. That is, the open-loop voltage gain is very high. But as the frequency increases and the open-loop gain rolls off, our assumptions begin to lose their validity. The most obvious proof of this exists beyond the point of intersection of the open- and closed-loop curves. In the region to the right of the intersection point, the open-loop gain is actually lower than our calculated closed-loop gain thus making it impossible for our circuit to deliver the desired amplification.

It is common to compute bandwidth in a circuit like that shown in Fig. 2.3 by applying the following equation:

$$\mathrm{bw} = \frac{f_{UG} R_I}{R_I + R_F} \quad (2.22)$$

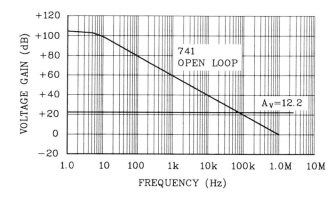

Figure 2.6 Frequency response of the standard 741 op amp.

Sec. 2.2 Inverting Amplifier

where f_{UG} is the unity gain frequency of the op amp. Substituting values and computing gives us the following:

$$bw = \frac{1 \times 10^6 \times 2.7 \times 10^3}{2.7 \times 10^3 + 33 \times 10^3}$$

$$= 75.6 \text{ khz}$$

The actual frequency response for the circuit shown in Fig. 2.3 is plotted in Fig. 2.7. This represents the circuit's real behavior. Two additional lines are superimposed on the plot for reference: the open-loop frequency response curve of the 741 op amp and the ideal gain curve of the circuit in Fig. 2.3.

Power supply rejection ratio. If the DC supply lines (V^+ and V^-) have noise, particularly high-frequency noise, these noise signals may affect the output signal. The degree to which the op amp is affected by the power supply noise is called the power supply rejection ratio. The manufacturer's data sheet normally expresses this parameter in microvolts per volt. To determine the magnitude of the noise signal on the output for a given amplitude of noise signal on the supply lines, we can use the following calculation:

$$v_{NO} = PSRR \; v_N \left(\frac{R_F}{R_i} + 1 \right) \quad (2.23)$$

where v_{NO}, v_N, R_F, R_I, and PSRR are the values of the output noise signal, the noise signal on the DC supply lines, the feedback resistor, the input resistor, and the power supply rejection ratio, respectively. For example, refer to Fig. 2.8.

The manufacturer's data sheet in App. 1 for a 741 op amp lists the power supply rejection ratio as ranging from 30 to 150 microvolts per volt. Thus, the worst-case effect on the output voltage for the circuit in Fig. 2.8 is computed with Eq. (2.23) as

$$v_{NO} = PSRR \; v_N \left(\frac{R_F}{R_i} + 1 \right)$$

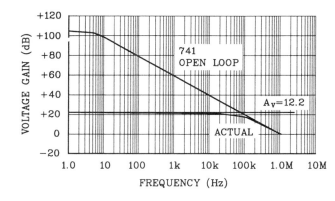

Figure 2.7 Actual frequency response of the circuit shown in Fig. 2.3.

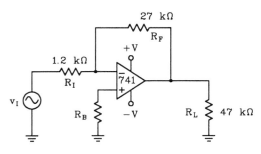

Figure 2.8 An inverting amplifier circuit used to demonstrate the effects of the power supply rejection ratio.

$$= 150 \times 10^{-6} \, v_N \left(\frac{27 \times 10^3}{1.2 \times 10^3} + 1 \right)$$

$$= v_N \times 0.003525$$

In other words, the amplitude of the power line noise (v_N) will be reduced by a factor of 0.003525. This means, for example, that if the DC supply lines had noise signals of 100 millivolts peak to peak, then we could anticipate a similar signal in the output with an amplitude of about

$$v_{NO} = v_N \times 0.003525$$

$$= 100 \times 10^{-3} \times 0.003525$$

$$= 352.5 \, \mu V$$

2.2.3 Practical Design Techniques

The following design procedures will enable you to design inverting op amp circuits for many applications. Although certain nonideal considerations are included in the design method, additional nonideal characteristics are described in Chap. 10.

To begin the design process, you must determine the following requirements based on the intended application:

1. voltage gain
2. maximum input current
3. frequency range
4. load resistance
5. maximum input voltage

As an example of the design process, design an inverting amplifier with the following characteristics:

1. voltage gain   12
2. maximum input current   250 microamps RMS

Sec. 2.2   Inverting Amplifier

3. frequency range   20 hertz to 2.5 kilohertz
4. load resistance   100 kilohms
5. maximum input voltage   500 millivolts RMS, zero-volt reference

Determine an initial value for R_I. The minimum value for R_I is determined by the maximum input voltage and the maximum input current and is computed with Ohm's Law as follows:

$$\boxed{R_I = \frac{v_I}{i_I}} \quad (2.24)$$

In this case, the calculations are

$$R_I = \frac{v_I}{i_I}$$
$$= \frac{500 \times 10^{-3}}{250 \times 10^{-6}}$$
$$= 2000 \, \Omega$$

As a general rule, you should avoid designing amplifiers with input resistances of less than 1000 ohms unless you have a specific need for it. In our present case, the computed minimum (2.0 kΩ) is greater than 1000 ohms, therefore, use the computed value.

Determine the value of R_F. R_F can be computed from the voltage gain equation, Eq. (2.6).

$$A_V = -\frac{R_F}{R_I}$$

or

$$R_F = A_V R_I$$

Note that the inversion sign has been omitted from the equation when computing a resistance value. For the present example, we compute R_F as follows:

$$R_F = A_V R_I$$
$$= 12 \times 2000$$
$$= 24 \, k\Omega$$

Determine the required unity gain frequency. The minimum unity gain frequency for the op amp can be estimated by applying Eq. (2.22). For our present case, we have

$$f_{UG} = \frac{bw(R_F + R_I)}{R_I}$$

$$= \frac{(2.5 \times 10^3)(24 \times 10^3 + 2 \times 10^3)}{2 \times 10^3}$$

$$= 32.5 \text{ kHz}$$

Since this is well below the 1.0 megahertz unity gain frequency of the 741, we should be able to use the 741 in this application (with regard to bandwidth).

Determine the minimum supply voltages. The minimum supply voltages are computed by simply insuring that the maximum expected output voltage swing is no greater than the $\pm V_{SAT}$ values. The maximum output swing can be found by using the basic equation, Eq. (2.1), for voltage gain.

$$A_V = \frac{v_{OUT}}{v_{IN}}$$

or

$$v_{OUT} = v_{IN} A_V$$

In our particular example, the maximum output voltage will be

$$v_{OUT} = v_{IN} A_V$$
$$= 500 \times 10^{-3} \times 12 \times 1.414$$
$$= 8.48 \text{ volts peak}$$

or

$$v_{OUT} = v_o(\max)$$
$$= 8.48 \times 2$$
$$= 16.96 \ v_{P-P}$$

Notice the multiplying factor 1.414 to convert our input voltage (given in RMS) to a peak or worst-case value. The manufacturer's data sheet in App. 1 indicates that the 741 op amp will produce at least a ± 12 volt output swing with a ± 15 volt supply voltage as long as the load resistor is at least 10 kilohms. Thus, we can infer that we have a worst-case internal voltage drop of 15 to 12 or 3 volts. This means that the minimum power supply voltage for our circuit must be higher than the maximum output voltage by the amount of the internal voltage drop (V_{INT}). That is

$$\boxed{\pm V_{MIN} = V_{OUT}(\max) + V_{INT}} \quad (2.25)$$

For our particular case, the minimum power supply voltages will be

Sec. 2.2 Inverting Amplifier

$$\pm V_{MIN} = V_{OUT}(max) + V_{INT}$$
$$= 8.48 + 3$$
$$= 11.48 \text{ V}$$

Anything greater than ± 11.48 volts for the DC supply will be adequate, therefore, choose the standard values of ± 15 volts for our application.

Determine the required slew rate. The required slew rate of the op amp is affected by the highest operating frequency and the maximum output voltage swing. In our present case, the highest input frequency has been specified as 2.5 kilohertz. The maximum peak-to-peak output voltage swing [$v_O(max)$] was previously computed as 16.96 volts. The minimum required slew rate for the op amp is determined by rearranging Eq. (2.11) to yield

$$\text{slew rate(min)} = \pi f(max) \, v_O(max)$$
$$= \pi \times 2.5 \times 10^3 \times 16.96$$
$$= 0.133 \text{ V}/\mu s$$

Since the slew rate of the 741 exceeds this minimum value, we can continue with our initial op amp selection. If this calculation indicates a higher requirement than our preliminary op amp selection can deliver, then another op amp must be selected which has a higher slew rate.

Calculate the value of compensation resistor (R_B). The compensating resistor (R_B) reduces the error in the output voltage caused by the voltage drops resulting from the op amp's input bias currents. To achieve maximum error reduction, we try to place equal resistances between both op amp input terminals and ground. If we were to apply Thevenin's Theorem to the inverting input circuit, we would see that resistors R_F and R_I are effectively in parallel. This means that the optimum value for R_B is simply the combined value of R_F and R_I in parallel.

$$\boxed{R_B = \frac{R_F R_I}{R_F + R_I}} \qquad (2.26)$$

For the present example, we compute R_B as follows:

$$R_B = \frac{R_F R_I}{R_F + R_I}$$
$$= \frac{24 \times 10^3 \times 2 \times 10^3}{24 \times 10^3 + 2 \times 10^3}$$
$$= 1.8 \text{ k}\Omega$$

The final schematic is shown in Fig. 2.9.

The actual behavior of the circuit is indicated in Fig. 2.10 by an oscilloscope display. The measured performance is compared to the design goals in Table 2.1.

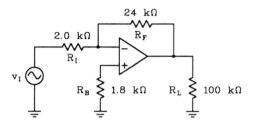

Figure 2.9 An inverting amplifier design.

TABLE 2.1

PARAMETER	DESIGN GOAL	MEASURED VALUE
Voltage gain	12	11.7–12
Frequency range	20 Hz–2.5 kHz	<20 Hz–>2.5 kHz

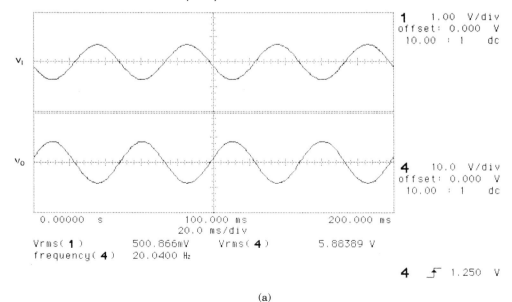

(a)

Figure 2.10 Oscilloscope displays showing the performance of the inverting amplifier shown in Fig. 2.9. (Test equipment courtesy of Hewlett-Packard Company)

Sec. 2.3 Noninverting Amplifier

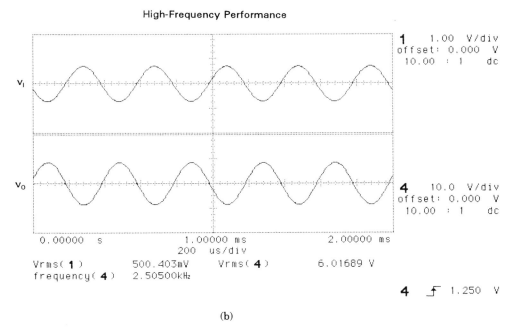

(b)

Figure 2.10 (continued)

2.3 NONINVERTING AMPLIFIER

2.3.1 Operation

Figure 2.11 shows the schematic diagram of a basic noninverting amplifier. As you might expect, the input signal is applied to the (+) or noninverting input. Resistor R_B is a compensation resistor similar to that described for the inverting amplifier. Since it has such a tiny current through it, ignore its effects for the immediate discussion.

Resistor R_F and resistor R_I form a voltage divider between the output terminal and ground. That portion of the output that appears across R_I will provide the input to the (−) input terminal. The input signal (v_I) supplies the voltage to the (+) input terminal. The difference between these to voltages (v_D) is amplified by the open-loop gain of the op amp. Recall that as long as the output of the op amp is in the linear range (i.e., not saturated), then the magnitude of v_D will be very near to zero volts. Since the (+) input terminal is equal to v_I, and since v_D is approximately zero, we can conclude that the voltage on the (−) input terminal must also be approximately equal to v_I. Recall that the source for the (−) input voltage is the output of the op amp. Now we can see that the output will go as high as necessary in order to develop enough voltage drop across R_I to equal v_I.

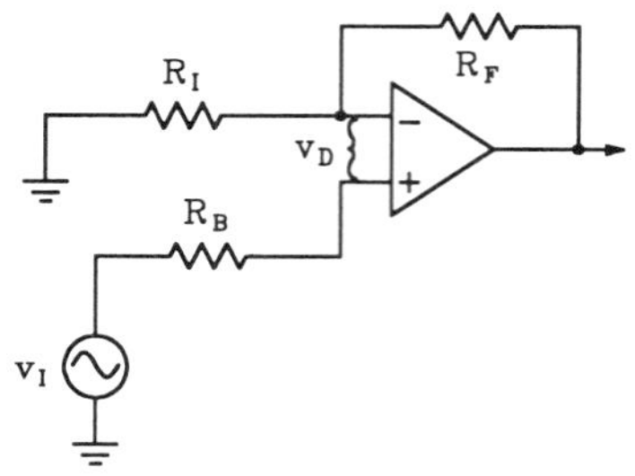

Figure 2.11 The basic noninverting amplifier circuit.

Suppose, for example, that the input voltage (v_I) made a sudden increase from zero volts to some positive level. At this first instant, the (+) input of the op amp would be positive and the (−) input would still be at its previous zero-volt level. The voltage v_D would now be amplified. Since the (+) input is more positive, the output would rise as quickly as possible in the positive direction. As the output goes positive, a portion is fed back through the R_F and R_I voltage divider to the (−) input. Since the (−) input is becoming more positive, the value of v_D is decreasing. That is, the two input terminal voltages are getting closer together. Finally, the output of the amplifier will stop going in the positive direction whenever the (−) input has come to within a few microvolts of the (+) input.

Now consider how high the output voltage had to go in order to bring the (−) input up to the same voltage as the (+) input. Can you see that it is strictly the values of the voltage divider R_F and R_I that determine the amount of output voltage change required? Thus, for a given input voltage change, the output will make a corresponding change. The magnitude of the change is the gain of the amplifier and is largely determined by the ratio of R_F to R_I. This action is explained with mathematics in the following section.

Additionally, note that as the input went positive, the output went positive. That is, the amplifier configuration is noninverting.

2.3.2 Numerical Analysis

Much of our analysis for the inverting amplifier is applicable to the noninverting amplifier circuit. Determine a method to enable us to compute the following circuit characteristics:

1. voltage gain
2. input impedance
3. input current requirement
4. maximum output voltage swing
5. slew-rate limiting frequency
6. maximum input voltage swing
7. output impedance
8. output current capability

Sec. 2.3 Noninverting Amplifier

9. bandwidth
10. power supply rejection ratio

For purposes of this discussion, let us analyze the noninverting amplifier circuit shown in Fig. 2.12.

Voltage gain. We know by inspection of the circuit in Fig. 2.12 that the voltage on the (+) input is approximately equal to v_I. That is, there is no significant voltage drop across R_B because the only current allowed to flow through R_B is the op amp bias current (ideally zero). We also know from previous discussions that the voltage between the (+) and (−) input terminals (v_D) is very near to zero volts. Thus, we may rightly conclude that the voltage on the (−) pin is approximately equal to the value of v_I.

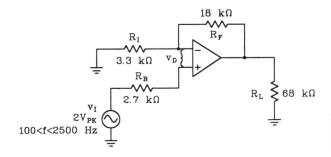

Figure 2.12 A noninverting amplifier circuit used for a numerical analysis example.

Ohm's Law can be used to compute the current through R_I as follows:

$$\boxed{i_{R_I} = \frac{v_I}{R_I}} \qquad (2.27)$$

For the circuit in Fig. 2.12, we have

$$i_{R_I} = \frac{v_I}{R_I}$$

$$= \frac{2}{3.3 \times 10^3}$$

$$= 606.1 \ \mu A \text{ peak}$$

Since negligible current flows into or out of the input of the op amp, we shall assume that all of the current flowing through R_I continues through R_F according to Kirchoff's Current Law. The voltage drop across R_F can be computed by applying Ohm's Law.

$$v_{R_F} = i_{R_F} \times R_F$$

$$= 606.1 \times 10^{-6} \times 18 \times 10^3$$

$$= 10.91 \text{ volts peak}$$

The output voltage can be determined through application of Kirchoff's Voltage Law. That is, we know the voltage on the ($-$) input is 2.0 volts peak. The output voltage will be greater than this by the amount of voltage drop across R_F. The output voltage is computed as

$$v_O = v_I + v_{R_F}$$
$$= 2 + 10.91$$
$$= 12.91 \text{ volts peak}$$

The voltage gain can be computed by the basic gain equation, Eq. (2.1) as shown

$$A_V = \frac{v_O}{v_I}$$
$$= \frac{12.91}{2}$$
$$= 6.45$$

Recall from an earlier discussion that the voltage gain of the circuit is largely determined by the ratio of R_F to R_I. Specifically, the low frequency or ideal voltage gain of the circuit can also be calculated with the following equation:

$$\boxed{A_V = \frac{R_F}{R_I} + 1} \tag{2.28}$$

In our case, the calculations are

$$A_V = \frac{R_F}{R_I} + 1$$
$$= \frac{18 \times 10^3}{3.3 \times 10^3} + 1$$
$$= 6.45$$

This latter method is the most common, but the former provides additional insight into circuit operation and the application of basic electronics principles.

The voltage can be expressed in decibels if desired as we did with inverting amplifiers. In our present example, the equivalent voltage gain expressed in decibels would be

$$A_V(\text{dB}) = 20 \log_{10} A_V$$
$$= 20 \log_{10} 6.45$$
$$= 16.2 \text{ dB}$$

Sec. 2.3   Noninverting Amplifier

It is important to note that the voltage gain computed in this section is the ideal closed-loop voltage gain of the circuit. The actual circuit gain will roll off as the input frequency is increased just as it did with inverting amplifiers. This effect is discussed as part of the discussion on bandwidth.

Input impedance. The input impedance of the noninverting amplifier circuit (refer to Fig. 2.12), is essentially equal to the input impedance of the (+) input terminal of op amp modified by the feedback effects. That is, the only current leaving the source must flow into or out of the op amp as bias current for the (+) input. The manufacturer's data sheet for a 741 is shown in App. 1. It indicates that the input resistance is at least 0.3 megohms and is typically about 2.0 megohms. Recall that this is the effective resistance between the two op amp inputs. By considering the output impedance to be near zero, we can sketch the equivalent circuit shown in Fig. 2.13(a).

Let us make the following substitution for the value of v_O:

$$v_O = v_I \left(\frac{R_F}{R_I} + 1 \right)$$

This of course comes from Eqs. (2.1) and (2.28). If we now apply Thevenin's Theorem to the portion of the circuit to the right of the dotted line, we will obtain the equivalent circuit shown in Fig. 2.13(b). Notice that the resistance in our equivalent circuit has the same voltage (2 V_{PK}) on both ends which produces a net voltage of zero. If there is no voltage, there will be no current so the effective input impedance is infinite. This represents the ideal condition.

In a real op amp circuit, the differential input voltage (v_D) is greater than zero and increases as the frequency increases. With reference to Fig. 2.13(b), as the input frequency increases, the two voltage sources become more and more unequal. This causes a difference in potential across the resistance in the circuit which in turn produces a current flow. This increasing current corresponds to a decreasing input impedance. Although the

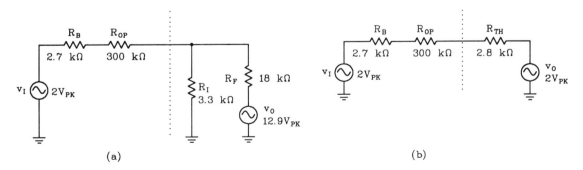

Figure 2.13 An equivalent circuit used to estimate the input impedance of the noninverting amplifier shown in Fig. 2.12.

actual input impedance is quite high and can normally be assumed to be infinite, it can be approximated by the following equation:

$$\text{input impedance} = Z_I = R_{OP} A_V \left(\frac{R_I}{R_F + R_I} \right) \quad (2.29)$$

where R_{OP} is the value of input resistance provided by the manufacturer, and A_V is the open-loop voltage gain of the op amp. For the circuit shown in Fig. 2.12, we can estimate input resistance at low frequencies as

$$Z_I = R_{OP} A_V \left(\frac{R_I}{R_F + R_I} \right)$$

$$= 300 \times 10^3 \times 200 \times 10^3 \left(\frac{3.3 \times 10^3}{18 \times 10^3 + 3.3 \times 10^3} \right)$$

$$= 9296 \text{ M}\Omega$$

If we had used the more typical value of 2.0 megohms for the op amps resistance (R_{OP}), we would have gotten a much higher value for input resistance. In either case, the actual effective input resistance is extremely high. This high input resistance is one of the primary advantages of the noninverting amplifier in many applications.

Input current requirement. The input current can be estimated by applying Ohm's Law to the input circuit as follows:

$$i_{IN} = \frac{v_I}{Z_I}$$

$$= \frac{2}{9.296 \times 10^9}$$

$$= 215 \text{ picoamperes peak}$$

Even this is a worst-case value. If we used the higher typical value for input resistance, we would have computed an even smaller value. For many, if not most, applications, this input current can be considered negligible. If it becomes necessary to consider this current, then additional considerations must be made since the exact value of input resistance varies considerably with temperature and frequency.

Maximum output voltage swing. As we found with the inverting amplifier, the output voltage of an op amp is limited by the $\pm V_{SAT}$ levels. For most applications utilizing a bipolar op amp, the saturation voltages can be estimated at about 2 volts less than the DC supply voltage. In the case of Fig. 2.12, we compute the maximum output swing, Eq. (2.10), as

$$v_O(\text{max}) = +V_{SAT} - (-V_{SAT})$$

Sec. 2.3 Noninverting Amplifier

$$= +13 - (-13)$$
$$= 26 \text{ V}$$

If a more accurate value is desired, the manufacturer's data sheet can be used to find a more precise value for the worst-case saturation voltage.

Slew-rate limiting frequency. The highest frequency that can be amplified without distorting the waveform because of the slew rate limitation of the op amp is given by Eq. (2.11).

$$f_{SRL} = \frac{\text{slew rate}}{\pi v_o(\text{max})}$$

$$= \frac{0.5/10^{-6}}{3.14 \times 26}$$

$$= 6.12 \text{ kHz}$$

If it is known for certain that the actual output swing will never be required to reach its limits, then the lower actual output swing can be used in place of $v_o(\text{max})$ in this calculation.

Maximum input voltage swing. The maximum input voltage swing is simply the highest input voltage that can be applied without driving the output past the saturation point. It is computed in the same manner as we did for the inverting amplifier.

$$v_I(\text{max}) = \frac{v_O(\text{max})}{A_V}$$

$$= \frac{26}{6.45}$$

$$= 4.03 \text{ volts peak-to-peak}$$

Since we are working with sinusoidal waveforms, we might choose to express this value as peak, Eq. (2.13), or RMS, Eq. (2.14), as shown

$$v_I(\text{peak}) = \frac{v_O(\text{max})}{2}$$

$$= \frac{4.03}{2}$$

$$= 2.015 \text{ volts peak}$$

also,

$$v_I(\text{RMS}) = v_I(\text{peak}) \times 0.707$$

$$= 2.015 \times 0.707$$
$$= 1.425 \text{ volts RMS}$$

If you attempt to amplify signals larger than 1.425 volts RMS, then the peaks on the output waveform will be flattened at the output saturation voltage limits.

Output impedance. You will recall from the analysis of the inverting amplifier that the effective output impedance decreases sharply from the open-loop value stated in the manufacturer's data sheets. The value of effective output impedance can be approximated as shown, Eq. (2.15)

$$r_O = \frac{(R_I + R_F) \times \text{output impedance (open loop)}}{A_{OL} R_I}$$

where A_{OL} is the open-loop gain of the op amp at a particular frequency. For the circuit in Fig. 2.12, the open-loop gain at 2500 hertz is computed with Eq. (2.16) as

$$A_{OL} = \frac{10^6}{2500}$$
$$= 400$$

The output impedance at 2500 hertz can then be estimated as

$$r_O = \frac{(R_I + R_F) \times \text{output impedance (open loop)}}{A_{OL} R_I}$$
$$= \frac{(3300 + 18000) \times 75}{400 \times 3300}$$
$$= 1.2 \text{ }\Omega$$

If we are concerned about the effect of r_O on the output on the output voltage, we can apply the voltage divider equation (developed in Sec. 2.2.2)

$$\text{load voltage} = v_{LOAD} = \frac{v_O R_L}{R_L + r_O}$$

Let us compute the actual load voltage in Fig. 2.12 at a frequency of 2500 hertz. First we compute, Eq. (2.1), the ideal output voltage as

$$v_O = v_I A_V$$
$$= 2.0 \times 6.45$$
$$= 12.9 \text{ volts peak}$$

We have already found the value of r_O at 2500 hertz to be 1.2 ohms. Using the voltage divider equation, Eq. (2.17), we can now determine the actual load voltage.

Sec. 2.3 Noninverting Amplifier

$$\text{load voltage} = v_{LOAD} = \frac{v_O R_L}{R_L + r_O}$$

$$= \frac{12.9 \times 68 \times 10^3}{68 \times 10^3 + 1.2}$$

$$= 12.899 \text{ volts peak}$$

The output impedance is so low relative to the value of load resistance that the output voltage is essentially unaffected. This is generally the case, but you can always be sure by performing this voltage divider calculation.

Output current capability. If the output of the op amp is short circuit protected (as in the 741), then the output current capability is limited by the maximum allowable drop in output voltage for the given application. This can be estimated with Ohm's Law as discussed in the preceeding section. Recall from our discussion of inverting amplifiers that the output must supply both load resistor current and the current through the feedback resistor. The feedback current is computed using Ohm's Law. For this particular circuit, the calculations are

$$i_F = \frac{v_O - v_I}{R_F}$$

$$= \frac{12.9 - 2}{18 \times 10^3}$$

$$= 605.6 \ \mu\text{A peak}$$

With no output current being supplied to the load, the output voltage stays at the expected v_O level, and the total output current is equal to i_F. As the load current is increased (load resistance decreased), the actual output voltage begins to drop due to the voltage divider action described in the previous section. Finally, if the load resistance is reduced all the way to zero ohms, the output current will be limited to the short circuit value. This value can be found in the data sheet (App. 1), and is 20 milliamps for the 741 device.

As the load resistance varies from infinity (open) to zero (short), the output current from the op amp varies from i_F to 20 milliamperes. The limiting factor is the amount of reduction that can be tolerated on the output voltage.

On an unprotected op amp, the value of load current plus the value of feedback current must be kept below the stated output current rating. If this value is not supplied in the data sheet, then it can be estimated by using the maximum power dissipation data; recall that *power = voltage × current*.

Bandwidth. The discussion of bandwidth presented for the inverting amplifer circuit is also applicable to the noninverting configuration. That is, as long as the circuit has no reactive components, the frequency response will extend all the way down to DC on the low-frequency end. We can estimate the high-frequency end of the frequency response by applying Eq. (2.22)

$$bw = \frac{f_{UG}R_I}{R_F + R_I}$$

$$= \frac{(1 \times 10^6)(3.3 \times 10^3)}{18 \times 10^3 + 3.3 \times 10^3}$$

$$= 155 \text{ kHz}$$

Recall that the open-loop gain of the op amp falls off rapidly as the input frequency is increased above a few hertz. As the open-loop gain value approaches the computed closed-loop gain value, the actual circuit gain also begins to drop. Thus, we begin to experience increased errors in our gain calculations as the frequency is increased.

For these equations to be valid, it is important that the op amp output voltage swing be small enough to avoid the effects of slew-rate limiting. The highest amplitude that can be amplified at a given frequency without experiencing the effects of slew-rate limiting is given as

$$\boxed{v_O(\text{max}) = \frac{\text{slew rate}}{\pi f}} \quad (2.30)$$

The slew rate is determined by the particular amplifier, f is the frequency of interest, and $v_O(\text{max})$ is the highest peak-to-peak amplitude in the output before slew-rate limiting begins to distort the signal. In the present case, if we were to try to operate at the upper cutoff frequency (155 kHz), then we would have to keep the output voltage below the value computed

$$v_O(\text{max}) = \frac{\text{slew rate}}{\pi f}$$

$$= \frac{0.5/10^{-6}}{3.14 \times 155 \times 10^3}$$

$$= 1.03 \text{ volts peak-to-peak}$$

Power supply rejection ratio. The power supply rejection ratio provides us with an indication of the degree of immunity the circuit has to noise voltages on the DC power lines. The change in output voltage (v_O) for a given change in DC power line noise voltage (v_N) is computed with Eq. (2.23)

$$v_{NO} = v_N \text{PSRR}\left(\frac{R_F}{R_I} + 1\right)$$

where v_O, v_N, R_F, R_I, and PSRR are the values of the output noise signal, the noise signal on the DC supply lines, the feedback resistor, the input resistor, and the power supply rejection ratio, respectively. The manufacturer's data sheet in App. 1 lists the power

Sec. 2.3   Noninverting Amplifier   69

supply rejection ratio (PSRR) as ranging from 30 to 150 microvolts per volt. The worst-case effect on the output voltage for the circuit in Fig. 2.12 then is

$$v_{NO} = v_N \text{PSRR}\left(\frac{R_F}{R_I} + 1\right)$$

$$= v_N \times 150 \times 10^{-6}\left(\frac{18 \times 10^3}{3.3 \times 10^3} + 1\right)$$

$$= v_N \times 0.000968$$

In other words, the amplitude of the power line noise (v_N) will be reduced by a factor of 0.000968. This means, for example, that if the DC supply lines had noise signals of 100 millivolts peak-to-peak, then we could anticipate a similar signal in the output with an amplitude of about

$$v_{NO} = v_N \times 0.000968$$

$$= 100 \times 10^{-3} \times 0.000968$$

$$= 96.8 \; \mu V$$

2.3.3 Practical Design Techniques

The following design procedures will enable you to design noninverting op amp circuits for many applications. Although certain nonideal considerations are included in the design method, additional nonideal characteristics are described in Chap. 10.

To begin the design process, the following requirements based on the intended application must be determined:

1. voltage gain
2. frequency range
3. load resistance
4. maximum input voltage

As an example of the design procedure, let us design a noninverting amplifier with the following characteristics:

1. voltage gain   8
2. frequency range   DC to 5 kilohertz
3. load resistance   27 kilohms
4. maximum input voltage   800 millivolts RMS

Determine an initial value for R_I.   There are endless combinations of R_F and R_I that will combine to produce the desired circuit voltage gain. The smaller the values of R_F and R_I the higher the value of feedback current. The feedback current subtracts from the maximum available output current. Thus, we want to avoid extremely small values.

The larger we make R_F and R_I the more the circuit operation is affected by certain nonideal characteristics. In general, neither resistor should be less than 1.0 kilohms nor more than 680 kilohms unless there is a compelling reason for doing so. With this rule of thumb in mind, we shall select R_I as 4.7 kilohms.

Determine the value of R_F. R_F can be computed from the voltage gain equation, Eq. (2.28)

$$A_V = \frac{R_F}{R_I} + 1$$

or

$$R_F = R_I(A_V - 1)$$

For the present design example we compute R_F as follows:

$$R_F = R_I(A_V - 1)$$
$$= 4700(8 - 1)$$
$$= 32.9 \text{ k}\Omega$$

We shall select the nearest standard value of 33 kilohms to use as R_F.

Determine the required unity gain frequency. You will recall from our discussions on bandwidth, that the error between the calculated or ideal gain and the actual gain increases as frequency increases. We can, however, estimate the required unity gain frequency by applying Eq. (2.22).

$$f_{UG} = \frac{bw(R_F + R_I)}{R_I}$$
$$= \frac{(5 \times 10^3)(33 \times 10^3 + 4.7 \times 10^3)}{4.7 \times 10^3}$$
$$= 40.1 \text{ kHz}$$

Thus, we must select an op amp that has minimum unity gain frequency of at least 40.1 kilohertz. Since the 741 has a 1.0 megahertz unit gain frequency, it should be adequate for this application with respect to bandwidth.

Determine the minimum supply voltages. The minimum supply voltages are computed by simply insuring that the maximum expected output voltage swing is no greater than the $\pm V_{SAT}$ values. The maximum output swing can be found by using the basic equation for voltage gain, Eq. (2.1).

Sec. 2.3 Noninverting Amplifier

$$A_V = \frac{v_{OUT}}{v_{IN}}$$

or

$$v_{OUT} = v_{IN} A_V$$

In our particular example, the maximum output voltage will be

$$v_{OUT} = v_{IN} A_V$$
$$= 800 \times 10^{-3} \times 1.414 \times 8$$
$$= 9.05 \text{ volts peak}$$

or

$$v_{OUT} = v_o(\text{max})$$
$$= 9.05 \times 2$$
$$= 18.1 \text{ volts peak-to-peak}$$

Notice the multiplying factor 1.414 to convert our input voltage (given in RMS) to a peak or worst-case value. The manufacturer's data sheet in App. 1 indicates that the 741 op amp will produce at least a ±12-volt output swing with a ±15-volt supply voltage and a load resistance of at least 10 kilohms. Thus, we can infer that we have a worst-case internal voltage drop of 15 to 12 or 3 volts. The minimum power supply voltage can be determined with Eq. (2.25)

$$\pm V_{MIN} = V_{OUT} + V_{INT}$$
$$= 9.05 + 3$$
$$= 12.05 \text{ V}$$

Anything greater than ±12.05 volts for the DC supply will be adequate so let us choose the standard values of ±15 volts for our application. Realize that this is a worst-case calculation, and a more typical internal drop would be 2 volts rather than 3 volts.

Determine the required slew rate. The minimum slew rate for the op amp is computed by transposing Eq. (2.11)

$$\text{slew rate(min)} = \pi f_{SRL} \, v_O(\text{max})$$
$$= 3.14 \times 5 \times 10^3 \times 18.1$$
$$= 0.284 \text{ V}/\mu s$$

Since the slew rate of the 741 exceeds this minimum value, we can continue with our initial op amp selection. If this calculation indicates a higher requirement than our preliminary op amp selection can deliver, then an another op amp must be selected which has a higher slew rate.

Calculate the value of compensation resistor (R_B). The compensating resistor (R_B) reduces the error in the output voltage caused by the voltage drops resulting from the op amp's input bias currents. As with the inverting configuration, we achieve maximum error reduction by inserting equal resistances between both op amp input terminals and ground. The resistance between the inverting input to ground is essentially equal to the parallel combination of R_I and R_F. This is easier to appreciate if you remember that the output impedance of an op amp is very low. For purposes of this analysis, assume that the output impedance is actually zero ohms. In this condition, one end of both R_I and R_F connect to ground, and the other ends connect to the inverting input terminal. Thus, they are effectively in parallel. The value of R_B is calculated as follows, Eq. (2.26):

$$R_B = \frac{R_F R_I}{R_F + R_I}$$
$$= \frac{33 \times 10^3 \times 4.7 \times 10^3}{33 \times 10^3 + 4.7 \times 10^3}$$
$$= 4.1 \text{ k}\Omega$$

We shall choose a standard value of 4.3 kilohms. The final schematic is shown in Fig. 2.14.

The actual performance of the circuit is indicated by the oscilloscope plots in Fig. 2.15. Additionally, Table 2.2 contrasts the measured performance with the original design goals.

A slight phase shift can be seen between input and output waveforms in Fig. 2.15. The effect is more pronounced as the input frequency is increased. For many applications, input/output phase relations are not important. In other applications they are critical. Chapter 10 discusses this issue in more detail.

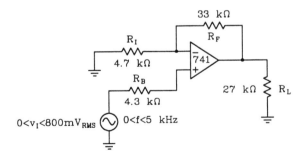

Figure 2.14 An example noninverting amplifier design.

TABLE 2.2

PARAMETER	DESIGN GOAL	MEASURED VALUES
Voltage gain	8	7.9–8.01
Frequency range	DC–5 kHz	DC–>5 kHz

Sec. 2.3 Noninverting Amplifier

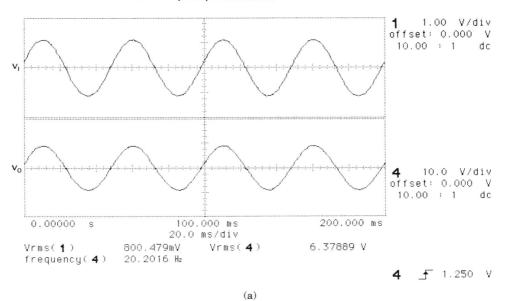

(a)

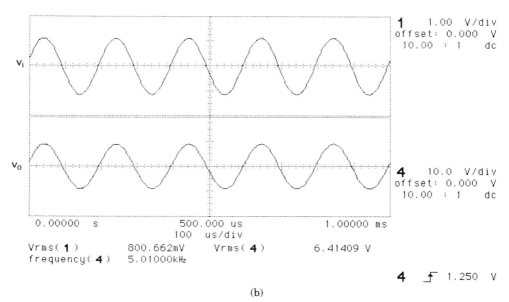

(b)

Figure 2.15 Oscilloscope displays showing the actual performance of the noninverting amplifier shown in Fig. 2.14. (Test equipment courtesy of Hewlett-Packard Company)

2.4 VOLTAGE FOLLOWER

2.4.1 Operation

A voltage follower circuit using an op amp is shown in Fig. 2.16. This is a very simple, but very useful, op amp configuration.

If you compare the voltage follower circuit to the noninverting amplifier previously discussed, you will see that R_I and R_F in the noninverting circuit have become infinity and zero, respectively, to form the follower circuit. Since there is no significant impedance in the path of the ($-$) input terminal, there is no need for the compensating resistor in the ($+$) terminal.

The voltage on the ($+$) input is equal to v_I because of the direct connection. Recall that v_D is approximately zero volts as long as the amplifier is not saturated. This means that the ($-$) input terminal will also be approximately equal to v_I. And, since the ($-$) pin is connected directly to the output, the output must also be equal to v_I. Since the output is essentially equal to the input at all times, the voltage gain is unity (i.e., one). The circuit is called a voltage follower since the output appears to follow or track the input voltage.

What value is a circuit that gives us an output voltage that is equal to the input? Well, although the voltage gain is only one, there are other very important reasons for using a voltage follower. One of the most important uses for the circuit is impedance transformation. By inspection, you can see that the input impedance is very high since the only current drawn from the source is the bias current for the ($+$) terminal. The output impedance, on the other hand, is quite low. As with the other configurations previously studied, the output impedance approaches an ideal value of zero. Therefore, the voltage follower circuit can interface a high impedance device or circuit to a lower impedance device or circuit. Although very little current is drawn from the source, a substantial current may be supplied to the load.

2.4.2 Numerical Analysis

The numerical analysis for the voltage follower is simpler than for previous circuits because of the lack of circuit complexity. Let us analyze the circuit shown in Fig. 2.16 and determine the following values:

1. voltage gain
2. input impedance

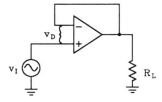

Figure 2.16 A basic voltage follower circuit.

Sec. 2.4 Voltage Follower

3. input current requirement
4. maximum output voltage swing
5. slew-rate limiting frequency
6. maximum input voltage swing
7. output impedance
8. output current capability
9. bandwidth
10. power supply rejection ratio

For purposes of the following analyses, let us assume that the op amp in Fig. 2.16 is a 741.

Voltage gain. The ideal voltage gain of a voltage follower circuit is always unity or one. This can be further demonstrated by applying the voltage gain equation, Eq. (2.28), presented for the noninverting amplifier circuit. Since R_F is now zero and R_I is infinity, our calculations become

$$A_V = \frac{R_F}{R_I} + 1$$

$$= \frac{0}{\infty} + 1$$

$$= 1$$

As with other amplifier configurations, the actual gain of the circuit falls off at high frequencies. This is further discussed along with bandwidth in a later section.

Input impedance. The input impedance of the voltage follower is ideally infinite since it is essentially the input resistance of the (+) input of the op amp modified by the effects of feedback. The value may be estimated by applying Eq. 2.29 with the quantity $R_I/(R_F + R_I)$ considered to be unity. Thus, for low frequencies (i.e., near DC), the circuit in Fig. 2.16 will have a minimum input impedance of

$$Z_I = R_{OP}A_V$$

$$= 0.3 \times 10^6 \times 200 \times 10^3$$

$$= 60{,}000 \text{ M}\Omega$$

If we had used typical values for R_{OP}, we would have gotten an even higher value for Z_{IN}. In any case, the value is so high that we can consider it as infinite for most applications.

Input current requirement. The input current for the circuit in Fig. 2.16 is only the bias current for the (+) input terminal. This is ideally zero, and for most applications may be neglected. If more precision is desired, then the manufacturer's data sheet in

App. 1 can be referenced. The data sheet indicates that the input bias current will be no higher than 500 nanoamperes with a more typical value being listed as 80 nanoamperes. Even though this current is temperature dependent, the absolute values are so small that they may be neglected in many applications.

Maximum output voltage swing. The maximum output voltage swing for the follower circuit is determined in the same manner, Eq. (2.10), as with preceding amplifiers. That is

$$v_o(\text{max}) = +V_{SAT} - (-V_{SAT})$$
$$= +13 - (-13)$$
$$= 26 \text{ V}$$

If a more accurate value is desired, the manufacturer's data sheet can be used to find a more precise value for the worst-case saturation voltage.

Slew-rate limiting frequency. As with the amplifier configurations discussed previously, the highest frequency that can be amplified with a full output voltage swing and no slew-rate limited distortion is computed as follows, Eq. (2.11):

$$f_{SRL} = \frac{\text{slew rate}}{\pi \ v_o(\text{max})}$$
$$= \frac{0.5/10^{-6}}{3.14 \times 26}$$
$$= 6.12 \text{ kHz}$$

If it is known for certain that the actual output swing will never be required to reach its limits, then the lower actual output swing can be used to compute the slew-rate limiting frequency.

Maximum input voltage swing. Since the amplifier has a voltage gain of one, the maximum input voltage swing is equal to the maximum output voltage swing. Thus, in the case of Fig. 2.16, we could have an input signal as large as ±13 volts without causing the amplifier to saturate. Again, if you plan to push the amplifier to its limits, then you should refer to the manufacturer's data sheet and select the worst-case output saturation voltage at the worst-case temperature. The computations, however, remain similar.

Output impedance. The output impedance of the voltage follower can be computed as follows:

$$r_o = \frac{\text{output impedance (open loop)}}{A_{OL}} \quad (2.31)$$

Sec. 2.4 Voltage Follower

where A_{OL} is the open-loop gain of the op amp at the specified frequency. You can determine the value of A_{OL} at the desired operating frequency as follows, Eq. (2.16):

$$A_{OL} = \frac{f_{UG}}{f_{IN}}$$

where f_{IN} is the specific input frequency being considered.

For the circuit in Fig. 2.16, the open-loop gain at 5 kilohertz, for example, is

$$A_{OL} = \frac{f_{UG}}{f_{IN}}$$

$$= \frac{10^6}{5000}$$

$$= 200$$

The output impedance then becomes, Eq. (2.31)

$$r_O = \frac{\text{output impedance (open loop)}}{A_{OL}}$$

$$= \frac{75}{200}$$

$$= 0.375 \ \Omega$$

As with most op amp circuits, the output impedance is so low relative to any practical load resistance that its effects may be ignored.

Output current capability. The total current flowing in or out of the output terminal of the op amp in Fig. 2.16 may be delivered directly to the load. That is, the feedback current is extremely small and can be disregarded in nearly all cases. As the load resistance varies from infinity (open) to zero (short), the output current from the op amp varies from zero to the short-circuit value of 20 milliamperes (given in the data sheet). The limiting factor is the amount of reduction that can be tolerated on the output voltage swing.

On an unprotected op amp, the value of load current must be kept below the stated output current rating. If this value is not supplied in the data sheet, then it can be estimated by using the maximum power dissipation data; recall that *power = voltage × current*.

Bandwidth. The bandwidth (i.e., the upper cutoff frequency) of a voltage follower circuit may be estimated by the following equation:

$$\boxed{bw = f_{UG}} \quad (2.32)$$

For the circuit shown in Fig. 2.16, we can compute the upper cutoff frequency and/or bandwidth as follows:

$$\text{bw} = f_{UG} = 1.0 \text{ MHz}$$

At lower frequencies, the voltage gain will be nearly equal to the calculated value of unity. As the frequency approaches the upper cutoff frequency, the voltage gain begins to decrease. Once the input frequency exceeds 1.0 megahertz (for a 741), the overall circuit gain will decrease dramatically.

Power supply rejection ratio. The change in output (v_O) voltage for a given change in DC power line noise voltage (v_N) is computed for the voltage follower with the following equation:

$$\boxed{v_{NO} = v_N \text{PSRR}} \quad (2.33)$$

where v_O, v_N, and PSRR are the values of the output noise signal, the noise signal on the DC supply lines, and the power supply rejection ratio, respectively. The manufacturer's data sheet in App. 1 lists the power supply rejection ratio (PSRR) as ranging from 30 to 150 microvolts per volt. The worst-case effect on the output voltage for the circuit in Fig. 2.16 then is

$$v_{NO} = v_N \text{PSRR}$$
$$= v_N \times 150 \times 10^{-6}$$

In other words, the amplitude of the power line noise (v_N) will be reduced by a factor of 0.000150. This means, for example, that if the DC supply lines had noise signals of 100 millivolts peak-to-peak, then we could anticipate a similar signal in the output with an amplitude of about

$$v_{NO} = v_N \times 0.000150$$
$$= 100 \times 10^{-3} \times 0.000150$$
$$= 15 \text{ }\mu\text{V peak-to-peak}$$

2.4.3 Practical Design Techniques

The design of a voltage follower circuit is fairly straightforward due to the lack of circuit complexity. Let us examine the design procedure by designing a voltage follower with the following characteristics:

1. input voltage range   100 to 500 millivolts RMS
2. frequency range   DC to 75 kilohertz
3. load resistance   4.7 kilohms
4. input resistance   greater than 100 kilohms
5. source impedance   1.8 kilohms

Sec. 2.4 Voltage Follower

Select the op amp. First, we must select an op amp that can provide unity gain up to the maximum input frequency. Thus, we will need to choose an op amp with a unity gain bandwidth of at least, Eq. (2.32)

$$f_{UG} = f_{MAX}$$
$$= 75 \text{ kHz}$$

Second, the slew rate of the op amp must be adequate to allow the required output voltage swing at the highest input frequency. The required slew rate is given by Eq. (2.11):

$$\text{slew rate(min)} = \pi f_{SRL} v_O(\text{max})$$
$$= 3.14 \times 75 \times 10^3 \times 500 \times 10^{-3} \times 1.414 \times 2$$
$$= 0.333 \text{ V}/\mu\text{s}$$

Since both the unity gain frequency and the slew rate requirements are within the limits of the 741 (see App. 1), let us choose this device for our design.

Select the power supply voltages. Now we must select a power supply voltage that is high enough to prevent saturation on the highest input voltage. The worst-case internal voltage drop on the output for a 741 is listed as 5 volts in App. 1 for load resistances between 2 and 10 kilohms. A more typical value is 2 volts. The minimum required power supply voltage can be determined as follows, Eq. (2.25):

$$\pm V_{MIN} = v_I(\text{peak}) + V_{INT}$$
$$= 500 \times 10^{-3} \times 1.414 + 5$$
$$= 5.707 \text{ V}$$

We shall choose a more standard value of ± 15 volts for our power supply voltages. The complete schematic of our voltage follower circuit is shown in Fig. 2.17.

Now let us check to be sure the 741 can supply the required current to our load without causing an appreciable voltage loss in our output. When the output voltage reaches its maximum voltage, the load current can be computed with Ohm's Law as

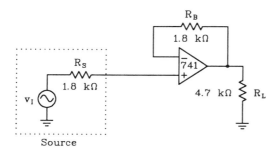

Figure 2.17 A voltage follower design which includes a compensation resistor (R_B).

$$i_L \text{(peak)} = \frac{v_O\text{(peak)}}{R_L}$$

$$= \frac{500 \times 10^{-3} \times 1.414}{4.7 \times 10^3}$$

$$= 150 \ \mu A$$

This should have negligible effect on the output voltage of the op amp since the 741 can supply significantly higher currents.

Figure 2.17 also illustrates the use of a compensating resistor R_B. You will recall from the previous amplifier designs that bias current in the op amp can cause output offsets due to the voltage drops across any resistances in line with the bias current. We minimize this offset by providing equal resistances in both (+) and (−) inputs. The resistance in the (+) input is simply the source resistance which was given as 1.8 kilohms. To minimize output errors, we insert an equal value R_B in the feedback loop. Note that no significant signal current flows through R_B. Therefore, the voltage gain is unaffected by the addition of R_B, and it remains constant at unity.

The actual performance of our voltage follower circuit is shown in Fig. 2.18 through the use of an oscilloscope plot. The measured performance is compared to the design goals in Table 2.3.

2.5 INVERTING SUMMING AMPLIFIER

2.5.1 Operation

Figure 2.19 shows the schematic diagram for an inverting summing amplifer. As indicated in the figure, the summing amplifer has several inputs. The circuit in Fig. 2.19 shows four inputs with the possibility of other inputs being indicated. Although the input sources are shown as DC signals (i.e., batteries) the circuit works equally well for AC signals or even a combination of AC and DC signals.

There are several ways to understand the operation of the inverting summing amplifier circuit. One simple method is an application of the Superposition Theorem. In this case, we consider the effects of each input signal one at a time with all other sources being set to zero. We know from our discussion of the basic inverting amplifier that the (−) input terminal is a virtual ground point. That is, unless the amplifier's output is saturated, the voltage on the (−) input will be within a few microvolts of ground potential. Thus, when we replace all but one source with a short (i.e., set them to zero volts) the associated input resistors essentially have a ground connection on both ends. That is, one end of each resistor is connected to ground through the temporary short that we inserted across the battery as part of the application of the Superposition Theorem. The opposite end of each input resistor is conneced to the (−) input which we know is a virtual ground point. Since all input resistors but one have ground potential on both sides, there will be no current flow through them and they can be totally disregarded for the remainder of our analysis.

Sec. 2.5 Inverting Summing Amplifier

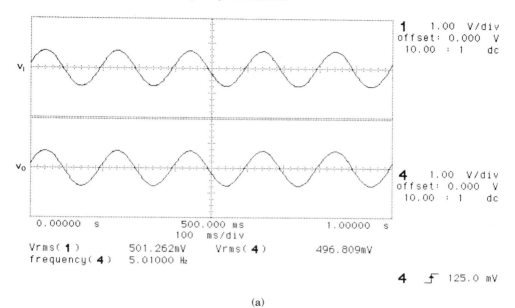

(a)

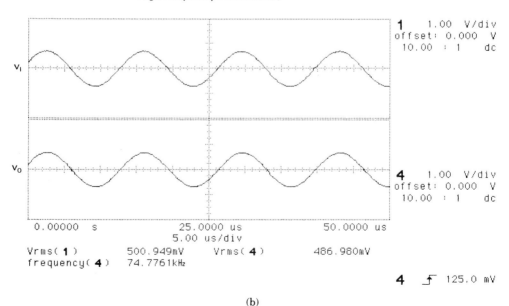

(b)

Figure 2.18 Oscilloscope displays showing the performance of the voltage follower circuit shown in Fig. 2.17. (Test equipment courtesy of Hewlett-Packard Company)

TABLE 2.3

PARAMETER	DESIGN GOAL	MEASURED VALUES
Input resistance	>100 kΩ	>100 kΩ
Voltage gain	1.0	0.97–0.99
Frequency range	DC–75 kHz	DC–>75 kHz

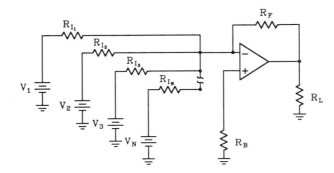

Figure 2.19 An inverting summing amplifier circuit.

By disregarding all input resistors and sources but one, we are left with a simple, single input, inverting amplifier circuit. We already know how this circuit works. Now compute voltage gain, input current, output voltage, and so on, for this single input, and perform a similar analysis for each of the other inputs one at a time. The actual output voltage of the circuit is the combination or sum of the voltages caused by the individual inputs.

One important point that should be recognized about the circuit shown in Fig. 2.19 is that the gains for each input signal are independent. That is, the ratio of R_F to R_{I_1} will determine the voltage gain that signal V_1 receives. V_2, on the other hand, is amplified by a factor established by the ratio of R_F and R_{I_2}. We can quickly conclude that the individual gains could be varied by changing the values of the input resistors while the gains of all signals could be changed simultaneously by varying the value of R_F. Consider, for example, that the circuit is being used as a microphone mixer. The signals from several microphones provide the inputs to the circuit. If the individual input resistors were variable, then they would adjust the amplitude (i.e., volume) of one microphone relative to another. If the feedback resistor were also variable, it would serve as a master volume control since it varies the amplification of all microphone signals but does not change the strength of one relative to another.

Resistor R_B is a compensating resistor and insures that both inputs of the op amp have similar resistances to ground. You will recall that this helps minimize problems caused by the op amp's bias currents.

Sec. 2.5   Inverting Summing Amplifier

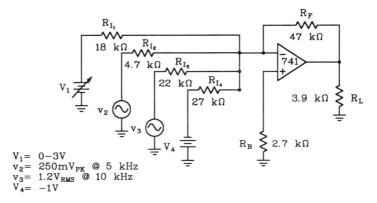

$V_1 = 0-3V$
$v_2 = 250\text{mV}_{PK}$ @ 5 kHz
$v_3 = 1.2\text{V}_{RMS}$ @ 10 kHz
$V_4 = -1V$

Figure 2.20 An inverting summing amplifier circuit used for a numerical analysis example.

2.5.2 NUMERICAL ANALYSIS

We shall now analyze the numerical performance of an inverting summing amplifier circuit. The circuit to be analyzed is shown in Fig. 2.20. Compute the following characteristics of the circuit:

1. Voltage gain of each input signal
2. Input impedance of each input signal
3. Input current requirement for each input signal
4. Maximum output voltage swing (total)
5. Maximum input voltage swing (individually)
6. Output impedance
7. Output current capability
8. Bandwidth
9. Slew rate limiting frequency

Voltage gain. The voltage gain for each input signal in Fig. 2.20 must be computed separately. Each gain, however, is computed in the same manner, Eq. (2.6) as a simple inverting amplifer circuit. That is

$$A_V = -\frac{R_F}{R_I}$$

where the minus sign is used to remind us of the phase inversion given to each signal. The individual voltage gains for the circuit in Fig. 2.20 are computed

$$A_{V_1} = -\frac{R_F}{R_{I_1}} = -\frac{47 \times 10^3}{18 \times 10^3} = -2.6$$

$$A_{V_2} = -\frac{R_F}{R_{I_2}} = -\frac{47 \times 10^3}{4.7 \times 10^3} = -10$$

$$A_{V_3} = -\frac{R_F}{R_{I_3}} = -\frac{47 \times 10^3}{22 \times 10^3} = -2.1$$

$$A_{V_4} = -\frac{R_F}{R_{I_4}} = -\frac{47 \times 10^3}{27 \times 10^3} = -1.7$$

Observe that each of these calculations is similar to our analysis on a single-input inverting amplifier and that the gains are independent of each other.

Input impedance. The input impedance seen by each input is equal to the value of input resistor on that particular input. That is, since each input resistor connects to a virtual ground point, its respective source sees the input resistor as the total input impedance. No calculations are required to determine the input impedance, we simply inspect the individual values of input resistors.

Input current requirement. Each source must supply the current for its own input. The amount of current can be determined by Ohm's Law and is simply the input voltage divided by the input resistance, Eq. (2.8). For the circuit shown in Fig. 2.20, we can compute the following values:

$$i_{I_1} = \frac{v_1}{R_{I_1}} = \frac{3}{18 \times 10^3} = 167 \ \mu A$$

$$i_{I_2} = \frac{v_2}{R_{I_2}} = \frac{250 \times 10^{-3}}{4.7 \times 10^3} = 53 \ \mu A \text{ peak}$$

$$i_{I_3} = \frac{v_3}{R_{I_3}} = \frac{1.2 \times 1.414}{22 \times 10^3} = 77 \ \mu A \text{ peak}$$

$$i_{I_4} = \frac{V_4}{R_{I_4}} = \frac{1}{27 \times 10^3} = 37 \ \mu A$$

In the case of V_1, a variable DC source, we computed the worst-case input current by using the maximum input voltage (3 V). Similarly, for the alternating voltage sources v_2 and v_3, we used peak values of input voltage. In each of these cases, the source must be capable of supplying the required current.

Maximum output voltage swing. The output voltage of the summing amplifier is limited by the $\pm V_{SAT}$ values. For the purposes of this analysis, we shall estimate the values of $\pm V_{SAT}$ to be 2 volts below the DC power supply values. The calculations, Eq. (2.10), to determine the maximum output voltage swing are

$$v_o(\text{max}) = +V_{SAT} - (-V_{SAT})$$

Sec. 2.5 Inverting Summing Amplifier

$$= +13 - (-13)$$
$$= 26 \text{ V}$$

As with previous circuits, we can utilize the data sheet supplied by the manufacturer if it becomes necessary to have a more accurate, or perhaps worst-case, value.

Maximum input voltage swing. The maximum input voltage swing of an amplifier is that voltage which causes the amplifier's output to reach saturation. Input voltages which exceed this limit will produce distorted (i.e., clipped) output signals. In the case of the summing amplifier, the situation is more complex than with previous, single-input amplifiers. That is, the instantaneous level of output is determined by the instantaneous values of input voltage on all inputs. First we shall consider each input separately to determine the maximum levels of an isolated input. The calculations, Eq. (2.1), are similar to those used with previous circuits. That is

$$\text{maximum input voltage swing} = \frac{\text{maximum output voltage swing}}{A_V}$$

where A_v is the voltage gain received by a particular input. The individual calculations are

$$V_1(\text{max}) = \frac{-13}{-2.6} = 5 \text{ volts DC}$$

$$v_2(\text{max}) = \frac{26}{10} = 2.6 \text{ volts peak-to-peak}$$

$$v_3(\text{max}) = \frac{26}{2.1} = 12.4 \text{ volts peak-to-peak}$$

$$V_4(\text{max}) = \frac{13}{-1.7} = -7.6 \text{ V}$$

Note that the negative and positive saturation limits were used as the maximum output "swing" for V_1 and V_4, respectively, since these two inputs are DC and will only be limited by one saturation barrier.

With reference to v_2 and v_3, we may want to express them in their peak and RMS forms to better compare them with the signals shown in Fig. 2.20. These conversions are

$$v_2(\text{peak}) = \frac{v_2(p-p)}{2} = \frac{2.6}{2} = 1.3 \text{ volts peak}$$

$$v_3(\text{RMS}) = \frac{v_3(p-p)}{2.828} = \frac{12.4}{2.828} = 4.39 \text{ volts RMS}$$

Since the maximum limits on all inputs (both DC and AC) are greater than the values listed on the schematic, we can assume that no single input can cause the amplifier output to saturate. However, two or more input signals may combine at some instant to drive the output to its saturation limit. Let us determine if this situation can occur in the circuit shown in Fig. 2.20. To perform this calculation, we want to determine the worst-case combination of input signals.

First observe that V_1 and V_4 are of opposite polarity and thus tend to reduce each other's effect in the output. A worst case would be when V_1 was zero or when V_1 was maximum (3 V DC). Let us evaluate them with Eq. (2.1) to determine the worst-case combination.

$$V_{O_1} = V_1 A_{V_1} = 3 \times (-2.6) = -7.8 \text{ volts DC}$$

$$V_{O_4} = V_4 A_{V_4} = -1 \times (-1.7) = +1.7 \text{ volts DC}$$

From these calculations we can see that if V_1 were reduced to zero, then V_4 would produce +1.7 volts in the output. On the other hand, if V_1 is set for maximum (3 V DC), then the net output voltage will be the difference between the V_1 produced and V_4 produced outputs. This worst-case output voltage is simply $-7.8 + 1.7$ or -6.1 volts.

Now we must consider the effects of the AC signals v_2 and v_3. The worst-case output condition will occur when these two inputs hit their peak values simultaneously and have the same polarity as V_1. The output voltages produced individually by v_2 and v_3 are

$$v_{O_2} = v_2 A_{V_2} = 250 \times 10^{-3} \times (-10) = -2.5 \text{ volts peak}$$

$$v_{O_3} = v_3 A_{V_3} = 1.2 \times 1.414 \times (-2.1) = -3.6 \text{ volts peak}$$

The net effect of V_1, v_2, v_3, and V_4 can be found by adding the individual output values (Superposition Theorem).

$$v_O = V_{O_1} + v_{O_2} + v_{O_3} + V_{O_4}$$
$$= -7.8 - 2.5 - 3.6 + 1.7$$
$$= -12.2$$

Since this worst-case value is less than our maximum output voltage limit (± 13 V typical), we should not have a problem. In extreme cases, however, we may still have a potential problem. Recall that the output limits of ± 13 volts were obtained by using typical performance values for the 741. If worst-case values are used, we will find that the limits fall to ± 10 volts under worst-case conditions. If this situation were to occur at the same time our inputs were all at their maximum values, we would drive the amplifier into saturation and produced a clipped output. If this is a serious concern for our particular application, we could reduce R_F slightly to prevent the combined signals from driving the output to saturation.

Output impedance. The output impedance of the summing amplifier can be estimated as follows:

Sec. 2.5   Inverting Summing Amplifier

$$\boxed{\text{output impedance (closed loop)} = r_o = \frac{\text{output impedance (open loop)}}{A_{OL} Y + 1}} \quad (2.34)$$

where A_{OL} is the open-loop gain of the op amp at the specified frequency, and Y is computed as follows:

$$\boxed{Y = \frac{R_{I_1} \| R_{I_2} \| R_{I_3} \| R_{I_4}}{(R_{I_1} \| R_{I_2} \| R_{I_3} \| R_{I_4}) + R_F}} \quad (2.35)$$

Now let us compute the output impedance for the circuit in Fig. 2.20. First we compute the value of the parallel combination of input resistors (R_X)

$$R_X = \frac{1}{\dfrac{1}{R_{I_1}} + \dfrac{1}{R_{I_2}} + \dfrac{1}{R_{I_3}} + \dfrac{1}{R_{I_4}}}$$

$$= \frac{1}{\dfrac{1}{18 \text{ k}} + \dfrac{1}{4.7 \text{ k}} + \dfrac{1}{22 \text{ k}} + \dfrac{1}{27 \text{ k}}}$$

$$= 2.85 \text{ k}\Omega$$

Next we use this value to compute the factor Y, Eq. (2.35)

$$Y = \frac{R_X}{R_X + R_F}$$

$$= \frac{2.85 \times 10^3}{2.85 \times 10^3 + 47 \times 10^3}$$

$$= 0.057$$

Next we determine the value of A_{OL} at the frequency of interest, Eq. (2.16). We shall use the worst-case value which occurs at the highest input frequency (10 kHz)

$$A_{OL} = \frac{f_{UG}}{f_{IN}}$$

$$= \frac{10^6}{10^4}$$

$$= 100$$

Finally, compute the estimated value of output resistance, Eq. (2.34)

$$r_o = \frac{\text{output impedance (open loop)}}{A_{OL} Y + 1}$$

$$= \frac{75}{100 \times 0.057 + 1}$$

$$= 11 \ \Omega$$

Since this value was computed at the highest input frequency (worst case), and since it is very low compared to the value of the load resistor, its effects on output voltage can be safely ignored.

Output current capability. The maximum value of load current occurs when the output reaches its highest instantaneous value. The maximum voltage was previously computed as 12.2 volts. The worst-case load current can be computed with Ohm's Law

$$i_L(\text{max}) = \frac{v_O(\text{max})}{R_L}$$

$$= \frac{12.2}{3900}$$

$$= 3.13 \text{ mA}$$

The output of the op amp must also supply the feedback current. In most applications, this current can be ignored since it is generally much smaller than the load current. Our present circuit is no exception. That is, we can see by inspection that the feedback path has over ten times as much resistance as the load.

The data sheet in App. 1 indicates that, even under worst-case conditions, the output can maintain at least 10 volts across a 2000 ohm load. By Ohm's Law we can conclude that this corresponds to an output current of

$$i_O = \frac{v_O}{R_L} = \frac{10}{2000} = 5 \text{ mA}$$

Of course the typical value of current is even higher. In any case, the current capability of the output clearly exceeds our requirements, and therefore, poses no problem. If our load resistor were smaller, then we could anticipate a reduced output voltage.

Bandwidth. For a meaningful discussion on bandwidth, one must consider the response of each input individually. When considered separately, we can estimate the bandwidth of any given input by applying the bandwidth equation, Eq. (2.22), used in previous analyses

$$\text{bw} = \frac{f_{UG} R_I}{R_F + R_I}$$

We know from earlier calculations that the bandwidth will decrease as the closed-loop gain is increased. Let us calculate the bandwidth for the input in Fig. 2.20 which has the highest gain. We have already determined the individual gains to be 2.6, 10, 2.1, and 1.7 for inputs V_1 through V_4. We shall compute the bandwidth for the v_2 input because

Sec. 2.5 Inverting Summing Amplifier

it has the highest gain. Incidentally, there would be very little point in computing the bandwidth for inputs V_1 and V_4 since these have DC signals applied. The bandwidth for the v_2 input is

$$\text{bw} = \frac{f_{UG}R_I}{R_F + R_I}$$

$$= \frac{(1 \times 10^6)(4.7 \times 10^3)}{(47 \times 10^3) + (4.7 \times 10^3)}$$

$$= 90.9 \text{ kHz}$$

A similar analysis could be made for input v_3 which has a computed gain of 2.1 and a maximum input frequency of 10 kilohertz. For large amplitude output signals, the slew rate will tend to restrict the operation to even lower frequencies. This is discussed in the following section.

Slew-rate limiting frequency. As discussed for previous amplifier configurations, the slew rate also limits the highest operating frequency for larger output voltage excursions. The slew-rate limiting frequency is found as follows, Eq. (2.11):

$$f_{SRL} = \frac{\text{slew rate}}{\pi \, v_O(\text{max})}$$

$$= \frac{0.5/10^{-6}}{3.14 \times 26}$$

$$= 6.12 \text{ kHz}$$

Thus, although the v_2 input was shown to have a 90.9 kilohertz bandwidth as established by the unity-gain frequency, the full-power upper limit is only 6.12 kilohertz. In the given application, however, the applied signal is only 5000 hertz so this should not hamper the operation of the circuit with respect to the v_2 input.

The v_3 input, on the other hand, operates at 10 kilohertz. This means that we could never get the full 26-volt swing in the output as a result of v_3 signals. The schematic indicates that the highest input voltage is 1.2 volts RMS. The gain for v_3 was previously computed as 2.1. The largest normal output swing from v_3 can be found by applying Eq. (2.1):

$$v_O(\text{max}) = V_3(\text{RMS}) \times 1.414 \times 2 \times A_V$$

$$= 1.2 \times 1.414 \times 2 \times 2.1$$

$$= 7.13 \text{ volts peak-to-peak}$$

The actual slew-rate limiting frequency for this input is then estimated with Eq. (2.11) as

$$f_{SRL} = \frac{\text{slew rate}}{\pi \, v_O(\text{max})}$$

$$= \frac{0.5/10^{-6}}{3.14 \times 7.13}$$

$$= 22.3 \text{ kHz}$$

In the given circuit, the slew rate should not interfere with the expected operation.

2.5.3 Practical Design Techniques

To illustrate the design method for an inverting summing amplifier circuit, let us design a three-input circuit with the following performance characteristics:

Input 1. 0 to 500 milivolts peak, at a frequency of 2.7 kilohertz. The source resistance is 1.0 kilohms, and the signal is to be amplified by a factor of -3.5.

Input 2. -2 to $+2$ volts DC. The source resistance is 0.75 ohms. The voltage is to pass through the circuit without amplification (i.e., inversion only).

Input 3. 0 to 3 volts RMS, at a frequency of 500 hertz. The source resistance is 50 ohms, and the signal is to be amplified by a factor of -2.

Recall that the minus signs preceding the gain factors tells us that the signal is inverted in the process of being amplified. The negative gains **do not** imply voltage reduction.

The output of the amplifier must drive a load resistance which varies from 10 to 50 kilohms.

Determine the worst-case input. Our first step is to determine which input to design first. If we choose the wrong one, then we will end up recalculating some of our values. The proper input can be identified by choosing the one that has the highest product of source resistance times voltage gain (absolute value). These calculations are shown for comparison.

Input 1. $1000 \times 3.5 = 3500$

Input 2. $0.75 \times 1 = 0.75$

Input 3. $50 \times 2 = 100$

Since input 1 had the highest gain-source-resistance product, we shall begin by selecting the input resistor for input 1.

Choose the value for the first input resistor. The source resistance and the input resistor are in series. Their sum in conjunction with R_F will determine the voltage gain of that input. In theory, there is no requirement to have a physical resistor for R_I. That is, the source resistance alone can serve as the input resistor. In practice, however, the source resistance is usually only an estimate and rarely a constant. Therefore, it is generally wise to include a separate resistor as R_I and to make this resistor large enough to minimize the effects of changes in the source resistance. The application must dictate the degree of stability needed, but in general, if the input resistor is ten times as large as the source resistance, then the effects of changes in the source resistance are reduced by about 90 percent. If greater protection is needed, then increase R_I accordingly.

For purposes of our sample design, let us choose R_{I_1} to be ten times the value of source resistance. The value of R_{I_1} then is computed as

$$R_{I_1} = 10 \times R_{S_1}$$
$$= 10 \times 1000$$
$$= 10{,}000 \; \Omega$$

Calculate the required feedback resistor (R_F). The feedback resistor is calculated by using a transposed version of the basic voltage gain equation, Eq. (2.6), for an inverting amplifier.

$$A_V = -\frac{R_F}{R_I}$$

or

$$R_F = -A_V R_I$$

In our particular circuit,

$$R_F = -A_V R_I$$
$$= -(-3.5) \times 10000$$
$$= 35 \; k\Omega$$

We shall choose the nearest standard (5% tolerance) value of 36 kilohms.

Compute the remaining input resistors. Values for each of the remaining input resistors can be calculated by using yet another transposed version of the basic voltage gain equation, Eq. (2.6).

$$A_V = -\frac{R_F}{R_I}$$

or

$$R_I = -\frac{R_F}{A_V}$$

Using this equation, we can now compute values for R_{I_2} and R_{I_3} as follows:

$$R_{I_2} = -\frac{R_F}{A_{V_2}}$$

$$= -\frac{36000}{-1}$$

$$= 36 \text{ k}\Omega$$

and

$$R_{I_3} = -\frac{R_F}{A_{V_3}}$$

$$= -\frac{36000}{-2}$$

$$= 18 \text{ k}\Omega$$

Compute the value of R_B. To minimize the effects of op amp bias currents, we want to make the value of R_B equal to the parallel combination of R_F and all of the input/source resistors.

$$\boxed{R_B = \frac{1}{\dfrac{1}{R_F} + \dfrac{1}{R_{I_1} + R_{S_1}} + \dfrac{1}{R_{I_2} + R_{S_2}} + \dfrac{1}{R_{I_3} + R_{S_3}}}} \quad (2.36)$$

In our present case, the value of R_B is computed as

$$R_B = \frac{1}{\dfrac{1}{36\text{k}} + \dfrac{1}{10\text{k} + 1\text{k}} + \dfrac{1}{36\text{k} + 0.75} + \dfrac{1}{18\text{k} + 50}}$$

$$= 4.95 \text{ k}\Omega$$

We shall select the nearest standard value of 5100 ohms.

Determine the required power supply voltages. The DC power supply voltages must be high enough to prevent saturation under the worst-case input conditions. Generally, the condition to be considered is when all inputs are at the maximum voltage at the same time. The worst-case output voltage, then, is computed by adding the output voltages caused by each of the individual inputs, Eq. (2.1).

$$v_{O_1}(\text{max}) = v_1(\text{max}) \times A_{V_1} = 500 \times 10^{-3} \times (-3.5) = -1.75 \text{ V}$$

$$v_{O_2}(\text{max}) = v_2(\text{max}) \times A_{V_2} = 2 \times (-1) = -2.0 \text{ V}$$

$$v_{O_3}(\text{max}) = v_3(\text{max}) \times A_{V_3} = 3 \times 1.414 \times (-2) = -8.48 \text{ V}$$

The worst-case output then will be

$$v_O(\text{max}) = v_{O_1}(\text{max}) + v_{O_2}(\text{max}) + v_{O_3}(\text{max})$$

Sec. 2.5 Inverting Summing Amplifier

$$= (-1.75) + (-2) + (-8.48)$$
$$= -12.2 \text{ V}$$

Unless the internal drop on the output of the selected op amp is unusually high, we should be able to use standard ±15 volt supplies. Suppose, for example, we decide to use a 741 op amp. The manufacturer's data sheet in App. 1 indicates that the op amp can deliver at least ±12 volts to a load ≥10 kilohms when ±15 volt supplies are used. We shall plan to use a 741 unless we encounter problems with bandwidth or slew rate (verified in subsequent sections).

Determine the required unity gain frequency. The minimum unity gain frequency for each input is computed with Eq. (2.22)

$$f_{UG_1} = \frac{bw(R_F + R_{I_1})}{R_{I_1}} = \frac{2700(36 \times 10^3 + 10 \times 10^3)}{10 \times 10^3} = 12.4 \text{ kHz}$$

$$f_{UG_2} = \text{Not applicable to DC inputs}$$

$$f_{UG_3} = \frac{bw(R_F + R_{I_3})}{R_{I_3}} = \frac{500(36 \times 10^3 + 18 \times 10^3)}{18 \times 10^3} = 1.5 \text{ kHz}$$

In all cases, the required minimum unity gain bandwidth is substantially below the 1.0 megahertz limit of a 741. Therefore, we shall initially plan to use a 741 in our design. If the minimum bandwidth requirement were greater than 1.0 megahertz, then we would have had to select a different op amp.

Determine the required slew rate. The minimum acceptable slew rate for the op amp is given by the following equation, Eq. (2.11):

$$\text{slew rate(min)} = \pi f_{SRL} v_O(\text{max})$$

Let us determine the minimum slew rate for each input

$$\text{slew rate(min) 1} = 3.14 \times 2.7 \times 10^3 \times 3.5 = 0.03 \text{ V}/\mu\text{s}$$

$$\text{slew rate(min) 2} = \text{DC input}$$

$$\text{slew rate(min) 3} = 3.14 \times 500 \times 16.97 = 0.027 \text{ V}/\mu\text{s}$$

In all cases, the required slew rate is substantially below the 0.5 volts per microsecond rating of the 741. Therefore, we shall select this device as our final choice.

The schematic of our design is shown in Fig. 2.21. The actual performance of the circuit is evident from the oscilloscope displays in Fig. 2.22. The measured performance is contrasted with the original design goals in Table 2.4.

94   Amplifiers   Chap. 2

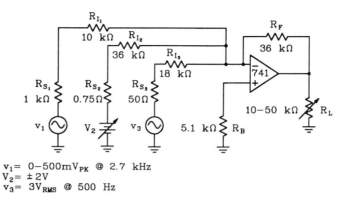

$v_1 = 0\text{-}500\,\text{mV}_{PK}$ @ 2.7 kHz
$V_2 = \pm 2\text{V}$
$v_3 = 3\text{V}_{RMS}$ @ 500 Hz

Figure 2.21 The final design of a three-input inverting summing amplifier circuit.

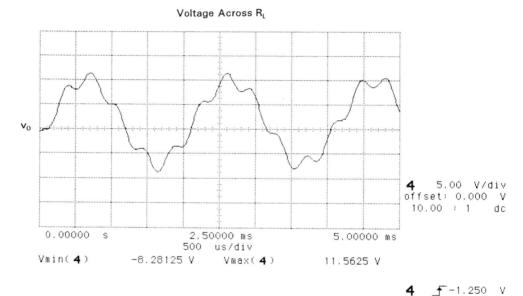

Figure 2.22 Oscilloscope displays showing the actual performance of the inverting summing amplifier shown in Fig. 2.21. (Test equipment courtesy of Hewlett-Packard Company)

TABLE 2.4

PARAMETER	DESIGN GOAL	MEASURED VALUE
Voltage gain 1	−3.5	−3.27
Voltage gain 2	−1.0	−1.0
Voltage gain 3	−2.0	−1.99

2.6 NONINVERTING SUMMING AMPLIFIER

2.6.1 Operation

Figure 2.23 shows a three input, noninverting summing amplifier circuit. Its operation is significantly more difficult to analyze than the inverting summing amplifier. In the present case, we will need to rely heavily on the use of Thevenin's Theorem to analyze the operation of the circuit. First, though, let us examine the fundamental theory of operation.

Although the network on the (+) input is somewhat difficult to analyze mathematically, we know intuitively that it must be equivalent to some value of voltage and some value of resistance. If we mentally replace the network on the (+) input with a simple voltage source and series resistance, we see that the circuit becomes a simple, familiar noninverting amplifier circuit. The gain of this equivalent circuit is determined by the ratio of R_F to R_I. So with the single exception of the network on the (+) input, analysis of the circuit is quite straightforward.

2.6.2 Numerical Analysis

Now let us analyze the circuit shown in Fig. 2.23 numerically. We will focus our efforts on the network associated with the (+) input terminal. If we can reduce this network to a simpler network consisting of a single voltage source and a single resistor, then we can analyze the rest of the circuit using the method presented for the simple noninverting amplifier.

To reduce the network on the (+) input, we shall apply Thevenin's Theorem in two stages. First, simplify V_1, V_2, and the associated resistors. Figure 2.24(a) shows the circuit divided between V_2 and V_3. Application of Thevenin's Theorem to the portion of the circuit on the left side of the break point, gives us a Thevenin voltage (V'_{TH}) of 2 volts and a Thevenin resistance (R'_{TH}) of 2.78 kilohms. This equivalent circuit is shown in Figure 2.24(b) reconnected to the original V_3/R_3 circuit.

If we apply Thevenin's Theorem to the partially simplified circuit in Fig. 2.24(b),

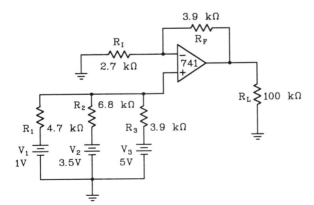

Figure 2.23 A three-input noninverting summing amplifier.

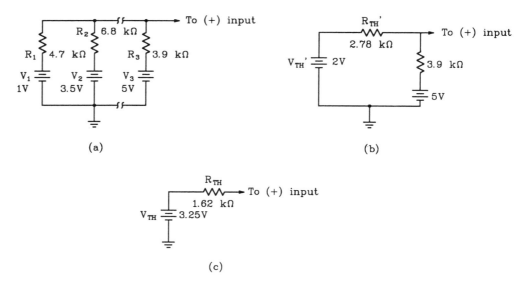

Figure 2.24 Thevenin's Theorem is used to simplify the summing network for the noninverting summing amplifier.

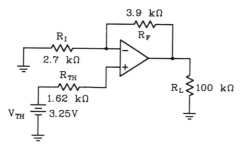

Figure 2.25 The summing network shown in Fig. 2.23 can be replaced by its Thevenin equivalent for analysis purposes.

we obtain the fully reduced equivalent circuit of Fig. 2.24(c). Thus, the network of resistors and voltage sources on the (+) input of the summing amplifier originally shown in Fig. 2.23, can be replaced by the Thevenin equivalent circuit shown in Fig. 2.24(c). This substitution is shown in Fig. 2.25.

We can now complete our analysis of the simplified circuit by applying techniques presented for the basic noninverting amplifier.

Voltage gain. The voltage gain of the circuit in Fig. 2.25 can be computed with the noninverting amplifier gain formula, Eq. (2.28).

$$A_V = \frac{R_F}{R_I} + 1$$

$$= \frac{3.9 \times 10^3}{2.7 \times 10^3} + 1$$

$$= 2.44$$

Sec. 2.7   AC Coupled Amplifier

Output voltage. The output voltage of the circuit in Fig. 2.25 can be determined by utilizing the basic gain equation, Eq. (2.1)

$$A_V = \frac{v_O}{v_I}$$

therefore,

$$\begin{aligned} v_O &= A_V v_I \\ &= 2.44 \times 3.25 \\ &= 7.93 \text{ V} \end{aligned}$$

2.6.3 Practical Design Techniques

The design of a noninverting summing amplifier like that shown in Fig. 2.23 is an involved process. Additionally, the resulting design is difficult to alter without affecting several parameters. Therefore, many designers who need a noninverting summing amplifier, will utilize an inverting summing amplifier followed by a simple inverting amplifier. This arrangement is much simpler to design, easier to modify, and costs little more to build. With this in mind, we will not explore the details for designing the generic noninverting summing amplifier. However, we will discuss the design of a special case which uses the same basic circuit when we study adder circuits in Chap. 9.

2.7 AC COUPLED AMPLIFIER

2.7.1 Operation

The term AC coupled identifies the fact that only AC signals are allowed to pass through the amplifier. DC and very low-frequency AC signals are blocked or at least severely attenuated. The concept of AC coupling is applicable to many different amplifier configurations. In the following discussion, we will consider the operation of the basic inverting and noninverting amplifier circuits when they are configured to be AC coupled. Most of the operation, analyses and design methods are similar to their DC coupled equivalents which have been previously covered in detail. Therefore, concentrate on those areas that are unique to the AC coupled circuit.

First let us examine the operation of the AC coupled inverting amplifier circuit shown in Fig. 2.26(a).

You will recall from basic electronics theory that a capacitor blocks DC and passes AC. More specifically, a capacitor's opposition to current flow (capacitive reactance) increases as the applied frequency decreases. As the input frequency in Fig. 2.26(a) decreases, the reactance of capacitors C_I and C_O both increase. As the reactance of C_I increases, the combined impedance of C_I and R_I also increases. Since the voltage gain of the inverting amplifier is determined by the ratio of the feedback resistor to the input resistance, and since the input resistance (actually the combined impedance of

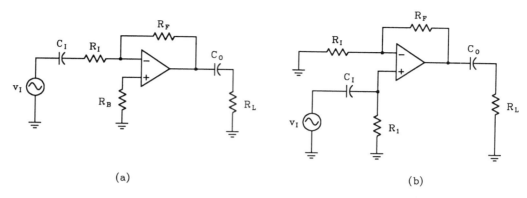

Figure 2.26 AC coupled versions of the basic inverting amplifier (a) and the basic noninverting amplifier (b) circuits.

C_I and R_I) is increasing, then we know that the amplifier gain must be decreasing.

Another way to view the operation of the AC coupled inverting amplifier is to consider that the output voltage of the amplifier is determined by the magnitude of feedback current. The feedback current that flows through R_F is identical to that which flows through R_I (ignoring the small bias current that flows in or out of the $(-)$ input terminal). The value of current flow through R_I is determined by the magnitude of the input voltage and the impedance of R_I and C_I in combination. As the frequency decreases toward DC, the input current, and therefore, the feedback current must decrease. This lowered feedback current causes a corresponding decrease in output voltage. Since the input voltage is constant but the output voltage is decreasing, we can conclude that the amplifier's gain is dropping as the frequency is lowered.

The output capacitor C_O also affects the frequency response of the circuit. Basically, the output resistance of the op amp, the load resistance and C_O form a series circuit across which the ideal output voltage is developed. That portion of the output voltage which appears across R_L is the final or effective output voltage of the circuit. The remaining voltage which is dropped across the internal resistance and across C_O is essentially lost. As the frequency in the circuit is decreased, the reactance of C_O increases. This causes a greater percentage of the output voltage to be dropped across C_O and leaves less to be developed across R_L. So the effects of C_O also cause the frequency response to drop off on the low end and, in fact, prohibits the passage of DC signals.

Resistor R_B helps compensate for the effects of op amp bias currents. Its value will generally be the same as the feedback resistor since the input resistor (R_I) is isolated by C_I for DC purposes.

The AC coupled noninverting amplifier circuit shown in Fig. 2.26(b) is nearly identical to its direct coupled counterpart that we discussed in an earlier section. Coupling capacitors C_I and C_O allow AC signals to be coupled in and out of the amplifier. Very low-frequency signals and DC in particular are not coupled through the capacitors and, are therefore, not allowed to pass through the amplifier. R_I and C_I form an RC coupling

Sec. 2.7   AC Coupled Amplifier

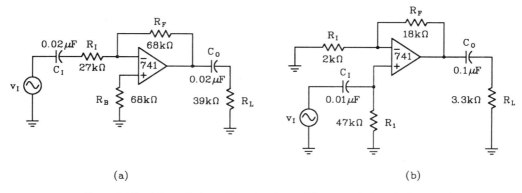

Figure 2.27 AC coupled amplifier circuits used for numerical analysis examples.

circuit on the input. That portion of the input signal which appears across R_I is actually amplified by the circuit.

2.7.2 Numerical Analysis

Most of the calculations for the basic direct coupled inverting and noninverting amplifier circuits apply to the AC coupled inverting and noninverting amplifier, respectively. For purposes of our present analyses, we will determine the following circuit parameters:

1. voltage gain
2. input impedance
3. bandwidth

We shall use the circuits shown in Fig. 2.27 for our numerical analysis example.

Voltage gain.   We shall compute the overall voltage gain of the inverting circuit, Fig. 2.27(a), by considering the gain to be made up of two parts. The first part of the gain will be determined by all components to the left of C_O. The second portion of the overall gain is determined by C_O and R_L. This latter part is actually a loss and tends to reduce the overall gain. Once these individual gains are computed, multiply them together to determine the overall voltage gain.

The voltage gain of the circuit to the left of C_O is computed basically the same way Eq. (2.6) as a direct coupled inverting amplifier. That is,

$$A_V = -\frac{R_F}{Z_I} \qquad (2.37)$$

The only modification to our original equation is that the denominator must also include the effects of C_I. Therefore, instead of dividing by R_I (as we did with the direct coupled circuit), we simply divide by the net impedance of R_I and C_I (i.e., Z). You will recall from basic electronics that the net impedance of a series RC circuit is computed with the following equation:

$$Z = \sqrt{R^2 + X_C^2}$$

where X_C is the capacitive reactance of C_I. We already know that the gain of an op amp varies with frequency, but now we have introduced an even more obvious frequency sensitive factor (X_C). Thus, when we speak of voltage gain, we must refer to a specific frequency in order to have a meaningful discussion. In most cases, we are interested in the lowest input frequency since this is where the capacitors will have their greatest effect (i.e., gain will be the lowest).

For the portion of the circuit in Fig. 2.27(a) left of C_O we can compute the voltage gain as shown. First, we need to determine the capacitive reactance with our basic electronics formula for X_C.

$$X_C = \frac{1}{2\pi fC}$$

For illustrative purposes, we shall assume an input frequency of 800 hertz. The first step, then, is to calculate the reactance of C_I at the frequency of interest.

$$X_{C_I} = \frac{1}{6.28 \times 800 \times 0.02 \times 10^{-6}} = 9.95 \text{ k}\Omega$$

Now we can compute the impedance of R_I and C_I

$$Z = \sqrt{R^2 + X_C^2}$$
$$= \sqrt{27000^2 + 9950^2}$$
$$= 28.8 \text{ k}\Omega$$

Substituting this into the voltage gain equation, Eq. (2.37), we can compute the gain of the circuit to the left of C_O.

$$A_{V_1} = -\frac{R_F}{Z_I}$$
$$= -\frac{68 \times 10^3}{28.8 \times 10^3}$$
$$= -2.4$$

Recall that the minus sign indicates a phase inversion but in no way implies a reduction in signal amplitude.

Our next step is to compute the effects of C_O and R_L. These two components form an RC voltage divider. You may apply your favorite circuit analysis method to determine the percentage of voltage that appears across R_L. This percentage is the effective "gain" of the RC coupling circuit. For our purposes, we shall use the following method which is based on the resistive voltage divider formula

Sec. 2.7 AC Coupled Amplifier

$$\boxed{A_V = \frac{R_L}{Z}} \qquad (2.38)$$

where Z is the net impedance of C_O and R_L.

First, we compute the reactance of C_O at the frequency of interest (800 Hz in this case)

$$X_C = \frac{1}{2\pi fC}$$

$$= \frac{1}{6.28 \times 800 \times 0.02 \times 10^{-6}}$$

$$= 9.95 \text{ k}\Omega$$

Next we compute the net impedance of R_L and C_O.

$$Z = \sqrt{R^2 + X_C^2}$$

$$= \sqrt{39000^2 + 9950^2}$$

$$= 40.2 \text{ k}\Omega$$

Finally, substituting this value into Eq. (2.38), we compute our gain as

$$A_{V_2} = \frac{R_L}{Z}$$

$$= \frac{39 \times 10^3}{40.2 \times 10^3}$$

$$= 0.97$$

That is to say, about 97 percent of the signal amplitude that appears at the output terminal of the op amp will be developed across R_L. The RC coupling circuit appears to be working well at 800 hertz since very little voltage is being lost across C_O.

The overall voltage gain for the circuit is found by multiplying these two gains as computed.

$$\boxed{A_V(\text{overall}) = A_{V_1} \times A_{V_2}} \qquad (2.39)$$

For our example, we compute overall voltage gain as

$$A_V(\text{overall}) = -2.4 \times 0.97 = -2.33$$

Notice that the method described for computing overall voltage gain does not include the effects of the variations in open-loop gain at different frequencies. Although this additional consideration could be included as we did with the direct coupled amplifier, it is not normally necessary since our calculations are accomplished at the lowest input frequency.

If you want to compute the gain at some relatively high frequency, then you should include the effects of reduce op amp internal gain.

Another point which you may wish to consider involves phase shift. In addition to the 180 degree phase shift provided by the op amp itself, the signal also receives a phase shift from the two RC networks. The mentioned calculations compute only the amplitude of the signal. If the phase is also an important consideration, then the same basic equations still apply but you can express the values as complex numbers. The final answer, then, will include not only the magnitude of the gain as computed, but it will also reveal the amount of phase shift given to the signal.

The voltage gain calculation for the noninverting circuit shown in Fig. 2.27(b) is similar, but will be considered as three separate gains that are multiplied together to find the overall gain. The three individual gains are

1. R_I/C_I network gain (actually a loss),
2. the R_L/C_O network gain (actually a loss),
3. the gain of the op amp circuit as determined by R_F and R_I.

The gains of the input and output RC circuits are computed the same way we computed the gain of the output RC circuit in Fig. 2.27(a). Let us first calculate the input RC circuit gain. Our initial step is to compute the reactance of C_I at the lowest frequency (assumed to be 800 Hz).

$$X_C = \frac{1}{2\pi fC}$$

$$= \frac{1}{6.28 \times 800 \times 0.01 \times 10^{-6}}$$

$$= 19.9 \text{ k}\Omega$$

Next, we find the net impedance of R_1 and C_1.

$$Z = \sqrt{R^2 + X_C^2}$$

$$= \sqrt{47000^2 + 19900^2}$$

$$= 51 \text{ k}\Omega$$

Finally, we compute the voltage gain (loss) of the RC network with Eq. (2.38).

$$A_{V_I} = \frac{R_1}{Z}$$

$$= \frac{47 \times 10^3}{51 \times 10^3}$$

$$= 0.922$$

Similar calculations for the output RC network of Fig. 2.27(b) can now be accomplished. First, we find the reactance of C_O.

Sec. 2.7 AC Coupled Amplifier

$$X_C = \frac{1}{2\pi f C}$$

$$= \frac{1}{6.28 \times 800 \times 0.1 \times 10^{-6}}$$

$$= 1.99 \text{ k}\Omega$$

Next, we compute the net impedance of R_L and C_O.

$$Z = \sqrt{R^2 + X_C^2}$$

$$= \sqrt{3300^2 + 1990^2}$$

$$= 3.85 \text{ k}\Omega$$

Finally, the effective voltage gain (loss) of the $R_L C_O$ network can be computed, Eq. (2.38).

$$A_{V_2} = \frac{R_L}{Z}$$

$$= \frac{3.3 \times 10^3}{3.85 \times 10^3}$$

$$= 0.857$$

The third portion of our overall gain calculation is the gain of the op amp circuit as determined by R_F and R_I. We compute this using the gain formula, Eq. (2.28) presented for the direct coupled amplifier.

$$A_{V_3} = \frac{R_F}{R_I} + 1$$

$$= \frac{18 \times 10^3}{2 \times 10^3} + 1$$

$$= 10$$

The effective overall gain at 800 hertz is found by multiplying the three individual gains, Eq. (2.39).

$$A_V(\text{overall}) = A_{V_1} \times A_{V_2} \times A_{V_3}$$

$$= 0.922 \times 0.857 \times 10$$

$$= 7.9$$

As with the AC coupled inverting amplifier, we have chosen to ignore the frequency dependent effects of open-loop op amp gain. This is generally a reasonable approach since our calculations are performed at the lowest input frequencies where the open-loop gain is the closest to its ideal value.

Input impedance. The input impedance for the AC coupled inverting amplifier circuit shown in Fig. 2.27(a) is equal to the net impedance of R_I and C_I. Recall that the ($-$) input of the op amp is a virtual ground point. The source, therefore, sees the input impedance offered by C_I and R_I. Since this is a frequency-dependent value, we must discuss input impedance at a particular frequency of interest. For purposes of our present discussion, let us compute the highest and lowest values for input impedance if the input frequency range is 800 hertz to 3 kilohertz. The following input impedance at 800 hertz is computed. First, we find the reactance of C_I at 800 hertz.

$$X_C = \frac{1}{2\pi fC}$$

$$X_C = \frac{1}{6.28 \times 800 \times 0.02 \times 10^{-6}}$$

$$= 9.95 \text{ k}\Omega$$

Now we can compute the impedance of R_I and C_I.

$$Z(\max) = \sqrt{R^2 + X_C^2}$$

$$= \sqrt{27000^2 + 9950^2}$$

$$= 28.8 \text{ k}\Omega$$

The minimum value for input impedance occurs at the highest input frequency. In most cases, the input impedance approaches the value of R_I. The computations, however, are similar to those as shown.

$$X_C = \frac{1}{2\pi fC}$$

$$= \frac{1}{6.28 \times 3000 \times 0.02 \times 10^{-6}}$$

$$= 2.65 \text{ k}\Omega$$

Now, we can compute the impedance of R_I and C_I.

$$Z(\min) = \sqrt{R^2 + X_C^2}$$

$$= \sqrt{27000^2 + 2650^2}$$

$$= 27.1 \text{ k}\Omega$$

The input impedance for the AC coupled noninverting amplifier circuit shown in Fig. 2.27(b) is essentially equal to the impedance offered by R_1 and C_1. Technically, the input impedance of the ($+$) input terminal appears in parallel with R_1. We can generally ignore this impedance, however, since it is usually an extremely high value. For our present example, we begin by finding the reactance of C_I at the lowest input frequency.

Sec. 2.7 AC Coupled Amplifier

$$X_C = \frac{1}{2\pi fC}$$

$$= \frac{1}{6.28 \times 800 \times 0.01 \times 10^{-6}}$$

$$= 19.9 \text{ k}\Omega$$

Next, we find the net impedance of R_1 and C_I.

$$Z(\text{max}) = \sqrt{R^2 + X_C^2}$$

$$= \sqrt{47000^2 + 19900^2}$$

$$= 51 \text{ k}\Omega$$

The minimum value for input impedance occurs at the highest input frequency as it did with the inverting circuit. In most cases, the input impedance approaches the value of R_1. The following computations, however, are shown:

$$X_C = \frac{1}{2\pi fC}$$

$$= \frac{1}{6.28 \times 3000 \times 0.01 \times 10^{-6}}$$

$$= 5.31 \text{ k}\Omega$$

Now, we can compute the impedance of R_1 and C_I.

$$Z(\text{min}) = \sqrt{R^2 + X_C^2}$$

$$= \sqrt{47000^2 + 5310^2}$$

$$= 47.3 \text{ k}\Omega$$

Bandwidth. Bandwidth can be defined as that range of frequencies which pass through a circuit with a voltage amplitude of at least 70.7 percent of the maximum output voltage. In other words, the range of frequencies between the two half-power points. These two frequencies can be readily determined in at least three ways:

1. numerical analyses involving higher mathematics,
2. computer-aided analysis,
3. direct measurements in the lab.

You may personally be able to employ all three of these methods, however, none of them are suitable for use in this reference book. Therefore, we will examine yet another, less direct, approach. Let us begin by making some observations. First, since the op amp has a frequency response that extends all the way to DC, the lower cutoff frequency will be unaffected by the op amp. That is, the input and output RC circuits will determine the lower cutoff frequency. Second, in a practical circuit, the upper cutoff frequency will be determined by the op amp itself. The RC circuits act as high-pass filters and will not restrict the gain at the higher input frequencies.

Calculation of the upper cutoff frequency was discussed in previous sections. We estimate it with Eq. (2.40)

$$f_U = \frac{f_{UG} R_I}{R_F + R_I} \quad (2.40)$$

In the case of Fig. 2.27(a), our ideal upper cutoff frequency is computed as

$$f_U = \frac{(1 \times 10^6)(27 \times 10^3)}{(68 \times 10^3) + (27 \times 10^3)} = 284 \text{ kHz}$$

Next, we calculate the lower cutoff frequency which is determined by the input and output RC networks. The cutoff frequency of each individual RC network is determined with the following equation:

$$f_C = \frac{1}{2\pi RC} \quad (2.41)$$

The lower cutoff frequency for the entire circuit is determined by the ratio of the cutoff frequencies for the RC circuits.

Let us now compute the lower cutoff frequency for the circuit in Fig. 2.27(a). First, we compute the individual cutoff frequencies for the two RC networks. The input circuit calculations, Eq. (2.41) are

$$f_{C_1} = \frac{1}{2\pi RC}$$

$$= \frac{1}{6.28 \times 27 \times 10^3 \times 0.02 \times 10^{-6}}$$

$$= 295 \text{ Hz}$$

A similar computation, Eq. (2.41) for the output RC circuit is

$$f_{C_2} = \frac{1}{2\pi RC}$$

$$= \frac{1}{6.28 \times 39 \times 10^3 \times 0.02 \times 10^{-6}}$$

$$= 204 \text{ Hz}$$

Computing the ratio of the two cutoff frequencies (using the higher frequency as the numerator) gives us the index needed for the lookup operation in Table 2.5.

Sec. 2.7 AC Coupled Amplifier

TABLE 2.5

	0.0	0.1	0.2	0.3	0.4	0.5	0.6	0.7	0.8	0.9
1	1.554	1.485	1.427	1.380	1.340	1.306	1.277	1.252	1.230	1.211
2	1.194	1.790	1.166	1.154	1.143	1.134	1.125	1.117	1.110	1.103
3	1.097	1.092	1.087	1.082	1.078	1.074	1.070	1.066	1.063	1.060
4	1.058	1.055	1.053	1.050	1.048	1.046	1.044	1.042	1.041	1.039
5	1.038	1.036	1.035	1.034	1.033	1.032	1.031	1.030	1.028	1.028
6	1.027	1.026	1.025	1.024	1.024	1.023	1.022	1.022	1.021	1.020
7	1.020	1.019	1.019	1.018	1.018	1.017	1.017	1.016	1.016	1.016
8	1.015	1.015	1.015	1.014	1.014	1.014	1.013	1.013	1.013	1.012
9	1.012	1.012	1.012	1.011	1.011	1.011	1.011	1.010	1.010	1.010
10	1.000	Use k = 1.0 for ratios greater than 10								

$$\text{index} = \frac{f_{C_1}}{f_{C_2}} \quad (2.42)$$

For our particular case, the index is computed as

$$\text{index} = \frac{295}{204} = 1.45$$

Finally, we use the lookup table shown in Table 2.5 to get our multiplying factor k. In this case, the value of k is about 1.32 (estimating the value between 1.34 and 1.306). The overall lower cutoff frequency can now be found by multiplying our factor k by the **higher** of the two individual cutoff frequencies.

$$f_L = f_C \times k \quad (2.43)$$

Thus, the lower cutoff frequency for the circuit in Fig. 2.27(a) is estimated as

$$f_L = 295 \times 1.32 = 389 \text{ Hz}$$

The approximate bandwidth of the circuit in Fig. 2.27(a) can now be expressed as Eq. (2.5).

$$\text{bw} = f_U - f_L$$
$$= (284 \times 10^3) - 389 \approx 283.6 \text{ kHz}$$

The bandwidth of the noninverting circuit shown in Fig. 2.27(b) is computed in the same way as we did for the inverting circuit. First, we estimate the upper cutoff frequency which is determined by the behavior of the op amp, Eq. (2.22).

$$f_U = \frac{f_{UG} R_I}{R_F + R_I}$$

$$= \frac{(1 \times 10^6)(2 \times 10^3)}{(18 \times 10^3) + (2 \times 10^3)}$$

$$= 100 \text{ kHz}$$

Next, we compute the individual cutoff frequencies for the input and output RC circuits, Eq. (2.41).

$$f_{C_1} = \frac{1}{2\pi RC}$$

$$= \frac{1}{6.28 \times 47 \times 10^3 \times 0.01 \times 10^{-6}}$$

$$= 339 \text{ Hz}$$

A similar computation for the output RC circuit is Eq. (2.41)

$$f_{C_2} = \frac{1}{2\pi RC}$$

$$= \frac{1}{6.28 \times 3.3 \times 10^3 \times 0.1 \times 10^{-6}}$$

$$= 483 \text{ Hz}$$

Computing the ratio of the two cutoff frequencies (using the higher frequency as the numerator) gives us the index, Eq. (2.42), needed for the lookup operation in Table 2.5:

$$\text{index} = \frac{f_{C_2}}{f_{C_1}} = \frac{483}{339} = 1.425$$

Using this value as the index into Table 2.5 gives us an approximate value of 1.33 for k. The lower cutoff frequency for the entire circuit in Fig. 2.27(b) can now be estimated with Eq. (2.43).

$$f_L = f_C \times k$$

$$= 483 \times 1.33$$

$$= 642 \text{ Hz}$$

The bandwidth of the circuit can now be estimated with Eq. (2.5)

Sec. 2.7 AC Coupled Amplifier

$$bw = f_U - f_L$$
$$= (100 \times 10^3) - 642$$
$$= 99.36 \text{ kHz}$$

Slew-rate limitations. The slew rate of the op amp will limit the upper cutoff frequency for high-amplitude output signals. The slew-rate limiting frequency is calculated in the same manner as described for previous amplifier configurations.

2.7.3 Practical Design Techniques

The design of either the inverting or the noninverting AC coupled amplifier is a relatively easy process. Following are the sequential steps:

1. Design the basic amplifier circuit according to the guidelines presented for the direct coupled inverting or noninverting amplifier circuits.
2. Compute the values for the input and output RC coupling components.

As an example, let us design a noninverting AC coupled amplifier that has the following characteristics:

1. midpoint voltage gain of 12
2. lower cutoff frequency of 500 hertz
3. upper cutoff frequency of at least 15 kilohertz
4. input impedance of at least 3000 ohms

Determine the value of R_I. We shall select R_I as 6.8 kilohms. Although this selection is somewhat arbitrary, we are keeping with the guidelines of choosing resistance between 1000 ohms and 680 kilohms. Additionally, the selection of R_I will have a major effect on the final input impedance since R_1 will be close to the same value as R_I, and it is R_1 that determines the input impedance of the amplifier. Thus, in order to meet the requirements for an output impedance of at least 3000 ohms, we must choose a value for R_I that is larger than 3000 ohms.

Determine the value of R_F. R_F can be computed from the voltage gain equation, Eq. (2.28)

$$A_V = \frac{R_F}{R_I} + 1$$

or

$$R_F = R_I(A_V - 1)$$

For the present design example, we compute R_F as follows:

$$R_F = 6800 \times (12 - 1) = 74.8 \text{ k}\Omega$$

We shall select the nearest standard value of 75 kilohms to use as R_F.

Determine the required unity gain frequency. We can compute the minimum unity gain frequency for our op amp with Eq. (2.22)

$$f_{UG} = \frac{\text{bw}(R_F + R_I)}{R_I} = \frac{(15 \times 10^3)(75 \times 10^3 + 6.8 \times 10^3)}{6.8 \times 10^3} = 180 \text{ kHz}$$

where f_{UG} is the minimum required unity gain frequency for the op amp and bw is the highest operating frequency. Thus, we must select an op amp that has a minimum unity gain frequency of at least 180 kilohertz. Since the 741 has a 1.0 megahertz unity gain frequency, it should be fine for our purposes. Now, let us determine the slew-rate requirement.

Determine the required slew rate. The minimum acceptable slew rate for the op amp is given by the following equation, Eq. (2.11):

$$\text{slew rate(min)} = \pi f_{SRL} v_O(\text{max})$$

In our case, let us assume that we want to deliver a full output swing (± 13 V) at the highest frequency (15 kHz). The minimum slew rate is computed as follows:

$$\text{slew rate(min)} = 3.14 \times 15 \times 10^3 \times 26 = 1.2 \text{ V}/\mu\text{s}$$

Since the 741 has a slew rate of 0.5 volts per microsecond, it will not be adequate for this application. There are many alternatives, but let us choose the MC1741SC op amp (App. 4). It has a unity gain frequency similar to the 741, but it offers a slew rate of 10 volts per microsecond.

Select R_1. R_1 is chosen to be the same value as the parallel combination of the feedback resistor (R_F) and the input resistor (R_I). This following value is computed:

$$R_1 = \frac{1}{\dfrac{1}{R_I} + \dfrac{1}{R_F}}$$

$$= \frac{1}{\dfrac{1}{6.8\text{k}} + \dfrac{1}{75\text{k}}}$$

$$= 6.23 \text{ k}\Omega$$

We shall use a standard value of 6.2 kilohms.

Sec. 2.7   AC Coupled Amplifier

Compute the value of C_I. To simplify subsequent calculations, we shall choose a value for C_I that produces a reactance which is much less than the resistance R_1. We shall design for a reactance of one-tenth R_1 or 620 ohms in this case at the lower cutoff frequency. The value for C_I is computed from the capacitive reactance equation.

$$X_C = \frac{1}{2\pi f C}$$

or

$$C = \frac{1}{6.28 f X_C}$$

In our present case

$$C_I = \frac{1}{6.28 \times 500 \times 620} = 0.514 \ \mu F$$

We shall select the next higher standard value of 0.56 microfarad.

Select the value of R_L. In many cases, R_L is the input resistance of a subsequent stage. In these cases, R_L is not selected; it is already defined by the nature of the problem. For our example design, we shall choose R_L as 27 kilohms. This is sufficiently large to eliminate any concerns of output loading on the op amp, but it is low enough to facilitate coupling to a subsequent op amp circuit if needed.

Compute the value of C_O. By following the guidelines given for the selection of the $R_1 C_I$ combination, we have assured ourselves that the lower cutoff frequency will be primarily determined by the $R_L C_O$ network. We use the fundamental formula for capacitive reactance to compute a value for C_O that produces a reactance which is equal to the value of R_L at the lower cutoff frequency (500 Hz).

$$C_O = \frac{1}{2\pi f X_C}$$

$$= \frac{1}{6.28 \times 500 \times 27 \times 10^3}$$

$$= 0.012 \ \mu F$$

The resulting circuit for our noninverting AC coupled amplifier is shown in Fig. 2.28. The actual performance of the circuit is reflected in the oscilloscope waveforms shown in Fig. 2.29. The design goals are compared to the measured performance in Table 2.6.

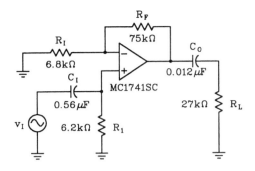

Figure 2.28 A noninverting AC coupled amplifier design.

TABLE 2.6

PARAMETER	DESIGN GOAL	MEASURED VALUES
Midpoint voltage gain	12	11.9
Frequency range	500 Hz–15 kHz	475 Hz–79.5 kHz

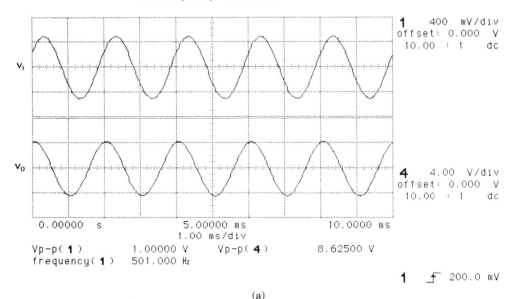

(a)

Figure 2.29 Oscilloscope displays showing the actual performance of the AC coupled amplifier shown in Fig. 2.28. (Test equipment courtesy of Hewlett-Packard Company)

Sec. 2.8 Current Amplifier

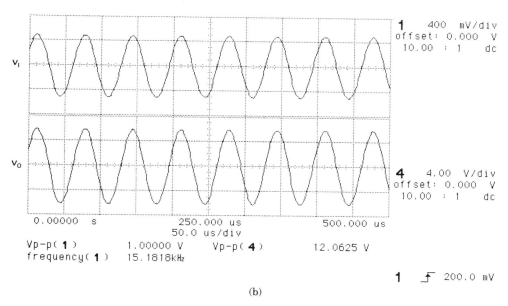

Figure 2.29 (continued)

2.8 CURRENT AMPLIFIER

2.8.1 Operation

Figure 2.30 shows the schematic diagram of a basic current amplifier. This circuit, as its name implies, accepts a current source as its input and delivers an amplified version of that current to the load. The load, in the case of Fig. 2.30, is not directly referenced to ground. A current source is normally designed to drive into a very low (ideally zero) impedance. In the case of the circuit in Fig. 2.30, the ($-$) input of the op amp is a virtual ground point. Thus the current source sees a very low-input resistance.

All of the current which leaves the source must flow through resistor R_2 since we know that no current flows in or out of the op amp input (except for bias current). The current flowing through R_2 produces a voltage drop that is determined by the value of R_2 (a constant) and the value of the input current. Once the circuit has been designed, the voltage drop across R_2 is strictly determined by the amount of input current (i_I). Notice that resistors R_2 and R_1 are essentially in parallel since R_2 is connected to a virtual ground point. Because the two resistors are in parallel, we know that the voltage across them must be the same. That is, the voltage across R_1 will be the same as the voltage across R_2 and is determined by the value of input current. The current through R_1 can be determined by Ohm's Law. If the value of R_1 is smaller than the value of R_2 (the normal case), then current i_1 will be proportionally larger than i_I (recalling that the voltages across the parallel resistors are equal).

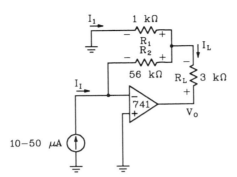

Figure 2.30 A basic current amplifier circuit.

Kirchoff's Current Law would show us that the current i_1 and i_1 must combine to produce the load current i_L. The value of i_L is strictly determined by the input current, but its value will be larger by the amount of current (i_1) flowing through R_1. Thus, we have current gain or current amplification. The larger we make i_1 as compared to i_1, the higher the current amplification. Examination of the circuit will confirm that the circuit can accept current of either polarity as long as the op amp is operating from a dual power supply.

2.8.2 Numerical Analysis

Now, analyze the current amplifier shown in Fig. 2.30 and compute the following values:

1. current gain
2. load current
3. range of acceptable input currents
4. maximum load resistance
5. input resistance
6. output resistance

Current gain. The current gain (A_I) can be initially described with the basic gain equation, Eq. (2.1).

$$A_I = \frac{\text{output current}}{\text{input current}} = \frac{i_L}{i_1}$$

Current i_L is composed of the two currents i_1 and i_1. That is,

$$i_L = i_1 + i_1$$

The voltage across R_1 is equal to the voltage across R_2 and is computed by Ohm's Law as

$$v_1 = v_2 = i_1 R_2$$

Sec. 2.8 Current Amplifier

The value of current (i_1) can also be computed by Ohm's Law as

$$i_1 = \frac{v_1}{R_1} = \frac{i_1 R_2}{R_1}$$

Substituting this into the equation for load current produces

$$i_L = i_1 + i_I$$

$$= \frac{i_1 R_2}{R_1} + i_I$$

Factoring i_1 gives us the equation for I_L

$$\boxed{i_L = i_1\left(\frac{R_2}{R_1} + 1\right)} \qquad (2.44)$$

In this form, it is easy to see that we do indeed have a current amplifier. That is, the input current (i_1) is multiplied by a constant ($R_2/R_1 + 1$) to produce the output or load current. The constant is the current gain of the circuit. That is

$$\boxed{\text{current gain} = A_I = \frac{R_2}{R_1} + 1} \qquad (2.45)$$

In the case of the circuit in Fig. 2.30, the current gain is calculated as shown

$$A_I = \frac{56 \times 10^3}{1 \times 10^3} + 1 = 57$$

It is especially important to note that the value of load current is independent of the value of load resistor. That is, the op amp circuit is acting as a current source.

Although many current sources are essentially DC (e.g., transducers), there may be an application requiring current amplification at higher frequencies. As the frequency of operation is increased, the actual current gain will begin to decrease from the low frequency value calculated. This effect is caused by the reduction in open-loop op amp gain as the input frequency is increased. The higher the value of current gain (A_I), the more significant the effects of op amp voltage gain variations.

Load current. The input current (i_1) for the circuit in Fig. 2.30 is indicated to be in the range of 10 to 50 microamps. The output current can be found by transposing the basic current gain equation, Eq. (2.1).

$$A_I = \frac{i_O}{i_I}$$

or

$$i_O = i_I A_I$$

In our case, the minimum load current is computed as

$$i_L(\min) = i_1(\min) \times A_1$$
$$= 10 \times 10^{-6} \times 57$$
$$= 570 \ \mu A$$

The highest load current is found in a similar manner as

$$i_L(\max) = i_1(\max) \times A_1$$
$$= 50 \times 10^{-6} \times 57$$
$$= 2.85 \ mA$$

Range of acceptable input currents. In order for the circuit in Fig. 2.30 to operate as a current source whose value is proportional to input current, it is essential that the output voltage (v_O) be less than the saturation voltage in either polarity. This then is the factor which restricts the range of acceptable input currents. The output voltage can be expressed as a sum of two voltages by using Kirchoff's Voltage Law.

$$v_O = v_{R_2} + v_{R_L}$$

Substituting current and resistance values (i.e., E = IR) produces

$$v_O = i_1 R_2 + i_1 \left(\frac{R_2}{R_1} + 1\right) R_L$$
$$= i_1 \left(R_2 + \frac{R_2 R_L}{R_1} + R_L\right)$$

To calculate the amount of input current needed to drive the output to saturation, we transpose this equation to find i_1 and substitute the value of V_{SAT} for v_O. For our example, let us assume that the saturation voltage is determined from the manufacturer's data sheet to be 13 volts. The maximum input current then is computed as

$$i_1(\max) = \frac{R_1 V_{SAT}}{R_1(R_2 + R_L) + R_2 R_L} \tag{2.46}$$

More specifically for our present circuit we have

$$i_1(\max) = \frac{1 \times 10^3 \times 13}{(1 \times 10^3)(56 \times 10^3 + 3 \times 10^3) + (56 \times 10^3 \times 3 \times 10^3)}$$
$$= 57.3 \ \mu A$$

The lower range is computed in a similar manner by using the other saturation limit. In most cases (i.e., balanced dual power supply circuits) the values of $\pm V_{SAT}$ will be the same. If this were true for the circuit in Fig. 2.30, we could then have a range of input currents that extended from -57.3 microamps to $+57.3$ microamps. The polarity of course telling us the direction of current flow.

Sec. 2.8   Current Amplifier   117

Maximum load resistance. Another way to view the preceding calculations is to consider a known range of input currents and a variable value for R_L. Again, the output voltage must be kept from reaching the saturation limits. We can transpose the equation, Eq. (2.46), for i_1 (maximum) to get the following result:

$$R_L(\text{max}) = \frac{V_{SAT}}{A_I i_1} - \frac{R_2}{A_I}$$

where i_1 is the highest expected input current. In the case of Fig. 2.30, we can determine the maximum value for the load resistance as

$$R_L(\text{max}) = \frac{13}{57 \times 50 \times 10^{-6}} - \frac{56 \times 10^3}{57} = 3.58 \text{ k}\Omega$$

Input resistance. Although the input resistance of the circuit in Fig. 2.30 is ideally zero, there may be applications which require us to know a more accurate value for the input resistance. The following equation can be used to estimate the input resistance of the current amplifier in Fig. 2.30:

$$\boxed{R_{IN} = \frac{R_2(R_X + R_L)}{R_X(1 + A_V) + R_L}} \quad (2.47)$$

where R_X is the resistance of R_1 and R_2 in parallel (i.e., $R_1 R_2/(R_1 + R_2)$) and A_V is the open-loop gain of the op amp at a particular frequency.

In the case of Figure 2.30, let us compute the input resistance for DC conditions. First, the open-loop gain at DC can be found in the data sheet (App. 1) to be at least 50,000. The value for R_X is calculated as

$$R_X = \frac{R_1 R_2}{R_1 + R_2}$$

$$= \frac{1 \times 10^3 \times 56 \times 10^3}{(1 \times 10^3) + (56 \times 10^3)}$$

$$= 982 \ \Omega$$

The input resistance can now be calculated, Eq. (2.47) as shown

$$R_{IN} = \frac{R_2(R_X + R_L)}{R_X(1 + A_V) + R_L}$$

$$= \frac{(56 \times 10^3)(982 + 3 \times 10^3)}{982(1 + 50000) + (3 \times 10^3)}$$

$$= 4.54 \ \Omega$$

As you might expect, the input resistance approaches the ideal value (to be driven by a current source) of zero ohms.

If the input frequency were higher than DC, then the input resistance would deviate more from the ideal value of zero. For example, if our input frequency were raised to 1 kilohertz, the input resistance would increase to about 226 ohms. Additionally, we would need to consider the effects of bandwidth and slew-rate limitations.

Output resistance. The output resistance of the circuit in Fig. 2.30 as viewed by the load resistor is ideally infinite since the circuit acts like a current source. A more accurate value for the output resistance can be computed with the following equation:

$$\boxed{R_O = A_V R_X} \quad (2.48)$$

where A_V is the open-loop voltage gain of the op amp at a particular frequency and R_X is the value of R_1 and R_2 in parallel. In the case of Fig. 2.30, let us estimate the output resistance at DC.

$$R_O = 50000 \times 982 = 49.1 \text{ M}\Omega$$

As evidenced in the equation, this value becomes less ideal as the frequency of operation is increased.

2.8.3 Practical Design Techniques

A practical current amplifier circuit can be designed by applying the equations discussed in the preceding paragraphs. Depending on the application, you will know some combination of the following parameters:

1. input current range
2. output current range
3. current gain
4. load resistance

For purposes of a design example, let us build a current amplifier that satisfies the following requirements:

1. output current of 10 milliamps (constant)
2. input current 500 microamps
3. load resistance < 500 ohms
4. available power supply is ±15 volts
5. use 741 op amp if practical

Compute the required current gain. The required current gain of our circuit can be computed with the basic current gain equation, Eq. (2.1)

$$A_I = \frac{i_O}{i_I}$$

Sec. 2.8 Current Amplifier

$$= \frac{10 \times 10^{-3}}{500 \times 10^{-6}}$$

$$= 20$$

Determine the maximum value for R_2. The maximum value for R_2 can be found by applying a transposed version of the equation we used for computing the maximum value for R_L.

$$R_L(\max) = \frac{V_{SAT}}{A_I i_1} - \frac{R_2}{A_I}$$

or

$$R_2(\max) = \frac{V_{SAT}}{i_1} - A_I R_L(\max)$$

$$= \frac{12}{500 \times 10^{-6}} - 20 \times 500$$

$$= 14 \text{ k}\Omega$$

We shall select a standard value less than this. For purposes of our design, we shall select a 12 kilohm resistor for R_2. Notice that we used the worst-case value of 12 volts as the saturation voltage for the op amp.

Compute the value of R_1. R_1 can be computed by applying a transposed version of the current gain equation, Eq. (2.45).

$$A_I = \frac{R_2}{R_1} + 1$$

or

$$R_1 = \frac{R_2}{(A_I - 1)}$$

$$= \frac{12 \times 10^3}{20 - 1}$$

$$= 632 \text{ }\Omega$$

We shall select the nearest standard value of 620 ohms. If it is essential to have a precise value of load current, then we would put a variable resistance in series with R_2.

The schematic of our completed design is shown in Fig. 2.31. The load resistor has been replaced with a zener diode to illustrate a possible application. By forcing the zener current to be a known value, we can measure the zener voltage and compute the zener resistance. As long as the effective zener resistance is below our established limit

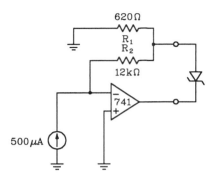

Figure 2.31 A current amplifier design being used to deliver a constant current to a zener load.

TABLE 2.7

PARAMETER	DESIGN GOAL	MEASURED VALUES
Input current	500 µA	500 µA
Output current	10 mA	9.97 mA
Current gain	20	19.9

for R_L, the circuit will work fine. The measured performance of the circuit is contrasted with the design goals in Table 2.7.

2.9 HI-CURRENT AMPLIFIER

2.9.1 Operation

A general purpose op amp can only supply a few milliamps to a load. The 741, for instance, has a short-circuit output current of 20 milliamps. Some applications require substantially higher currents. The circuit in Fig. 2.32 illustrates one common method for increasing the available current at the output of an op amp circuit. This technique is illustrated for a simple inverting amplifier, but is applicable to most voltage amplifier circuits.

The majority of the circuit operation is identical to that discussed with reference to the basic inverting amplifier circuit and will not be repeated here. Recall that resistors R_F and R_I determine the voltage gain of the circuit. Resistor R_B is to compensate for op amp bias currents. Potentiometer R_P has been added to the basic inverting amplifier. This will be used to force an offset in the output.

The output voltage of the op amp appears on the base of Q_1. A similar voltage appears on the emitter of Q_1 which is connected as a voltage follower. The actual voltage on the emitter is less than the base voltage by a small amount (0.7 volts). In general, the output voltage across R_L is essentially the same as the output voltage of the op amp. The output current of the op amp provides base current to the transistor. The load current,

Sec. 2.9 Hi-Current Amplifier

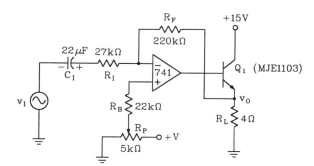

Figure 2.32 An inverting voltage amplifier with additional current amplification.

on the other hand, is provided by the transistor's emitter current. You will recall that transistor emitter current and base current are related by a current gain factor called β, which can range from about 20 up to several thousand. With the circuit connected as shown, the load can draw β times as much current as the basic op amp supplies.

In order for the circuit to operate properly, the base voltage on Q_1 must always be positive with reference to ground since the load is returned to ground. Ordinarily, an AC signal applied to a split-supply op amp would produce bipolar signals on its output. To prevent this situation, we adjust R_P to establish a positive bias on the base of Q_1 that is approximately equal to half of the positive saturation voltage. The output signal then can swing from near zero up to near saturation. Recognize that this swing is only half of the swing available with previous amplifiers and represents a disadvantage of the circuit shown in Fig. 2.32. An alternative is to return the load to the negative 15-volt supply (emitter bias), or use a single-supply op amp.

One very important characteristic of the circuit in Fig. 2.32 is that output voltage is unaffected by changes in V_{BE}. We know V_{BE} is approximately 0.7 volts for silicon transistors, but we also know it changes with temperature and varies from one transistor to another. By connecting the feedback resistor (R_F) to the emitter rather than directly to the output of the op amp, we include the base-emitter junction in the feedback circuit. Changes in the base-emitter voltage are now effectively compensated by the op amp.

Since the (+) input on the op amp is at some positive level, the (−) pin will also be at a similar level. If the input signal were centered on zero volts, this could cause an undesired DC offset in the output. Capacitor C_1 is included to isolate the DC level on the (−) pin from the DC level associated with the signal source. If it is sufficiently large, it has no effect on the gain calculations in the circuit. If its reactance at the lowest input frequency exceeds one tenth the resistance of R_I then the gain of the circuit should be computed in the same manner as the AC coupled amplifiers discussed in a previous section.

2.9.2 Numerical Analysis

Let us now analyze numerically the behavior of the circuit shown in Fig. 2.32. Appendix 1 shows the data sheet for the 741 op amp, and App. 2 shows the specifications for an MJE1103 transistor. First, we know from our basic transistor theory that the impedance looking into the base will be approximately equal to β times the resistance in the emitter circuit. Thus, the op amp sees the load resistance as

$$R'_L = \beta R_L \qquad (2.49)$$

In the case of Fig. 2.32, the emitter resistance appears as

$$R'_L = 750 \times 4 = 3000 \ \Omega$$

We can now refer to the 741 data sheet and determine the worst-case saturation voltage when using a ±15 volt supply and driving a 3000 ohm load. This value is listed as 10 volts.

The transistor data sheet indicates that the base-emitter voltage drop is 2.5 volts or less. Thus, the highest voltage (worst-case) that we can expect at the load is

$$v_O(\text{peak}) = V_{SAT} - V_{BE} \qquad (2.50)$$

For the circuit in Fig. 2.32, we have

$$v_O(\text{peak}) = 10 - 2.5 = 7.5 \text{ V}$$

With Ohm's Law, we can calculate the maximum instantaneous (i.e., peak) load current as

$$i_L(\text{peak}) = \frac{v_O(\text{peak})}{R_L}$$

$$i_L(\text{peak}) = \frac{7.5}{4}$$

$$= 1.88 \text{ A}$$

The op amp must supply a current that is smaller than load current by a factor of β. That is,

$$i_O(\text{op amp}) = \frac{i_L}{\beta} \qquad (2.51)$$

In our particular case, the calculations are

$$i_O(\text{op amp}) = \frac{1.88}{750} = 2.51 \text{ mA}$$

This current is well within the range of op amp output currents even though the load current itself is nearly two amps.

In order to insure that we have a maximum symmetrical swing for the output signal, we will establish a positive DC offset in the output. The value of this offset should be midway between the two extremes. One extreme is the value of V_{SAT} or +10 volts. The other extreme is the minimum turn on voltage for Q_1 and is 2.5 volts. The DC level on the output of the op amp then must be

Sec. 2.9 Hi-Current Amplifier

$$V_{BIAS}(\text{output}) = \frac{V_{SAT} + V_{BE}}{2} \qquad (2.52)$$

In the present case, we have

$$V_{BIAS}(\text{output}) = \frac{10 + 2.5}{2} = 6.25 \text{ V}$$

The maximum voltage swing at the output of the op amp will be

$$v_O(\text{op amp}) = V_{SAT} - V_{BE}$$
$$= 10 - 2.5$$
$$= 7.5 \text{ volts peak-to-peak}$$

We will get this same swing at the load, but the DC level will be reduced by the amount of V_{BE}.

If the output of the op amp is allowed to go more positive than $+V_{SAT}$ (estimated here as $+10$ volts), the waveform will be clipped on its positive peaks. This clipped waveform would also appear across the load.

If the output of the op amp is allowed to go below the minimum turn on voltage for Q_1 (estimated here as $+2.5$ volts), the waveform will be clipped on its negative peaks. The load voltage will also have a clipped waveform.

If the amplitude of the input signal remains fixed, but the DC offset voltage in the output of the op amp is changed, then similar waveform clipping can occur. That is, if the instantaneous value of the combined AC and DC voltages on the output of the op amp go more positive than $+V_{SAT}$ or less positive than the turn on voltage for Q_1, the output waveform will be distorted.

The sketch below clarifies the relationships between the output waveforms (op amp and load), the bias level and the clipping levels.

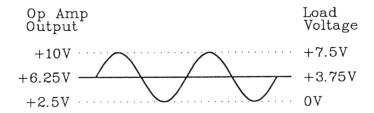

In order to establish the 6.25 volts DC bias at the output of the op amp, we need to adjust potentiometer R_p to the necessary level. Although this would have to be done in a lab environment, we can compute the required value of voltage at the (+) input of the op amp. Since capacitor C_1 acts as an open to DC, the op amp is essentially configured as a voltage follower with reference to the DC offset voltage at potentiometer R_P. Therefore to obtain a 6.25 volt offset in the output, we will need a 6.25 volt offset on the noninverting (+) input of the op amp.

We have already determined that the maximum output swing is 7.5 volts peak-to-peak. We can determine the maximum input swing before distortion by applying the voltage gain formula, Eq. (2.6), for an inverting amplifier.

$$A_V = -\frac{R_F}{R_I}$$

$$= -\frac{220 \times 10^3}{27 \times 10^3}$$

$$= -8.15$$

This can now be used with the basic voltage gain equation, Eq. (2.1) to compute the maximum allowable input swing.

$$A_V = \frac{v_O}{v_I}$$

or

$$v_I(\text{max}) = \frac{v_O(\text{max})}{A_V}$$

$$= \frac{7.5}{8.15}$$

$$= 0.92 \text{ volts peak-to-peak}$$

This can be more conveniently discussed as an RMS value so we shall convert it using our basic electronics conversion factor.

$$v_I(\text{RMS}) = 0.707 \left(\frac{\text{peak-to-peak}}{2}\right)$$

$$= 0.707 \left(\frac{0.92}{2}\right)$$

$$= 0.325 \text{ volts RMS}$$

If we drive the amplifier with a signal greater than 0.325 volts RMS we can expect clipping to occur in the output.

Sec. 2.9   Hi-Current Amplifier   125

The input impedance of the circuit is approximately equal to the value of R_I. We can apply Ohm's Law to compute the current supplied by the AC input source under maximum input voltage conditions.

$$i_I(\text{peak}) = \frac{v_I(\text{peak})}{R_{IN}}$$

$$= \frac{0.46}{27 \times 10^3}$$

$$= 17 \;\mu\text{A}$$

The usefulness of the circuit should become very apparent after this last calculation. A signal source delivering a peak current of 17 microamps is driving a load resistance which requires 1.88 amps peak current.

2.9.3 Practical Design Techniques

Much of the design procedure was covered in our numerical analysis discussion in the preceding section. Let us now design a high-current amplifier that will perform according to the following:

1. input voltage of 1.0 volt RMS,
2. input resistance > 10 kilohms,
3. input frequency range between 10 hertz and 2.0 kilohertz,
4. load resistance of 50 ohms,
5. ±15 volt supplies are to be used,
6. 741 op amp is to be used if practical.

Select the output transistor. There are basically five transistor parameters that must be reviewed to select a transistor:

1. forward current transfer ratio (h_{FE}) or current gain (β),
2. base-emitter voltage drop (V_{BE}),
3. emitter-collector breakdown voltage,
4. maximum collector current,
5. power dissipation.

In some cases, the frequency characteristics of the transistor must be evaluated, but in most cases the transistor performance exceeds that of the op amp and can be ignored.

In our case, we need an emitter-collector breakdown voltage greater than 15 volts. The exact collector current will be computed later, but we need to estimate a worst-case value so that we can select the transistor. For this purpose, we can assume that the entire

+15 volts of the supply are felt across the 50 ohm load. Ohm's Law then tells us the value of load current.

$$i_L(\text{estimate}) = \frac{V_{CC}}{R_L}$$

$$= \frac{15}{50}$$

$$= 300 \text{ mA}$$

The actual collector current will be less than this, but this is a good value to use for initial transistor selection.

Now, we need to determine the required current gain (β) of the transistor. In data sheets, this is generally labeled as h_{FE}. We can again make a rough estimate for purposes of transistor selection. If we divide the load current computed above by half of the short-circuit output current of the op amp, we will have a good place to start. The short-circuit current for a 741 is listed as 20 milliamps. The following computation gives us the minimum value of β that our transistor should have.

$$\boxed{\beta(\text{min}) = \frac{i_L(\text{estimate})}{0.5 I_{SC}}} \quad (2.53)$$

In the present case, we have

$$\beta(\text{min}) = \frac{300 \times 10^{-3}}{0.5 \times 20 \times 10^{-3}} = 30$$

The power dissipation of the transistor can be estimated with the following equation:

$$\boxed{P_D(\text{estimate}) = \frac{V_{CC}^2}{4R_L}} \quad (2.54)$$

where V_{CC} is the positive supply voltage. For the present case, the estimated power dissipation of the transistor is

$$P_D(\text{estimate}) = \frac{V_{CC}^2}{4R_L}$$

$$= \frac{15^2}{4 \times 50}$$

$$= 1.125 \text{ W}$$

By scanning a transistor data book (selector guides in particular), a transistor that satisfies these requirements can be found. For illustration purposes, let us select a 2N3440 transistor. The data sheet for this common device is presented as App. 9. Its critical parameters are listed

Sec. 2.9 Hi-Current Amplifier

1. β or h_{FE} — 160
2. base-emitter voltage drop (V_{BE}) — 1.3 volts (maximum)
3. emitter-collector breakdown voltage — 250 volts (minimum)
4. maximum collector current — 1.0 amps
5. power dissipation — 10 watts
6. thermal resistance, junction to case (θ_{JC}) — 17.5 °C/W
7. thermal resistance, junction to air (θ_{JA}) — 175 °C/W
8. maximum junction temperature — 200 °C

These values exceed our rough, worst-case requirements. Now let us extend our estimate to include the determination of a heat sink (see App. 10 for a more complete discussion). We shall assume that 50 °C will be the highest expected ambient temperature. The required thermal resistance (θ_{JA}) can be estimated as follows:

$$\theta_{JA}(\text{req'd}) = \frac{T_J(\text{max}) - T_A}{P_D}$$

$$= \frac{200 - 50}{1.125}$$

$$= 133.33 \text{ °C/W}$$

Since the required value of θ_{JA} is greater than the transistor's θ_{JC}, this transistor can be used for this application. However, since the required value of thermal resistance (θ_{JA}) is less than the (θ_{JA}) for the transistor, a heat sink will be needed to insure safe operation. The required thermal resistance (θ_{SA}) of the heat sink can be estimated as follows:

$$\theta_{SA} = \theta_{JA}(\text{req'd}) - \theta_{JC} - \theta_{CS}$$

$$= 133.33 - 17.5 - 2$$

$$= 113.8 \text{ °C/W}$$

The case-to-sink resistance was estimated as 2 °C/W. It should be easy to find or make a satisfactory heat sink for this application. There are many transistor/heat sink combinations that are adequate for a given application. Final selection must include cost and availability considerations.

Determine the maximum output voltage of the op amp. The maximum op amp output voltage is simply V_{SAT}. Appendix 1 lists the data sheet for a 741. The worst-case saturation voltage is listed as 10 volts for resistive loads of 2 to 10 kilohms. If the load on the op amp is over 10 kilohms, the saturation voltage is listed as 12 volts minimum. The load as seen by our op amp is computed by applying a basic transistor equation, Eq. (2.49).

$$R'_L = \beta R_L$$

$$= 160 \times 50$$
$$= 8.0 \text{ k}\Omega$$

Thus the maximum available voltage at the output of the op amp will be considered to be 10 volts. In practice, it will likely be higher.

Determine the minimum output voltage of the op amp. The lower limit on op amp output voltage is determined by the V_{BE} value of the transistor. The worst-case value for the 2N3440A is given as 1.3 volts. Thus, our op amp output voltage can swing as low as 1.3 volts without fear of clipping.

Determine the required bias voltage at the output. The output of the op amp should be biased half way between its two limits (V_{SAT} and V_{BE}). This is computed as shown, Eq. (2.52):

$$V_{BIAS}(\text{output}) = \frac{V_{SAT} + V_{BE}}{2}$$
$$= \frac{10 + 1.3}{2}$$
$$= 5.65 \text{ V}$$

Determine the maximum AC swing at the output. The output of the op amp is centered at the bias level and can swing between V_{SAT} and V_{BE}. The RMS value of output voltage is computed as

$$\boxed{v_O(\text{RMS}) = 0.707 \left(\frac{V_{SAT} - V_{BE}}{2} \right)} \qquad (2.55)$$

For the present circuit, we have

$$v_O(\text{RMS}) = 0.707 \left(\frac{10 - 1.3}{2} \right) = 3.08 \text{ V}$$

Compute the required voltage gain. The required voltage gain of the amplifier circuit is determined by applying the basic voltage gain equation, Eq. (2.1).

$$A_V = \frac{v_O}{v_I}$$
$$= -\frac{3.08}{1.0}$$
$$= -3.08$$

Note the negative value simply indicates an inversion.

Sec. 2.9 Hi-Current Amplifier 129

Determine the value of R_I. The value of input resistor is chosen to establish the required input resistance of the circuit. In our case, anything greater than 10 kilohms should suffice. Let us choose a standard value of 18 kilohms.

Calculate the value of R_F. R_F is calculated with the inverting amplifier gain equation, Eq. (2.6).

$$A_V = -\frac{R_F}{R_I}$$

or

$$\begin{aligned}R_F &= -A_V R_I \\ &= -(-3.08 \times 18 \times 10^3) \\ &= 55.4 \text{ k}\Omega\end{aligned}$$

We shall select a standard value of 56 kilohms.

Determine the required unity gain frequency. The required unity gain frequency is computed in a manner similar to that in previous discussions, Eq. (2.22).

$$\begin{aligned}f_{UG} &= \frac{bw(R_F + R_I)}{R_I} \\ &= \frac{(2 \times 10^3)(56 \times 10^3 + 18 \times 10^3)}{18 \times 10^3} \approx 8.2 \text{ kHz}\end{aligned}$$

Select a value for R_P. The value of the potentiometer R_P is essentially arbitrary. As it is made smaller, its power rating requirement becomes higher, and the current draw from the supply becomes greater. If R_P is made excessively large, then the effects of bias currents which flow through R_P are more pronounced. As a guideline, select R_P to be approximately equal to $R_I/10$, but consider 1.0 kilohm to be the minimum practical value. In our case, we have

$$\begin{aligned}R_P &= \frac{R_I}{10} \\ &= \frac{18 \times 10^3}{10} \\ &= 1.8 \text{ k}\Omega\end{aligned}$$

We shall choose a standard value of 2 kilohms.

Compute the value of R_B. The optimum value for R_B varies as the wiper arm of R_P is moved. However, the preceeding method for selecting R_P reduces this dependency.

We shall compute the value of R_B needed when the wiper arm of R_P is at midpoint. R_B is computed as shown.

$$R_B = \frac{R_F R_I}{R_F + R_I} - \frac{R_P}{2} \qquad (2.56)$$

In our particular case, we compute R_B as

$$R_B = \frac{56 \times 10^3 \times 18 \times 10^3}{56 \times 10^3 + 18 \times 10^3} - \frac{2 \times 10^3}{2} = 12.6 \text{ k}\Omega$$

We shall choose a standard value of 12 kilohms.

Compute the value of C_I. The purpose of the input coupling capacitor is to isolate the DC levels between the signal source and the ($-$) pin on the op amp. It should be selected to have a reactance of less than one tenth of R_I at the lowest input frequency (10 Hz in this case). The calculations are

$$C_I = \frac{1}{2\pi f X_C}$$

$$= \frac{1}{6.28 \times 10 \times 1.8 \times 10^3}$$

$$= 8.85 \text{ }\mu\text{F}$$

We shall choose a standard value of 10 microfarad.

Bandwidth and slew-rate considerations. Since our application requires only modest performance, neither slew rate nor bandwidth limitations should pose problems. If the application were more demanding, then these restrictions would have to be considered. The methods described for previous amplifier circuits can be utilized to evaluate the effects of bandwidth and slew-rate limitations.

The schematic of our completed design is shown in Fig. 2.33. Actual performance of the circuit is indicated by the oscilloscope display shown in Fig. 2.34.

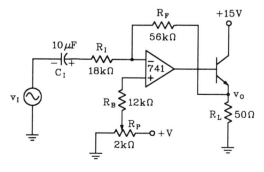

Figure 2.33 A high-current amplifier design.

Sec. 2.9 Hi-Current Amplifier

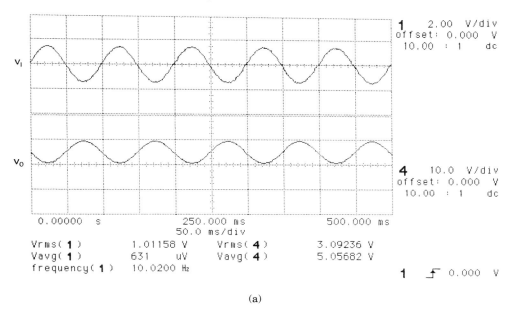

(a)

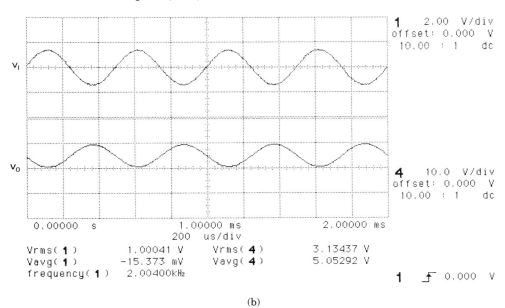

(b)

Figure 2.34 Oscilloscope displays showing the actual performance of the high-current amplifier shown in Fig. 2.33. (Test equipment courtesy of Hewlett-Packard Company)

2.10 TROUBLESHOOTING TIPS FOR AMPLIFIER CIRCUITS

In order for an amplifier to operate properly, it must be biased in its linear range of operation. That is, the output must be between the two saturation limits with no signal applied. Many, if not most, of the problems encountered when troubleshooting op amps configured as linear amplifiers result in the output being driven to one of the saturation limits. Your task, then, is to recognize the symptoms and to locate the defective component.

If the amplifier circuit is properly designed (i.e., capable of achieving the desired performance), then you can generally diagnose the problem by comparing the actual behavior to ideal op amp behavior. The following are two critical characteristics to remember when troubleshooting amplifier circuits utilizing op amps:

1. the output should be between the saturation limits,
2. the differential input voltage (v_D) should be very near zero.

2.10.1 Basic Troubleshooting Concepts

When troubleshooting any type of circuit, it is important to use a logical, systematic technique. Although there are several accepted methods, the following sequence of activities is a common and effective procedure:

1. observation
2. signal injection/tracing
3. voltage measurements
4. resistance measurements

Observation. This is probably the most important step in the process if done effectively. Observation means more than just looking at the circuit. It includes all of the following actions:

1. Interrogate the owner, user or operator for clues regarding how the trouble developed.
2. Operate the user controls and observe the behavior for clues.
3. Use your senses. Do you see any visible damage? Do you smell burned components? Do you hear suspicious sounds? Can you feel a component that is hotter than normal?
4. Be alert to similarities between observable symptoms on the defective unit and the symptoms of previously diagnosed circuits.

Many problems can be identified during the observation stage. How many of us have "successfully" traced a malfunction throughout a complex circuit until we located a

Sec. 2.10 Troubleshooting Tips for Amplifier Circuits

suspected switch or variable resistor on the schematic? Then when we physically locate the suspected component on the system it turns out to be a front panel control! Had we applied this procedure faithfully, we could have reduced our efforts dramatically.

Signal injection/tracing. All electronic circuits can be diagnosed to some extent by signal injection, signal tracing or a combination of the two methods. The underlying goal for this process is to reduce the number of possible culprits from a set consisting of every component in the system down to a smaller set consisting of only a few components.

Signal injection requires us to inject a known, good signal at some point in the circuit and observe the effects. If the circuits which utilize this signal then appear to operate normally, we can infer that the malfunction is located ahead of our injection point. We then move our injection point closer to the source of the trouble and inject another signal. Again, the behavior of the subsequent circuits will provide guidance as to our next injection point. Two common types of test equipment for signal injection are signal generators and logic pulsers.

Signal tracing is similar in concept except we put a known good signal at the input and verify (trace) its presence throughout the circuit. If we lose the signal (or it becomes distorted) at a certain point, then we can infer that the trouble lies ahead of the monitored point. The oscilloscope and logic probe are two common types of signal tracing equipment.

Both signal tracing and signal injection can be enhanced by using the split-half method of troubleshooting. By selecting your injection or monitor point to be approximately half way through the suspected range of components, each measurement effectively reduces the number of possible components by half.

Voltage measurements. Voltage measurements normally occur after you have isolated the problem down to a particular stage consisting of up to perhaps 10 components. The voltage checks contrasted with normal values should result in the narrowing of suspects down to one or two possibilities. Distinction between the signal tracing and voltage measuring phases often becomes blurred when using an oscilloscope. The concept remains valid, however.

Resistance measurements. Resistance checks are performed last since accurate measurements often require desoldering of a component. Desoldering is not only time consuming, but it also risks damage to an expensive printed circuit board in many cases. The resistance checks are done to verify that you have in fact located the defective component. Component testers can also be used at this point if available and appropriate.

2.10.2 Specific Techniques for Op Amps

Through observation and signal injection/tracing the technician can normally isolate the problem down to a specific circuit. For purposes of this discussion, we shall assume the problem has been isolated to an amplifier circuit built around an op amp. The following sequence of activities will normally isolate the defective component:

134   Amplifiers   Chap. 2

1. verify the power supply voltage on the op amp,
2. measure output voltage,
3. measure the differential input voltage (v_D),
4. compare the results of steps 2 and 3. If the results violate basic theory (e.g., noninverting input is more positive than inverting input, but output is negative) then the op amp is probably bad. If no basic theory principles are violated, check the following:
   a. correct input (especially the DC level),
   b. feedback path,
   c. input path.

Your most powerful troubleshooting tool when diagnosing op amp circuits is a solid grasp of the basic theory of operation. Although the performance of op amps can deteriorate in some ways, it is far more common for the device to exhibit catastrophic failure.

REVIEW QUESTIONS

1. A certain amplifier has a voltage gain of 100. Express this gain in decibels.
2. Suppose the amplifier circuit shown in Fig. 2.3, is altered to have the following values:

$$R_I = 39 \text{ k}\Omega$$
$$R_F = 470 \text{ k}\Omega$$
$$R_B = 36 \text{ k}\Omega$$
$$R_L = 68 \text{ k}\Omega$$

What is the voltage gain of the circuit with the new values? What happens to the voltage gain if R_L is decreased to 27 kilohms?

3. Refer to the amplifier circuit described in Ques. 2. Compute the input impedance of the circuit. Does the input impedance change if R_L is reduced to 27 kilohms?
4. If a 741 op amp were powered by a ±15 volt supply, what is the largest voltage swing that can be guaranteed on the output if the load is 18 kilohms? Repeat this question for a 2.0 kilohm load.
5. A certain op amp application requires a ±10 volt RMS output voltage swing and operates at a maximum sinewave frequency of 21.5 kilohertz. What is the minimum slew rate for the op amp that will allow the signal to pass without substantial slew rate distortion?
6. A simple noninverting amplifier (similar to Fig. 2.12) has the following component values:

Op amp	741
R_I	4.7 kΩ
R_F	68.kΩ
R_L	18 kΩ
R_B	4.3 kΩ

Compute the small signal bandwidth of the amplifier (ignore slew-rate considerations).

Chap. 1 Review Questions   135

7. The value of C_I in Fig. 2.26(a) is the primary factor which sets the upper cutoff frequency. (True or False) Explain why or why not.
8. If capacitor C_O in Figure 2.26(b) becomes open, what effect will this have on circuit operation to the left of capacitor C_O?
9. If capacitor C_I in Fig. 2.26(b) becomes open, what happens to the DC voltage on the output pin of the op amp? What happens to the DC voltage across R_L?
10. Refer to Fig. 2.32. As the wiper arm of R_P is moved to the right, what happens to the average current through R_L?
11. Refer to Fig. 2.32. If the wiper arm of R_P is moved too far to the left, the output waveform will start to clip. Explain which peak (positive or negative) is clipped and why?
12. Refer to Fig. 2.21. What happens to the average (i.e., DC) current through R_L if R_{I_1} becomes open?
13. What is another name for a noninverting amplifier with a voltage gain of one?
14. Can a standard 741 op amp be used to amplify a 33 kilohertz signal if the desired voltage gain is 5, and the maximum peak output voltage swing is 11 volts? Explain your answer.
15. While troubleshooting the circuit shown in Fig. 2.27(a), you discover that the voltage on the inverting ($-$) pin of the op amp is approximately zero volts (with a normal signal applied at the input). If you think this is normal, explain why. If you think it is abnormal, what is the most likely defect?

CHAPTER 3

VOLTAGE COMPARATORS

3.1 VOLTAGE COMPARATOR FUNDAMENTALS

A voltage comparator circuit compares the values of two voltages and produces an output to indicate the results of the comparison. The output is always one of two values (i.e, the output is digital). Suppose, for example, we have two voltage comparator inputs labeled A and B. The circuit can be designed so that if input A is a more positive voltage than input B, the output will go to $+V_{SAT}$. Similarly, if input A is less positive than input B, the output will go to $-V_{SAT}$. In general, the voltage comparator circuit accepts two voltages as inputs and produces one of two distinct output voltages depending on the relative values of the two inputs.

During the preceeding discussion, we were careful not to consider what happens when the two input voltages are equal. In a simple voltage comparator, this condition can produce indeterminate operation. That is, the output may be at either of the two normal output voltage levels or, more probably, oscillating between the two output levels. This erratic behavior is easily overcome by adding positive feedback to the comparator. With positive feedback, the circuit has hysteresis. In the simple comparator circuit, output switching occurs when the two input voltages are equal. Hysteresis causes the circuit to have two different switching points. This important concept will be explained in greater detail in Sec. 3.3.

Voltage comparator circuits are widely used in analog-to-digital converter applications and for various types of alarm circuits. In the case of the alarm application, one input of the comparator is controlled by the monitored signal (e.g., the voltage produced by a pressure transducer). The second input to the comparator is connected to a reference voltage representing the safe level. If the pressure in the device being monitored exceeds the safe limit, the comparator output will change states and sound an alarm. Figure 3.1 illustrates a voltage comparator circuit used in conjuction with a pressure sensor and a

Sec. 3.2   Zero-Crossing Detector   137

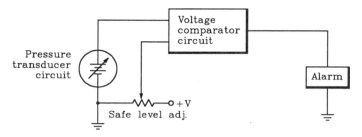

Figure 3.1   The voltage comparator is often used in alarm applications. In this figure, the alarm sounds when the voltage from the pressure sensor exceeds the voltage set by the potentiometer.

potentiometer. If the pressure being monitored exceeds a certain prescribed value, the voltage generated by the pressure sensor exceeds the preset voltage on the potentiometer. This causes the output voltage to change states and to sound the alarm.

3.2 ZERO-CROSSING DETECTOR

3.2.1 Operation

Figure 3.2 shows the schematic of a simple inverting voltage comparator being used as a zero-crossing detector. That is, the output of the comparator switches every time the input signal passes through (i.e., crosses) zero volts. As simple as it is, this circuit has practical applications.

One way to view the operation of the circuit in Fig. 3.2 is to consider it to be an open-loop amplifier. That is, with no feedback, the gain of the amplifier is simply the open-loop gain of the op amp itself. Since this gain value is very high (at least at low frequencies), we know the output will be driven to either $+V_{SAT}$ or $-V_{SAT}$ if the input is more than perhaps one microvolt or so above or below ground potential.

Input resistor R_I partially determines the input impedance for the circuit. Resistor R_B helps compensate for the effects of input bias currents. That is, by making R_B and R_I equal in value, then the voltage drops caused by the $(+)$ and $(-)$ bias currents will be approximately equal, thereby, tending to cancel their effects. R_B also contributes to the input impedance.

Since the output voltage is at one of the two saturation levels at all times (except during the short switching time), the circuit essentially converts the sinewave input into a square wave. The resulting square wave will have the same frequency as the input, but the amplitude will always swing between $\pm V_{SAT}$ regardless of the value of input voltage. Figure 3.3 illustrates the relationship between the input and output waveforms.

The output of a voltage comparator switches between two voltage limits ($\pm V_{SAT}$ in the case of the simple comparator in Fig. 3.2). In a real op amp, it takes a small but definite amount of time for the output to switch between the two voltage levels. The maximum rate at which the output can change states is called the slew rate of the op amp

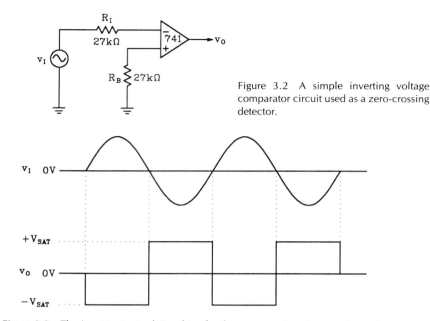

Figure 3.2 A simple inverting voltage comparator circuit used as a zero-crossing detector.

Figure 3.3 The input/output relationships for the zero-crossing detector shown in Fig. 3.2.

and is specified in the manufacturer's data sheet. If the input frequency is too high (i.e., changes too quickly), then the output of the op amp cannot change fast enough to keep up with the input. The initial effects of slew rate become evident by nonideal rise and fall times on the output waveshape. If the frequency continues to increase, the rise and fall times—which are established by the slew rate—become a significant part of the output waveform. Figure 3.4 illustrates the effects of slew rate on the output of the simple comparator circuit.

3.2.2 Numerical Analysis

For purposes of numerical analysis of the circuit shown in Fig. 3.2, let us assume the following input signal characteristics:

1. input frequency   1.8 kilohertz
2. input voltage   3 volts RMS
3. input reference   zero volts

Minimum input impedance. The input impedance is established by R_I, R_B, and the differential input resistance of the op amp. These are all effectively in series. The input resistance of the 741 (R_D) is listed in the manufacturer's data sheet (App. 1) as at least 300 kilohms. The minimum total input resistance then is computed as

Sec. 3.2 Zero-Crossing Detector

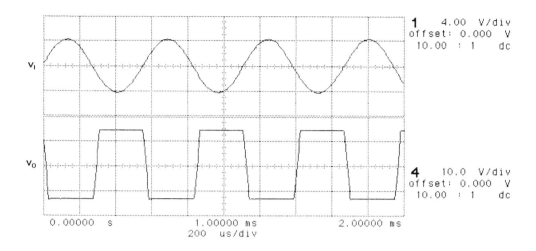

Figure 3.4 Oscilloscope displays show the effect of slew rate on the output of a zero-crossing detector circuit. (Test equipment courtesy of Hewlett-Packard Company)

$$R_{IN} = R_I + R_D + R_B \quad (3.1)$$

In this particular case

$$R_{IN} = 27k + 300k + 27k = 354 \text{ k}\Omega$$

It is important to note that without feedback, the $(-)$ input does **not** behave as a virtual ground.

Maximum input current. The maximum input current can be calculated by application of Ohm's Law

$$i_I(\text{max}) = \frac{v_I}{R_{IN}(\text{min})}$$

$$= \frac{3}{354 \times 10^3}$$

$$= 8.47 \text{ }\mu\text{A}$$

$$i_I(\text{peak}) = i_I(\text{RMS}) \times 1.414$$

$$= 8.47 \times 10^{-6} \times 1.414$$

$$= 12 \text{ }\mu\text{A}$$

Output voltage. The limits of the output voltage in the circuit shown in Fig. 3.2 are simply the values of $\pm V_{SAT}$. For ± 15 volt power supplies and a load resistance of greater than 10 kilohms, the manufacturer's data sheet in App. 1 lists a minimum output swing of ± 12 volts. If we loaded the circuit with a resistance of less than 10 kilohms, then we could expect the output levels to decrease.

3.2.3 Practical Design Techniques

For low-frequency noncritical applications, the simple circuit shown in Fig. 3.2 can be very useful. For purposes of a design example, let us develop a circuit to satisfy the following requirements:

1. input voltage   500 millivolts peak
2. input frequency   1 to 10 kilohertz
3. input reference   zero volts
4. maximum input current   1.2 milliamps
5. minimum output voltage ± 10 volts
6. load resistance   68 kilohms

Op amp selection. We must select an op amp that can satisfy the output voltage requirements, survive the input voltage swings, and respond to the input frequencies. The manufacturer's data sheet in App. 1 confirms that a 741 is capable of delivering a ± 10 volt output. More specifically, the minimum output voltage with ± 15 volt supplies and a load of greater than 10 kilohms is ± 12 volts.

Additionally, the data sheet indicates that input voltage levels may be as high as the value of supply voltage. So far, the 741 seems like a good choice. Now, let us consider the frequency effects.

The open-loop voltage comparator application requires the output voltage of the op amp to change from one extreme to the other. This change requires a finite amount of time. For DC or low-frequency applications, it is generally an insignificant amount of time. As the input frequency increases, however, the switching time becomes a greater portion of the total time for one alternation of the input signal. In the extreme case, if the input alternation were shorter than the time required for the output to change states, then the comparator would cease to function properly. That is, the output voltage would not have time to reach its limits.

It is the slew rate of the op amp that determines the maximum rate of change in the output voltage. The minimum acceptable rate of change is determined by the application. For purposes of example and a good rule of thumb, let us design our circuit to have rise and fall times of no greater than 10 percent of the time for an alternation of the input signal. For our present design, the highest input frequency was specified as 10 kilohertz. The time for one alternation can be calculated from our basic electronics theory as

Sec. 3.2 Zero-Crossing Detector

$$t(\text{alternation}) = \frac{t(\text{period})}{2} \qquad (3.2)$$

where $t(\text{period}) = 1/\text{frequency}$. In our case,

$$t(\text{period}) = \frac{1}{\text{frequency}} = \frac{1}{10 \times 10^3} = 100 \ \mu s$$

and

$$t(\text{alternation}) = \frac{100 \times 10^{-6}}{2} = 50 \ \mu s$$

Our maximum rise and fall times then will be computed as

$$t_R(\text{max}) = t_F(\text{max}) = 0.1 \times t(\text{alternation}) \qquad (3.3)$$

In the present case, we have

$$t_R(\text{max}) = t_F(\text{max}) = 0.1 \times 50 \times 10^{-6} = 5 \ \mu s$$

The minimum acceptable slew rate for our op amp can be computed with the following equation:

$$\text{slew rate(min)} = \frac{+V_{SAT} - (-V_{SAT})}{t_R(\text{max})} \qquad (3.4)$$

In our present example, the minimum acceptable slew rate is computed as

$$\text{slew rate(min)} = \frac{12 - (-12)}{5 \times 10^{-6}} = 4.8 \times 10^6 \ \text{V/sec}$$

It is common to divide this result by 10^6 and express the slew rate in terms of volts per microsecond. In our case

$$\frac{4.8 \times 10^6 \ \text{volts/second}}{10^6} = 4.8 \ \text{V}/\mu s$$

The slew rate for a 741 op amp is listed in the data sheet as 0.5 volts per microsecond. This is clearly too slow for our application. If we used the 741, our output signal would look more like a triangle wave than a square wave. Appendix 4 shows the data for another alternative.

The MC1741SC op amp should satisfy the voltage specifications of our design. Additionally, the minimum slew rate is listed as 10 volts per microsecond. We shall use the MC1741SC for our design.

Select R_I and R_B. For many applications, resistor R_I can be omitted. Its primary purpose is to help establish the input resistance of the circuit. The required value of R_I can be computed with the following equation:

$$\boxed{R_I = \frac{R_{IN} - R_D}{2}} \quad (3.5)$$

where R_{IN} is the minimum required input resistance and R_D is the minimum differential resistance of the op amp. If the computation yields a negative value for R_I, then both R_I and R_B can be omitted from the design.

The value of input resistance is often explicitly stated in the design criteria. If not, it can be computed with Ohm's Law. In our case, the minimum input resistance is

$$R_{IN}(\min) = \frac{v_I(\max)}{i_I(\max)}$$

$$= \frac{500 \times 10^{-3}}{1.2 \times 10^{-3}}$$

$$= 416.7 \; \Omega$$

If the differential resistance of the op amp is not directly listed in the manufacturer's data sheet, we can estimate it with Ohm's Law as follows:

$$\boxed{R_D = \frac{\text{maximum input voltage}}{\text{maximum bias current}}} \quad (3.6)$$

More specifically,

$$R_D = \frac{30}{500 \times 10^{-9}} = 60 \; M\Omega$$

The required value of R_I then is computed with Eq. (3.5) as

$$R_I = \frac{R_{IN} - R_D}{2}$$

$$= \frac{416.7 - 60 \times 10^6}{2} \approx -30 \; M\Omega$$

Since the result is negative, we can safely omit R_I and R_B. In cases where R_I is required, then R_B should be selected to be the same value.

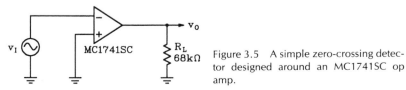

Figure 3.5 A simple zero-crossing detector designed around an MC1741SC op amp.

Sec. 3.2 Zero-Crossing Detector

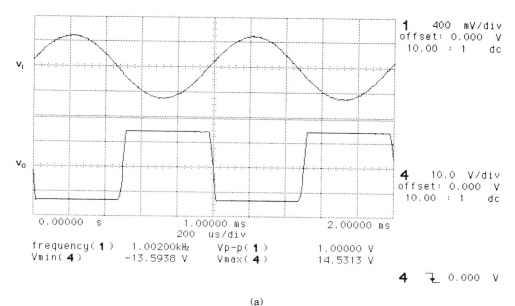

(a)

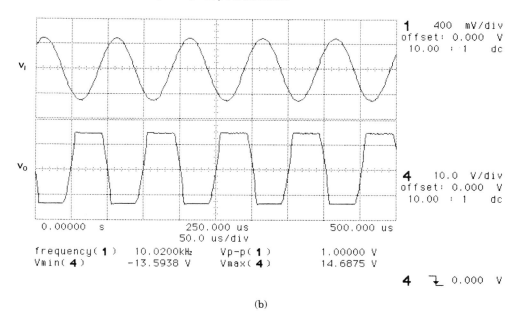

(b)

Figure 3.6 Oscilloscope displays showing the actual performance of the circuit in Fig. 3.5. (Test equipment courtesy of Hewlett-Packard Company)

Figure 3.5 shows the resulting design. The oscilloscope displays in Fig. 3.6 reveal the actual performance of the circuit.

3.3 ZERO-CROSSING DETECTOR WITH HYSTERESIS

3.3.1 Operation

Figure 3.7 shows the schematic diagram of a zero-crossing detector with hysteresis. At first glance, the configuration may resemble a basic amplifier circuit similar to those discussed in Chap. 2. A more careful examination, however, will reveal that the feedback is applied to the (+) input terminal. That is, the circuit is using positive feedback.

Resistors R_F and R_1 form a voltage divider for the output voltage. That portion of the output voltage which appears across R_1 is felt on the (+) input of the op amp. This voltage establishes what is called the threshold voltage. When the output is positive, the voltage on the (+) input is called the positive or upper threshold voltage. The voltage on this same input when the output is negative is called the lower or negative threshold voltage. For the circuit in Fig. 3.7, these two threshold levels will be the same magnitude but opposite polarity. In some circuits, it is desirable to have different values and/or different polarities for the upper and lower thresholds.

To examine the operation of the circuit, let us assume that the input voltage is at its most negative value and that the output is driven to its positive saturation level. A portion of the positive output voltage will be developed across R_1 and appear on the (+) input. This is our upper threshold voltage. As long as the input voltage is below the value of the upper threshold voltage, the circuit will remain in its present condition (positive saturation).

Once the input voltage exceeds (i.e., becomes more positive than) the upper threshold voltage, the (−) pin becomes more positive than the (+) pin of the op amp. Basic op amp operation will tell us that the amplifier will produce a negative output voltage. In our case, the output will drive all the way to its negative saturation limit.

Once the output has reached its negative limit, the (+) input now has a different voltage. It now has the negative or lower threshold voltage on it. The circuit will remain

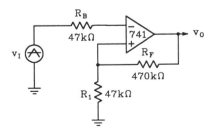

Figure 3.7 Positive feedback adds hysteresis to the zero-crossing detector.

Sec. 3.3 Zero-Crossing Detector With Hysteresis

in this stable condition as long as the input voltage is more positive than the lower threshold voltage.

An important point to notice is the voltages at which output switching occurs. When the input signal is rising, the switching point is determined by the upper threshold voltage. When the signal returns to a lower voltage, however, the output does not switch states as the upper threshold voltage is passed. Rather, the input voltage must go all the way down to below the lower threshold before the output will change states.

The difference between the two threshold voltages is called the hysteresis. The amount of hysteresis is determined by the values of the voltage divider consisting of R_F and R_1 and by the levels of output voltage ($\pm V_{SAT}$ in the present case).

Positive feedback makes the circuit more immune to noise. To understand this effect, consider the drawings in Fig. 3.8 which represent a dual-ramp input voltage with superimposed noise pulses. The output voltage waveform (v_O) for a circuit without hysteresis shows a response every time the combination of input voltage and noise crosses zero. For the circuit with hysteresis, however, once the output has changed states, the combination of input ramp and noise voltage must extend all the way to the opposite threshold voltage before the output will show a response.

Resistor R_B partially determines the input resistance of the circuit in conjunction with R_I and the differential input resistance of op amp. Additionally, R_1 helps to compensate for the effects of input bias current. In many cases, R_B can be omitted.

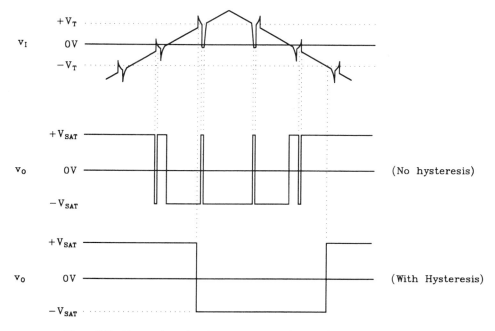

Figure 3.8 Hysteresis makes the zero-crossing detector less sensitive to noise.

3.3.2 Numerical Analysis

Let us now analyze the circuit shown in Fig. 3.7. We shall determine the following values:

1. input resistance
2. upper threshold voltage
3. lower threshold voltage
4. hysteresis
5. maximum frequency of operation

Input resistance. The input resistance for the inverting zero-crossing detector shown in Fig. 3.7 consists of R_B, R_1 and the differential input resistance (R_D) of the op amp in series. Appendix 1 lists 300 kilohms as the minimum value for R_D. Thus we have, Eq. (3.1),

$$R_{IN} = R_1 + R_D + R_B$$
$$= 47k + 300k + 47k$$
$$= 394 \text{ k}\Omega$$

The actual input resistance will likely be even higher.

Upper threshold voltage. The upper threshold voltage is the value of voltage that appears across R_1 when the output is at its maximum positive level. In our present circuit, the maximum output is essentially $+V_{SAT}$. We shall use the lowest value given in App. 1 (10 volts) for $+V_{SAT}$. The value of threshold voltage is computed by applying the basic voltage divider formula.

$$\boxed{V_{R_1} = \left(\frac{R_1}{R_1 + R_F}\right)V_T} \tag{3.7}$$

where V_T is the total voltage across the series resistors R_1 and R_2. In our case, we have

$$V_{UT} = 10\left(\frac{47 \times 10^3}{47 \times 10^3 + 470 \times 10^3}\right) = 0.91 \text{ V}$$

This calculation reveals a significant limitation for this circuit. The value of threshold voltage is directly affected by the value of V_{SAT}. The value of V_{SAT}, however, is far from constant. Appendix 1 indicates that V_{SAT} can vary from a low value of 10 volts (with a heavy load) to as high as 14 volts when lightly loaded. The resulting variation in threshold voltage may be objectionable in some applications. If so, the feedback voltage can be regulated (e.g., by using a pair of zener diodes) for more consistant performance.

Sec. 3.3 Zero-Crossing Detector With Hysteresis

Lower threshold voltage. The lower threshold voltage for the circuit shown in Fig. 3.7 is computed using the same method, Eq. (3.7), discussed for the upper threshold. The value is computed as

$$V_{LT} = -V_{SAT}\left(\frac{R_1}{R_1 + R_F}\right)$$

$$= -10\left(\frac{47 \times 10^3}{47 \times 10^3 + 470 \times 10^3}\right)$$

$$= -0.91 \text{ V}$$

The lower threshold suffers from the same variations described for the upper threshold. Also notice that in this particular circuit, the upper and lower threshold voltages are equal in magnitude and positioned on either side of zero volts. Other circuits may have dissimilar magnitudes for the two thresholds. Additionally, the thresholds are not necessarily centered around zero in all detector circuits.

Hysteresis. The hysteresis of the circuit shown in Fig. 3.7 is simply the difference between the two threshold voltages. That is

$$\boxed{V_H = V_{UT} - V_{LT}} \qquad (3.8)$$

In the present case, we have

$$V_H = +0.91 - (-0.91) = 1.82 \text{ V}$$

The higher the value of hysteresis, the more noise immunity offered by the circuit. In the present case, once the input voltage has crossed one of the threshold levels, it would take a noise pulse on the input of the opposite polarity and with a magnitude of at least 1.82 volts before the output would respond.

In the case of equal $\pm V_{SAT}$ voltages, the hysteresis may be computed directly with the following equation:

$$\boxed{V_H = 2V_{SAT}\left(\frac{R_1}{R_1 + R_F}\right)} \qquad (3.9)$$

Maximum frequency of operation. Since the output voltage of the zero-crossing detector switches between two extreme voltages, the upper frequency limit is more appropriately determined by considering the effects of slew rate rather than the falling amplification. You will recall that the slew rate of an op amp limits the rate of change of output voltage. For purposes of this calculation, we shall determine the highest operating frequency which allows the output to switch fully between $\pm V_{SAT}$. If we exceed this frequency, then the output amplitude will begin to diminish. The reduced output voltage will produce a similar reduction in threshold voltages and the hysteresis voltage.

Appendix 1 lists the slew rate of a 741 op amp as 0.5 volts per microsecond. The output must change from one saturation level to the other during the time for half of the input period (assuming a symmetrical output signal). For purposes of worst-case design, let us assume that the saturation voltages are at their highest magnitudes (listed as ± 14 volts in App. 1). The minimum time required to switch between these two limits is computed as shown

$$t_S(\min) = \frac{+V_{SAT} - (-V_{SAT})}{\text{slew rate}} \tag{3.10}$$

In the case of equal magnitudes of $\pm V_{SAT}$ voltages, this can be expressed as

$$t_S(\min) = \frac{2V_{SAT}}{\text{slew rate}} \tag{3.11}$$

In our present case, the minimum switching time is determined as shown

$$t_S(\min) = \frac{2 \times 14}{0.5 \text{ V}/\mu\text{s}} = 56 \text{ }\mu\text{s}$$

Since this corresponds to half of the period of the highest input frequency, we can determine the upper frequency as shown

$$f(\max) = \frac{1}{2t_S(\min)} \tag{3.12}$$

In our case, we have

$$f(\max) = \frac{1}{2 \times 56 \text{ }\mu\text{s}} = 8.9 \text{ kHz}$$

This represents a worst-case situation. It should be noted, however, that the output waveform under these extreme conditions will more closely resemble a triangle waveform than a square wave. Whether or not this is objectionable is totally dependent on the application. In cases where output rise and fall times must be short compared to the pulse width, the following equation can be used to determine the highest operating frequency for a particular ratio (ρ) of switching time (t_S) to stable time (t_P).

$$f(\max) = \frac{\rho}{2t_S(\rho + 1)} \tag{3.13}$$

In the case of Fig. 3.7, we have already computed t_S as 56 microsecond. Now suppose we want the switching times (rise and fall) to be one tenth (0.1) of the stable time (t_P). This establishes our ratio ρ as 0.1. The highest frequency is then computed as

$$f(\max) = \frac{\rho}{2t_S(\rho + 1)}$$

Sec. 3.3 Zero-Crossing Detector With Hysteresis

$$= \frac{0.1}{2 \times 56 \times 10^{-6}(0.1 + 1)}$$
$$= 811.7 \text{ Hz}$$

If the input waveform is such that the output will not be symmetrical, then t_S establishes the shortest (either positive or negative) alternation of the output waveform. The highest frequency of operation, however, will be obtained when the output waveform is symmetrical.

3.3.3 Practical Design Techniques

Let us now design a zero-crossing detector circuit similar to that shown in Fig. 3.7. We shall design to achieve the following:

1. upper threshold   +0.5 volts
2. lower threshold   −0.5 volts
3. hysteresis   1.0 volts
4. minimum input resistance   680 kilohms
5. highest operating frequency   15 kilohertz
6. maximum ratio (ρ) of switching time to stable time   0.2
7. power supply voltages   ±15 volts

Determine the required slew rate. In this type of application, slew rate is probably the most critical parameter with regard to op amp selection. The slew rate must be high enough to allow the output to switch between saturation levels within the allowed switching time (t_S). The switching time is computed by using a transposed version of the $f(\text{max})$ equation. That is

$$\boxed{t_S = \frac{\rho}{2(\rho + 1) \times f(\text{max})}} \qquad (3.14)$$

For the present circuit, we have

$$t_S = \frac{0.2}{2(0.2 + 1) \times 15 \times 10^3} = 5.56 \text{ }\mu\text{s}$$

For purposes of op amp selection, we can assume that the output swings between the two power supply limits. That is, assume that $\pm V_{SAT} = \pm V_{CC}$. The required slew rate can then be computed as

$$\boxed{\text{slew rate(min)} = \frac{2V^+}{t_s}} \qquad (3.15)$$

For the present circuit, we have

$$\text{slew rate(min)} = \frac{2 \times 15}{5.56 \times 10^{-6}} = 5.4 \text{ V}/\mu s$$

Select an op amp. Appendix 1 indicates that the slew rate for a 741 is only 0.5 volts per microsecond which clearly eliminates this device as an option since we will require a slew rate of at least 5.4 volts per microsecond. Appendix 4, however, shows that the MC1741SC op amp has a minimum slew rate of 10 volts per microsecond. This will satisfy our present requirements nicely. It is also compatible with our power supply requirements. We shall build our design around the MC1741SC op amp.

Determine R_F and R_1. The ratio of R_F and R_1 is dependent on the ratio of V_{SAT} voltage to hysteresis voltage. Appendix 4 indicates that the unloaded output swing will be approximately ± 13 volts. We use this value for our computations. If the circuit were expected to drive a greater load (smaller load resistor), then the output will be correspondingly smaller. The ratio of R_F to R_1 is computed as follows:

$$\boxed{\frac{R_F}{R_1} = \frac{2V_{SAT}}{V_H} - 1} \qquad (3.16)$$

More specifically, for the present circuit we have

$$\frac{R_F}{R_1} = \frac{2 \times 13}{1} - 1 = 25$$

There are many combinations of R_F and R_1 that will produce a 25:1 ratio. We shall select R_1 and calculate the value of R_F. If possible, we generally want both resistors in the range of 1.0 kilohm to 1.0 megohms although these do not represent absolute limits. Additionally, the series combination of R_D and R_1 should equal or exceed the desired input resistance for the circuit. In our present case, App. 4 lists R_D as 1.0 megohms typical. Since this exceeds our required input resistance, we have considerable freedom on the value of R_1. For purposes of this example, we select R_1 to be 4.7 kilohms. Having done this, we can now compute R_F by simply multiplying R_1 by the R_F/R_1 ratio.

$$\boxed{R_F = \left(\frac{R_F}{R_1}\right) R_1} \qquad (3.17)$$

And, in the present case

$$R_F = 4.7 \times 10^3 \times 25 = 118 \text{ k}\Omega$$

We select a standard value of 120 kilohms.

Sec. 3.3 Zero-Crossing Detector With Hysteresis

Determine the value of R_B. R_B is intended to compensate for the effects of bias currents. Its value is chosen to be equal to the parallel combination of R_F and R_1. That is,

$$\boxed{R_B = R_F \| R_1} \quad (3.18)$$

For the present circuit, we have

$$R_B = \frac{120 \times 10^3 \times 4.7 \times 10^3}{120 \times 10^3 + 4.7 \times 10^3} = 4.53 \text{ k}\Omega$$

We select a standard value of 4.7 kilohms for R_B. This completes the design of the simple zero crossing detector circuit. The schematic is shown in Fig. 3.9. The circuit performance

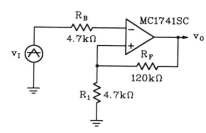

Figure 3.9 A zero-crossing detector designed for 1.0 volts hysteresis and operation up to 15 kilohertz.

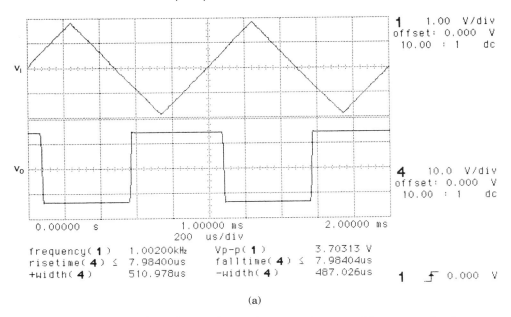

(a)

Figure 3.10 Waveforms showing the actual circuit performance of the zero-crossing detector shown in Fig. 3.9. (Test equipment courtesy of Hewlett-Packard Company) (continued)

152 Voltage Comparators   Chap. 3

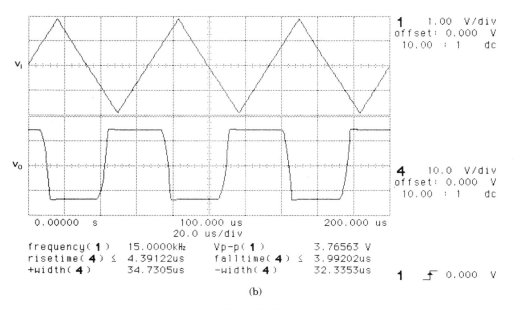

(b)

Figure 3.10b

TABLE 3.1

	DESIGN GOAL	MEASURED VALUES
Upper threshold	+0.5 V	+0.54 V
Lower threshold	−0.5 V	−0.51 V
Hysteresis	1.0 V	1.05 V
Maximum switching ratio	0.2	0.126

is shown by the oscilloscope displays shown in Fig. 3.10. The original design goals are contrasted with the measured performance in Table 3.1.

3.4 VOLTAGE COMPARATOR WITH HYSTERESIS

3.4.1 Operation

Figure 3.11 shows the schematic diagram of an inverting voltage detector with hysteresis. The operation of this circuit is similar to the zero crossing detector discussed in a previous section, but the upper and lower thresholds are on either side of a reference voltage (V_{REF})

Sec. 3.4 Voltage Comparator With Hysteresis

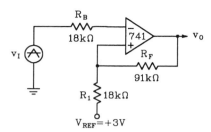

Figure 3.11 A voltage comparator with hysteresis.

rather than either side of zero. The reference voltage can be either positive or negative. It is interesting to note, that if the reference is zero volts then the circuit is identical to the zero crossing detector previously discussed.

To begin the discussion, let us assume that the input voltage is at its most negative value, and that the output of the op amp is driven to its $+V_{SAT}$ level. The $+V_{SAT}$ output is divided between R_F and R_1 in normal voltage divider fashion. The voltage appearing across R_1 plus the value of the reference voltage determines the voltage on the (+) input terminal. This is the upper threshold voltage. The circuit will remain in this condition as long as the input voltage is below the voltage on the (+) terminal.

Now suppose, the input voltage is allowed to exceed the upper threshold voltage which is present on the (+) input. If this happens, the output will quickly go to the $-V_{SAT}$ level. R_F and R_1 will divide the negative output voltage. The portion across R_1 plus the value of the reference voltage determines the voltage on the (+) input terminal. This is the lower threshold voltage. The circuit will remain in this stable condition until the input voltage falls below the negative threshold voltage.

In many practical comparator circuits, the reference voltage is provided by a zener diode (See, for example, Fig. 3.12.) or other voltage regulator circuit. Resistor R_B is included to compensate for the effects of op amp bias current.

3.4.2 Numerical Analysis

We now analyze the circuit shown in Fig. 3.11 to determine the following:

1. input resistance
2. upper threshold voltage
3. lower threshold voltage
4. hysteresis
5. maximum frequency of operation

Input resistance. The input resistance for the inverting voltage comparator shown in Fig. 3.11 consists of R_B, the differential input resistance (R_D) of the op amp, and the parallel combination of R_1 and R_F all in series. Appendix 1 lists the 300 kilohms as the minimum value for R_D. Thus, we have Eq. (3.1) modified to include R_F

$$R_{IN} = R_1 \| R_F + R_D + R_B$$
$$= 18k \| 91k + 300k + 18k$$

$$= 15k + 300k + 18k$$
$$= 333 \text{ k}\Omega$$

The actual input resistance will generally be even higher.

Upper threshold voltage. The upper threshold voltage is the value of voltage that appears across R_1 when the output is at its maximum positive level plus the value of the reference voltage. Appendix 1 lists a minimum value of 10 volts for $+V_{SAT}$. The value of threshold voltage is computed by applying the basic voltage divider formula to R_F and R_1 and then adding the result to the reference voltage (V_{REF}).

$$\boxed{V_{UT} = +V_{SAT}\left(\frac{R_1}{R_1 + R_F}\right) + V_{REF}} \tag{3.19}$$

For this particular circuit, we have

$$V_{UT} = 10\left(\frac{18 \times 10^3}{18 \times 10^3 + 91 \times 10^3}\right) + 3 = 4.65 \text{ V}$$

As with the zero-crossing detector previously discussed, the threshold voltages vary with V_{SAT}. If this variation is objectionable, then the output can be regulated by zeners as discussed in Sec. 3.6.

Lower threshold voltage. The lower threshold voltage for the circuit shown in Fig. 3.11 is computed using the same method discussed for the upper threshold. The value is computed as

$$\boxed{V_{LT} = -V_{SAT}\left(\frac{R_1}{R_1 + R_F}\right) + V_{REF}} \tag{3.20}$$

Or, more specifically

$$V_{LT} = -10\left(\frac{18 \times 10^3}{18 \times 10^3 + 91 \times 10^3}\right) + 3 = 1.35 \text{ V}$$

The lower threshold suffers from the same variations described for the upper threshold. Also, notice that in this particular circuit, the upper and lower threshold voltages are both above zero volts but equally spaced on either side of the reference voltage. The thresholds in some circuits are not equally spaced around the reference.

Hysteresis. The hysteresis of the circuit shown in Fig. 3.11 is simply the difference, Eq. (3.8), between the two threshold voltages. That is

$$V_H = V_{UT} - V_{LT}$$
$$= +4.65 - 1.35$$
$$= 3.3 \text{ V}$$

Sec. 3.4 Voltage Comparator With Hysteresis

Recall that the value of hysteresis primarily determines the noise immunity offered by the circuit. In the present case, once the input voltage has crossed one of the threshold levels, it would take a noise pulse on the input of the opposite polarity and with a magnitude of at least 3.3 volts before the output would respond.

The hysteresis in this circuit may also be computed directly with Eq. (3.9).

Maximum frequency of operation. The upper frequency of operation is limited in the same manner as the zero crossing detector discussed previously. This frequency is estimated with Eq. (3.12)

$$f(\text{max}) = \frac{1}{2t_S(\text{min})}$$

where t_S is computed with Eq. (3.10) as shown

$$t_S(\text{min}) = \frac{+V_{SAT} - (-V_{SAT})}{\text{slew rate}}$$

Or, in the usual case of symmetrical power supplies, we can simply use Eq. (3.11) as follows:

$$t_S(\text{min}) = \frac{2V_{SAT}}{\text{slew rate}}$$

In our present case, let us determine the minimum switching time with Eq. (3.11)

$$t_S(\text{min}) = \frac{2V_{SAT}}{\text{slew rate}}$$

$$= \frac{2 \times 10}{0.5 \text{ V}/\mu s}$$

$$= 40 \ \mu s$$

Substituting this value into the maximum frequency formula, Eq. (3.12), gives us

$$f(\text{max}) = \frac{1}{2t_S(\text{min})}$$

$$= \frac{1}{2 \times 40 \times 10^{-6}}$$

$$= 12.5 \text{ kHz}$$

This represents a worst-case situation for a symmetrical output waveform. As noted in the zero-crossing circuit, the output waveform under these extreme conditions will more closely resemble a triangle waveform than a square wave. In cases where output rise and fall times must be short compared to the pulse width, Eq. (3.13) can be used to determine the highest operating frequency for a particular ratio (ρ) of switching time (t_S) to stable time (t_P). That is

$$f(\max) = \frac{\rho}{2t_S(\rho + 1)}$$

where

$$\rho = \frac{t_S}{t_P}$$

In the case of Fig. 3.11, we have already computed t_S as 40 microseconds. Now suppose we want the switching times (rise and fall) to be one eighth (0.125) of the stable time (t_P). This establishes our ratio ρ as 0.125. The highest frequency is then computed with Eq. (3.14) as

$$f(\max) = \frac{0.125}{2 \times 40 \times 10^{-6}(0.125 + 1)} = 1.39 \text{ kHz}$$

3.4.3 Practical Design Techniques

We now design a voltage comparator circuit with hysteresis. We design it to provide the following performance:

1. upper threshold — -4.25 volts
2. lower threshold — -7.75 volts
3. hysteresis — 3.5 volts
4. minimum input resistance — 100 kilohms
5. highest operating frequency — 60 hertz
6. maximum ratio (ρ) of switching time to stable time — 0.1
7. power supply voltages — ± 15 volts

Determine the required slew rate. The slew rate must be high enough to allow the output to switch between saturation levels within the allowed switching time (t_S). The switching time is computed with Eq. (3.14) as follows:

$$t_S = \frac{\rho}{2(\rho + 1) \times f(\max)}$$

$$= \frac{0.1}{2 \times 60(0.1 + 1)}$$

$$= 758 \text{ μs}$$

For purposes of op amp selection, we can assume that the output swings between the two power supply limits. That is, assume that $\pm V_{SAT} = \pm V_{CC}$. The required slew rate can then be computed with Eq. (3.15) as

$$\text{slew rate(min)} = \frac{2V^+}{t_S}$$

Sec. 3.4 Voltage Comparator With Hysteresis

$$= \frac{2 \times 15}{758 \times 10^{-6}}$$

$$= 0.04 \text{ V}/\mu\text{s}$$

These calculations assume a 50 percent duty cycle on the output waveform. If the output is asymmetrical, then the shortest allowable alternation is given as

$$t_{MIN} = t_S \left(\frac{\rho + 1}{\rho} \right)$$

Select an op amp. Appendix 1 indicates that the slew rate for a 741 is 0.5 volts per microsecond which exceeds our requirement of at least 0.04 volts per microseconds. The power supply requirements are also compatible with our stated design requirements. We shall build our design around the 741 op amp.

Determine R_F and R_1. The ratio of R_F and R_1 is dependent on the ratio of V_{SAT} voltage to hysteresis voltage. Appendix 1 indicates that the lightly loaded ($R_L > 10k\Omega$) output swing will be typically ± 14 volts. We shall use this value for our computations. If the circuit were expected to drive a greater load (smaller load resistor and/or feedback network), then the output would be correspondingly smaller. The ratio of R_F to R_1 is computed with Eq. (3.16).

$$\frac{R_F}{R_1} = \frac{2V_{SAT}}{V_H} - 1$$

$$= \frac{2 \times 14}{3.5} - 1$$

$$= 7$$

We select R_1 and calculate the value of R_F. Since the input resistance of the 741 far exceeds our minimum input resistance requirement, we can select R_1 to be any convenient value. For purposes of this example, we select R_1 to be 33 kilohms. Having done this, we can now compute R_F by applying Eq. (3.17) where the ratio (R_F/R_1) is known.

$$R_F = \left(\frac{R_F}{R_1} \right) R_1$$

$$= 7 \times 33k$$

$$= 231 \text{ k}\Omega$$

We shall select a standard value of 240 kilohms.

Determine the value of R_B. R_B is intended to compensate for the effects of bias currents. Its value is chosen, Eq. (3.18), to be equal to the parallel combination of R_F and R_1. That is

$$R_B = R_F \| R_1$$
$$= \frac{240 \times 10^3 \times 33 \times 10^3}{240 \times 10^3 + 33 \times 10^3}$$
$$= 29 \text{ k}\Omega$$

We shall select a standard value of 30 kilohms for R_B.

Calculate V_{REF}. A simple way to calculate the required value of V_{REF} is to apply Eq. (3.21).

$$\boxed{V_{REF} = V_{LT} + \frac{V_H}{2}} \qquad (3.21)$$

where V_{LT} and V_H are the lower threshold and hysteresis voltages, respectively. In our present case, we can determine V_{REF} as shown

$$V_{REF} = -7.75 + \frac{3.5}{2} = -6.0 \text{ V}$$

For purposes of illustration, let us derive this reference voltage from a zener diode network across the -15 volt supply. As long as the equivalent (Thevenin) resistance of the zener circuit is small compared to R_F and R_1 (the usual case) it will not affect our previous selection of components.

Appendix 5 shows the data sheet for a family of zener diodes. One of the listed devices is the 1N5233B which is a 6.0 volt, 1/2 watt zener diode. The maximum zener current can be estimated by applying the power formula

$$\boxed{I_Z(\text{max}) = \frac{P_Z}{V_Z}} \qquad (3.22)$$

where P_Z and V_Z are the power and voltage ratings of the zener. In the case of the 1N5233B, the maximum current is

$$I_Z(\text{max}) = \frac{0.5}{6} = 83.3 \text{ mA}$$

This establishes the upper limit of zener current and even this must be derated for temperatures above 25 degrees centigrade. The zener test current (I_{ZT}) is listed as 20 milliamps. This is generally a good quiescent current choice.

The series current limiting resistor for the zener regulator can be computed as follows:

$$\boxed{R_S = \frac{V_{CC} - V_Z}{I_Z}} \qquad (3.23)$$

Sec. 3.4   Voltage Comparator With Hysteresis

In the present case, we have

$$R_S = \frac{15 - 6}{20 \times 10^{-3}} = 450 \ \Omega$$

We shall choose the standard value of 470 ohms. This completes our design. The final schematic is shown in Fig. 3.12. The waveforms in Fig. 3.13 reveal the performance of the circuit. The original design goals are contrasted with the measured performance in Table 3.2.

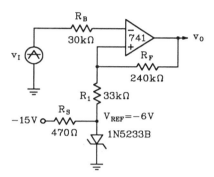

Figure 3.12 A voltage comparator design using a zener diode as the reference voltage.

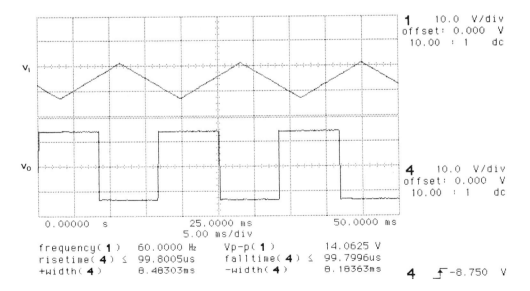

Figure 3.13 Waveforms showing the performance of the circuit shown in Fig. 3.12.

TABLE 3.2

	DESIGN GOAL	MEASURED VALUE
Upper threshold	-4.25 V	-4.25 V
Lower threshold	-7.75 V	-7.813 V
Hysteresis	3.5 V	3.56 V
Maximum switching ratio	0.02	0.012

3.5 WINDOW VOLTAGE COMPARATOR

3.5.1 Operation

A basic op amp window detector circuit is shown in Fig. 3.14. This circuit is essentially a dual comparator circuit and produces a two-state output which indicates whether or not the input voltage (v_I) is between the limits (i.e., within the window) established by the $\pm V_{REF}$ voltages. It is frequently used to sound an alarm or signal a control circuit when a measured variable (v_I) goes outside of a preset range. The reference voltages in Fig. 3.14 are established by two zener diode circuits.

To examine the operation of the circuit, let us start by assuming the input voltage is within the window. That is, the input voltage is less than $+V_{REF}$ and greater than $-V_{REF}$. Under these conditions, the outputs of both op amps will be driven to the $+V_{SAT}$ level. This reverse biases the two isolation diodes (1N914) and allows the output (v_O) to rise to +15 volts indicating an "in window" condition. (Note: If the positive saturation levels of the op amps is sufficiently low, then the isolation diodes will not be reverse biased, but the output will still be at its most positive level.)

Now suppose the input either exceeds $+V_{REF}$ or falls below $-V_{REF}$. In these cases, the output of one of the two op amps will go to the $-V_{SAT}$ level and forward bias its associated isolation diode. This will cause the output of the circuit (v_O) to be pulled to -15 volts (ideally). In practice, the output voltage will be equal to the negative saturation level *plus* the forward voltage drop of the conducting isolation diode. This negative level indicates an "out of window" condition.

3.5.2 Numerical Analysis

Let us now analyze the behavior of the circuit shown in Fig. 3.14 in greater detail. We shall determine the following characteristics:

1. $+V_{REF}$
2. $-V_{REF}$
3. output voltage (v_O)

Sec. 3.5 Window Voltage Comparator

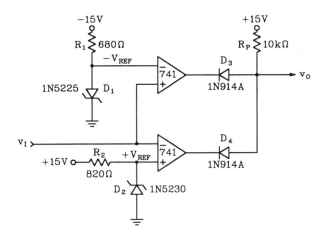

Figure 3.14 A window detector is used to determine whether the input voltage (v_I) is within the limits of $\pm V_{REF}$.

$+V_{REF}$. The value of $+V_{REF}$ is established by the 1N5230 zener diode. Appendix 5 indicates that this is a 4.7 volt zener. The $+V_{REF}$ is therefore approximately $+4.7$ volts. If desired, we can determine the amount of zener current with Eq. (3.23) (transposed)

$$I_Z = \frac{+V_{CC} - V_Z}{R_2}$$

$$= \frac{15 - 4.7}{820}$$

$$= 12.6 \text{ mA}$$

$-V_{REF}$. The 1N5225 zener diode is used to establish the $-V_{REF}$ source. Appendix 5 lists the 1N5225 as a 3.0 volt zener. Its current can be calculated with Eq. (3.23) (transposed) as

$$I_Z = \frac{+V_{CC} - V_Z}{R_1}$$

$$= \frac{15 - 3}{680}$$

$$= 17.6 \text{ mA}$$

Output voltage (v_O). The upper limit of v_O occurs when both of the isolation diodes (1N914) are reverse biased or effectively open. This means the pull-up resistor (R_P) has essentially no current flow, and therefore, no voltage drop across it. Since R_P drops no voltage under these conditions, the output will be at a $+15$ volt level. As mentioned previously, if $+V_{SAT}$ is sufficiently low, then the isolation diode will not be reverse biased and the output voltage (v_O) will be less than $+15$ volts ($+V_{SAT} + V_F$, where V_F is the forward drop of the diode).

If either of the op amp outputs are forced to their $-V_{SAT}$ level, then the associated isolation diode (1N914) will be forward biased. Appendix 1 indicates that the $-V_{SAT}$ will be about -11 volts. The current through R_P can now be estimated as follows:

$$I_{R_P} = \frac{V_{CC} - (-V_{SAT}) - V_F}{R_P} \qquad (3.24)$$

where V_F is the forward voltage drop of the isolation diode (typically 0.7 volts). In our present case, we can compute I_{R_P} as

$$I_{R_P} = \frac{15 - (-11) - 0.7}{10 \times 10^3} = 2.53 \text{ mA}$$

The actual output voltage (v_O) under these conditions is determined as follows:

$$v_O = V_{CC} - I_{R_P} R_P \qquad (3.25)$$

And, for the present circuit

$$v_O = 15 - (2.53 \times 10^{-3} \times 10 \times 10^3) = -10.3 \text{ V}$$

This same result can be obtained by applying Kirchoff's Voltage Law. That is

$$v_O = -V_{SAT} + V_F$$
$$= -11 + 0.7$$
$$= -10.3 \text{ V}$$

3.5.3 Practical Design Techniques

Let us now design a window detector to meet to following specifications:

1. upper window limit $+10$ volts
2. lower window limit: $+7.5$ volts
3. power supply ± 15 volts
4. input frequency   0 to 100 hertz

Select the op amp. Since the circuit is being driven by a very low-frequency source, the high-frequency characteristics of the op amp are unimportant to us. The DC stability of the op amp is more important in circuits like this and will be determined by the requirements of the specific application being considered. If the switching speed of the device is important for a particular application that has a higher input frequency, then you would do well to select an op amp that is specifically designed for fast comparator applications.

For purposes of our present example, let us choose the 741.

Sec. 3.5 Window Voltage Comparator

Select the zener diodes. Appendix 5 lists a family of zener diodes. The 1N5236 and 1N5240 devices will satisfy the requirements for our lower and upper reference voltages of 7.5 volts and 10 volts, respectively.

Calculate the zener current limiting resistors. Unless we have some reason to do otherwise (e.g., ultra-low-current designs), we can use the zener test current as the design value. Appendix 5 lists 20 milliamps as the test current for both diodes. Basic circuit theory, Eq. (3.23), allows us to compute the values of current limiting resistors.

$$R_1 = \frac{V_{CC} - V_{REF}}{I_Z}$$

$$= \frac{15 - 7.5}{20 \times 10 - 3}$$

$$= 375 \, \Omega$$

We shall choose a standard value of 390 ohms. In a similar manner

$$R_2 = \frac{V_{CC} - V_{REF}}{I_Z}$$

$$= \frac{15 - 10}{20 \times 10 - 3}$$

$$= 250 \, \Omega$$

We choose the next higher standard value of 270 ohms. Note that in this particular case, both zeners will be connected across the positive supply voltage since we require both references to be positive.

Select R_P. The correct value of R_P is determined by two primary considerations

1. current capability of the driving op amps,
2. type of circuit or device being driven.

Since we have no information regarding the driven circuit, let us select R_P such that the op amp output current is limited to 5 milliamps. This calculation, Eq. (3.24) is based on Ohm's Law as follows:

$$R_P = \frac{V_{CC} - (-V_{SAT}) - V_F}{I_{R_P}}$$

$$= \frac{15 - (-11) - 0.7}{5 \times 10^{-3}}$$

$$= 5.06 \, k\Omega$$

Let us select the next higher standard value of 5.1 kilohms.

Select the isolation diodes. The requirements for the isolation diodes are not stringent. The two primary parameters that need to be considered are

1. reverse voltage breakdown,
2. maximum forward current.

The maximum current is the same as the value of op amp output current. In our case, we have designed this to be 5 milliamps. The maximum reverse voltage is approximately (ignoring diode drops) equal to the difference between $+V_{SAT}$ of one op amp and $-V_{SAT}$ of the other. That is

$$V_{PRV} = +V_{SAT} - (-V_{SAT}) \tag{3.26}$$

More specifically

$$V_{PRV} = 11 - (-11) = 22 \text{ V}$$

Appendix 6 shows the characteristics for a 1N914A diode. Most any diode will satisfy our modest requirements. The 1N914A is a standard, low-cost diode that we can use for this application.

This completes the design of the window detector circuit. The final schematic is shown in Fig. 3.15. Figure 3.16 shows the actual waveforms produced by the cir-

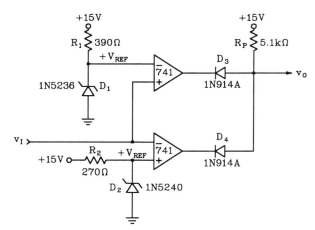

Figure 3.15 A window detector designed for a lower limit of +7.5 volts and an upper limit of +10 volts.

TABLE 3.3

	DESIGN GOAL (volts)	MEASURED VALUE (volts)
Upper window limit	+10	+10
Lower window limit	+7.5	+7.5

Sec. 3.6 Voltage Comparator With Output Limiting

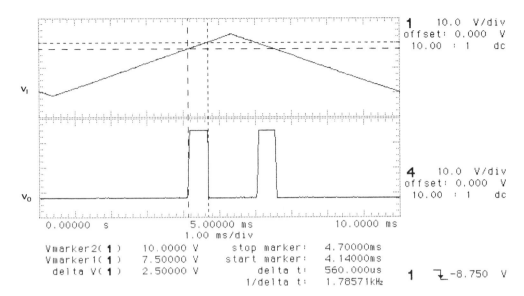

Figure 3.16 Oscilloscope displays showing the performance of the circuit in Fig. 3.15. (Test equipment courtesy of Hewlett-Packard Company)

cuit. Table 3.3 compares the original design goals with the measured circuit performance.

3.6 VOLTAGE COMPARATOR WITH OUTPUT LIMITING

3.6.1 Operation

The voltage comparator shown in Fig. 3.17 is configured to be noninverting. Additionally, the output uses two zener diodes (D_2 and D_3) to limit the swing of v_O. The zener pair along with resistor R_2 acts like a bidirectional, biased limiter circuit.

D_1 and current limiting resistor R_S establish the reference voltage. Feedback resistor R_F in conjunction with R_I establish hysteresis for the circuit. Finally, R_1 serves to compensate for the effects of op amp bias current.

For purposes of discussion, let us assume that the input is well below the upper threshold voltage. Since this is a noninverting circuit, we know that the output of the op amp will be driven to the $-V_{SAT}$ level. The zener pair in the output circuit along with R_2 regulate the $-V_{SAT}$ voltage to a value established by D_3. This reduced and regulated voltage appears as v_O.

Resistors R_F and R_I form a voltage divider that appears between the regulated v_O and the changing input voltage. The circuit will remain in this stable condition as long as the (+) input of the op amp remains lower than the reference voltage on the (−) input.

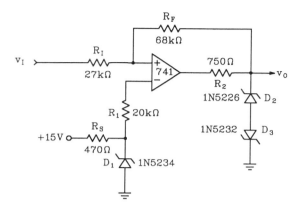

Figure 3.17 A voltage comparator with output limiting.

As the input voltage increases, the voltage on the (+) input also increases. Once the (+) input goes above the voltage on the (−) input even momentarily, the output of the op amp will go toward the $+V_{SAT}$ level. This rising potential, through R_F, further increases the potential on the (+) input pin. With the output of the op amp at the $+V_{SAT}$ level, diode D_2 establishes the value of voltage at v_O. Again the circuit will remain in this state until the input falls below the voltage on the (−) terminal. It is important to note that since the rising output has increased the potential on the (+) input, the actual input voltage (v_I) will have to go to a much lower level to cause the circuit to switch states. This effect is, of course, the very nature of hysteresis.

If the input voltage now decreases to a level which causes the voltage on the (+) pin to fall below the voltage on the (−) pin, then the circuit will switch back to its original state.

3.6.2 Numerical Analysis

Now let us extend our analysis of Fig. 3.17 to calculate the following:

1. upper threshold voltage
2. lower threshold voltage
3. hysteresis
4. all zener currents
5. output voltage limits (v_O^+ and v_O^-)

Upper threshold voltage. The upper threshold voltage can be found by applying Kirchoff's Law and basic circuit theory to the resistor network R_I and R_F. Our knowledge of op amp operation tells us that no substantial current enters or leaves the (+) pin. Therefore, $i_1 = i_2$ in Fig. 3.18. At the instant v_I reaches the upper threshold, the junction of R_F and R_I will just equal V_{REF}. This is so labeled on Fig. 3.18.

Using Ohm's Law, we can write expressions for the values of i_1 and i_2

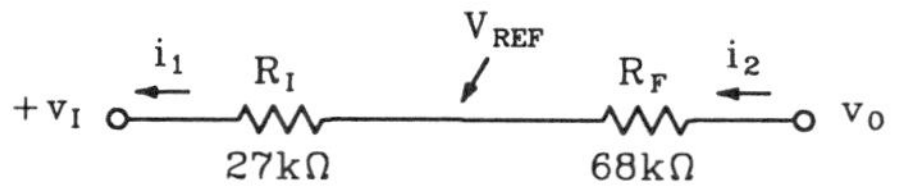

Sec. 3.6 Voltage Comparator With Output Limiting

Figure 3.18 Basic circuit theory can be used to compute the upper threshold voltage of the circuit in Fig. 3.17.

$$i_1 = \frac{v_I - V_{REF}}{R_I}$$

$$i_2 = \frac{V_{REF} - v_O^-}{R_F}$$

If we equate these two currents, we get

$$i_1 = i_2$$

$$\frac{v_I - V_{REF}}{R_I} = \frac{V_{REF} - v_O^-}{R_F}$$

Some algebraic manipulation gives us the expression for v_I at the moment it reaches the upper threshold

$$v_I = V_{UT} = \frac{R_I(V_{REF} - v_O^-)}{R_F} + V_{REF} \qquad (3.27)$$

In the present example, v_O^- will equal the voltage of D_3 (-5.6 volts as listed in App. 5) during this period of time. The reference voltage is 6.2 volts (see App. 5). Substituting values enables us to calculate the value of the upper threshold voltage

$$V_{UT} = 27 \times 10^3 \left(\frac{6.2 - (-5.6)}{68 \times 10^3} \right) + 6.2 = 10.9 \text{ V}$$

This value could be made slightly more accurate by including the effects of the forward voltage drop of D_2 (about 0.7 volts). That is, v_O^- will equal the voltage of D_3 plus the forward voltage drop of D_2 or -6.3 volts. If this effect is included, the threshold is computed to be 11.2 volts.

Lower threshold voltage. A similar application of basic circuit theory when the output is at the $+V_{SAT}$ level and the input is approaching the lower threshold voltage yields the following expression for the lower threshold voltage

$$V_{LT} = V_{REF} - \frac{R_I(v_O^+ - V_{REF})}{R_F} \qquad (3.28)$$

Recognizing that v_O^+ will be equal to the voltage of D_2 (3.3 volts) during this time, we can calculate the value of lower threshold voltage

$$V_{LT} = 6.2 - \frac{27 \times 10^3 (3.3 - 6.2)}{68 \times 10^3} = 7.35 \text{ V}$$

If the forward voltage drop of D_3 is included in the calculation, the threshold voltage will be computed as 7.07 volts.

Hysteresis. Hysteresis is simply the difference between the two threshold voltages. In our present case, hysteresis is computed, Eq. (3.8), as shown

$$V_H = V_{UT} - V_{LT}$$
$$= 10.9 - 7.35$$
$$= 3.55 \text{ V}$$

If the diode drops are included, the hysteresis will be computed as 4.13 volts.

Zener currents. The current through D_1 can be computed, Eq. (3.23), as follows:

$$I_{D_1} = \frac{+V_{CC} - V_{D_1}}{R_S}$$
$$= \frac{15 - 6.2}{470}$$
$$= 18.7 \text{ mA}$$

The reverse current through D_3 is computed with the following expression:

$$\boxed{I_{D_3} = \frac{-V_{SAT} - V_{D_3} + 0.7}{R_2}} \quad (3.29)$$

Continuing with the calculations, we get

$$I_{D_3} = \frac{-11 - (-5.6) + 0.7}{750} = -6.27 \text{ mA}$$

Note that the minus sign simply indicates direction and is of no significance to us at this time. D_2 current is computed in a similar manner as shown

$$\boxed{I_{D_2} = \frac{+V_{SAT} - V_{D_2} - 0.7}{R_2}} \quad (3.30)$$

Substituting values for the present circuit gives us

$$I_{D_2} = \frac{11 - 3.3 - 0.7}{750} = 9.33 \text{ mA}$$

The power dissipation of the zeners can be found by applying the basic power equation $P = IE$ where I and E are the current and voltages associated with a particular zener. In our particular case

$$P_{D_1} = 18.7 \times 10^{-3} \times 6.2 = 116 \text{ mW}$$

Sec. 3.6 Voltage Comparator With Output Limiting

$$P_{D_2} = 6.27 \times 10^{-3} \times 5.6 = 35.1 \text{ mW}$$

$$P_{D_3} = 9.33 \times 10^{-3} \times 3.3 = 30.8 \text{ mW}$$

Output voltage limits. The output voltage swing was essentially determined in a prior step. The upper excursion is established by the zener voltage of D_2 plus the forward voltage drop of D_3. In our case

$$v_O^+ = V_{D_2} + 0.7 = 3.3 + 0.7 = 4.0 \text{ V}$$

The lower limit of v_O is computed as

$$v_O^- = V_{O_3} - 0.7 = -5.6 - 0.7 = -6.3 \text{ V}$$

3.6.3 Practical Design Techniques

We shall now design a voltage comparator with output limiting that satisfies the following specifications:

1. upper output voltage (v_O^+)   +5.0 volts
2. lower output voltage (v_O^-)   −4.0 volts
3. upper threshold voltage   +2.0 volts
4. lower threshold voltage   +0.8 volts
5. power supply   ±15 volts
6. op amp   741

These specifications (i.e., input and output requirements) would normally be dictated by the application.

Compute hysteresis voltage. Our first step will be to calculate the required hystersis voltage. This is very simply the difference between the two threshold voltages, Eq. (3.8).

$$V_H = V_{UT} - V_{LT}$$
$$= 2 - 0.8$$
$$= 1.2 \text{ V}$$

Compute R_F and R_I. The ratio of R_F to R_I is determined by the ratio of the output voltage swing to the hysteresis voltage. That is

$$\boxed{\frac{R_F}{R_I} = \frac{v_O^+ - v_O^-}{V_H}} \tag{3.31}$$

In our design example, the required R_F/R_I ratio is computed as shown

$$\frac{R_F}{R_I} = \frac{5 - (-4)}{1.2} = 7.5$$

We shall now select R_I and compute R_F. Let us select a value of 10 kilohms for resistor R_I. The feedback resistor R_F can now be computed, Eq. (3.17) as shown

$$R_F = \left(\frac{R_F}{R_I}\right) R_I$$
$$= 7.5 \times R_I$$
$$= 7.5 \times 10 \times 10^3$$
$$= 75 \text{ k}\Omega$$

The factor 7.5 in this equation is simply the R_F/R_I ratio previously computed.

Select the output zener diodes. The voltage rating of the two zener diodes is determined by the stated output voltage swing. That is

$$\boxed{V_{D_2} = v_O^+ - 0.7} \qquad (3.32)$$

Substituting values gives us

$$V_{D_2} = 5 - 0.7 = 4.3 \text{ V}$$

Similarly, the voltage rating for D_3 is computed as,

$$\boxed{V_{D_3} = v_O^- + 0.7} \qquad (3.33)$$

Values for the present circuit are

$$V_{D_3} = -4 + 0.7 = -3.3 \text{ V}$$

The power ratings for the zeners are not critical but must be noted for subsequent calculations. By referring to App. 5, we can select a 1N5229 and a 1N5226 for diodes D_2 and D_3, respectively. We also observe that both of these diodes are rated at 500 milliwatts.

Compute R_2. Resistor R_2 is a current limiting resistor for the zener diodes. First, we shall compute, Eq. (3.22), the maximum allowable currents through each of the zeners.

$$I_{D_2} = \frac{P_{D_2}}{V_{D_2}}$$
$$= \frac{0.5}{4.3}$$
$$= 116 \text{ mA}$$

Sec. 3.6 Voltage Comparator With Output Limiting

Similarly,

$$I_{D_3} = \frac{P_{D_3}}{V_{D_3}}$$

$$= \frac{0.5}{3.3}$$

$$= 152 \text{ mA}$$

Since these both exceed the short-circuit output current of the 741, we shall have to limit the current well below the maximum amount. Let us plan to limit the current to 5 milliamps.

Next, we compute the minimum values for R_2 in order to limit the diode currents to the desired value by applying Ohm's Law. The minimum value as dictated by D_2 is found, Eq. (3.23) to be

$$R_2(1) = \frac{+V_{SAT} - V_{D_2}}{I_{D_2}}$$

$$= \frac{13 - 4.3}{0.005}$$

$$= 1.74 \text{ k}\Omega$$

Similarly, the value required to limit the current through D_3 is computed as

$$R_2(2) = \frac{-V_{SAT} - V_{D_3}}{I_{D_3}}$$

$$= \frac{-13 - (-3.3)}{-0.005}$$

$$= 1.94 \text{ k}\Omega$$

The larger of these two values (1.94 kilohms) sets the lower limit on R_2. We shall select the next higher standard value of 2 kilohms for R_2.

Compute R_1. Resistor R_1 helps to compensate for the effects of op amp bias current. Its value is computed as the parallel combination of R_F and R_I.

$$\boxed{R_1 = R_F \| R_I = \frac{R_F R_I}{R_F + R_I}} \tag{3.34}$$

In this particular case, we get

$$R_1 = \frac{75 \times 10^3 \times 10 \times 10^3}{75 \times 10^3 + 10 \times 10^3} = 8.82 \text{ k}\Omega$$

We shall select the nearest standard size of 9.1 kilohms for R_1.

Select the reference zener diode. The required reference voltage can be determined by using the following equation:

$$V_{REF} = \frac{V_{UT}R_F + R_I v_O^-}{R_I + R_F} \qquad (3.35)$$

In our present design, the required reference is determined as follows:

$$V_{REF} = \frac{2 \times 75 \times 10^3 + 10 \times 10^3 \times (-4)}{10 \times 10^3 + 75 \times 10^3} = 1.29 \text{ V}$$

A brief scan of App. 5 reveals that it may be very difficult to locate a 1.3 volt zener. Let us rely on our knowledge of basic semiconductors to discover an alternative. Recall that a forward biased silicon diode has about 0.6 to 0.7 volts and remains fairly constant. We can obtain the equivalent of a 1.3 volt zener by using two series silicon diodes. Appendix 6 lists the data for 1N914A diodes. A 1N914A diode will have about 0.64 volts across it with a forward current of 0.25 milliamps. Similarly, this same diode will have about 0.74 volts across it with a forward current of 1.5 milliamps. Let us select 1N914A diodes for our application and establish a forward current of about 0.5 milliamps.

Determine the value of R_S. The purpose of resistor R_S is to limit the current through reference diode D_1. In our case, it will limit the current through two series 1N914A diodes. The value of R_S is computed, Eq. (3.23), as follows:

$$R_S = \frac{V^+ - V_{REF}}{I_{REF}}$$

where I_{REF} is the specified current through the reference diode. For our design, R_S is calculated as shown

$$R_S = \frac{15 - 1.3}{0.5 \times 10^{-3}} = 27.4 \text{ k}\Omega$$

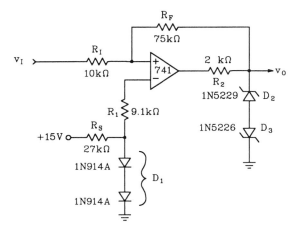

Figure 3.19 Final design for a comparator with output limiting.

Sec. 3.6 Voltage Comparator With Output Limiting

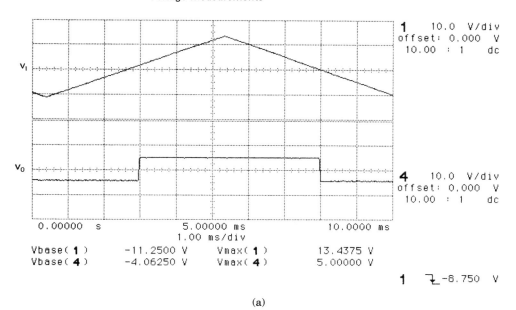

(a)

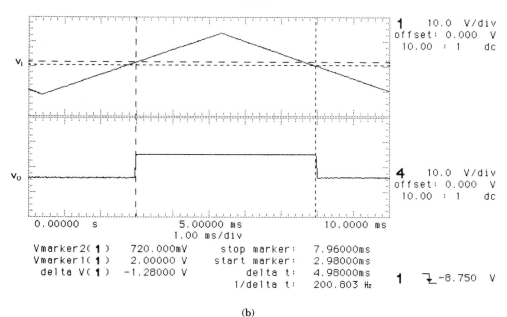

(b)

Figure 3.20 Oscilloscope displays showing the behavior of the comparator in Fig. 3.19. (Test equipment courtesy of Hewlett-Packard Company)

TABLE 3.4

	DESIGN GOAL (volts)	MEASURED VALUE (volts)
Output voltage (+)	+5.0	+5.0
Output voltage (−)	−4.0	−4.06
Upper threshold voltage	+2.0	+2.0
Lower threshold voltage	+0.8	+0.77

We shall choose the standard value of 27 kilohms for R_S.

This completes the design of our voltage comparator with output limiting. The final schematic is shown in Fig. 3.19. The actual performance of the circuit is shown in Fig. 3.20 by means of oscilloscope displays. The design goals are contrasted with the measured circuit values in Table 3.4.

3.7 TROUBLESHOOTING TIPS FOR VOLTAGE COMPARATORS

Comparator circuits are generally some of the easier op amp circuits to troubleshoot provided you pay close attention to the symptoms and keep the basic theory of operation in mind at all times. If the circuit worked properly at one time (i.e., it does not have design flaws), then the symptoms of the malfunction will normally fall into one of the following categories:

1. output is driven to one extreme ($\pm V_{SAT}$) regardless of the input signal,
2. switching levels (input or output or both) are wrong.

Output saturated. It is always a good first check to verify the power supply voltages. A missing supply can cause the output to go to the opposite extreme.

If the power supplies are both correct, then compare the voltage readings on the (+) and (−) inputs of the op amp. If the polarity on the two inputs periodically switches (i.e., one input becomes more positive than the other and then later changes such that it is less positive than the second input), then the op amp is a likely suspect. That is, the inputs **directly on the op amp** are telling the device to switch, and the op amp has the correct power source, yet the output remains in saturation. The op amp is the most probable trouble.

If, on the other hand, the (+) and (−) input terminal measurements reveal that one of the inputs is always more positive than the other, then the op amp is not being told to change states. In this case, you should check the input signal for proper voltage levels. Pay particular attention to any DC offset signals that may be present. A DC offset

Sec. 3.8 Nonideal Considerations

at the input can shift the entire operation so far off center that the input signal cannot cause the op amp to switch.

If the input signal is correct, verify the proper voltage on the reference input. If this is incorrect then the problem lies in the reference circuit (i.e., voltage divider, zener diode, etc.).

If both the input signal and the reference voltages are correct but one of the input pins continues to be more positive than the other at all times, measure the output of the op amp (particularly in circuits with output limiting). Although the output is at an extreme voltage, determine if the extreme voltage is one of the expected levels (e.g., a proper zener voltage) or some higher voltage. If the level is incorrect (i.e., too high) then suspect one of the zener diodes in the output.

Incorrect switching levels. Although there are many things that can cause minor shifts in switching levels (e.g., component value drifts), some are more probable than others. If the circuit has adjustable components such as a variable reference voltage, suspect this first. If the variable components are properly adjusted but the problem remains, suspect any solid state components other than the op amp (e.g., zener diodes).

The zeners can be checked for proper operation by measuring the voltage across them. A forward-biased zener will drop about 0.7 volts. A reverse-biased zener should have a voltage drop that is approximately equal to its rated voltage. Keep in mind, that zeners are not precision devices. For example, a 5.6 volt zener that drops 6 volts is probably not defective.

As a last resort, verify the resistance values. Resistor tolerances in a low-power circuit of this type do not frequently present problems.

3.8 NONIDEAL CONSIDERATIONS

For many comparator applications, slew rate is the primary nonideal parameter that must be considered. This limitation was discussed in earlier sections along with methods for determining the effects of a finite slew rate. Additionally, the zener diodes become less ideal as the input frequency is increased.

Throughout the earlier sections of this chapter, it was assumed that the op amp changed states whenever the differential input voltage passed through zero. The input bias current for the op amp, however, can cause the actual switch point to be slightly above or below zero. This problem is minimized by keeping the resistance between the ($-$) input to ground equal to the resistance between the ($+$) input and ground. This was the primary purpose of resistor R_B in the circuits presented previously. If a greater measure of switching accuracy is required, then resistor R_B can be made variable and adjusted to cause precise switching.

Input offset voltage is another nonideal op amp parameter that can affect the switching points of the comparator. The effect of a nonzero input offset voltage can be cancelled by adjusting R_B as described or by utilizing the offset null terminals. Appendix 4 illustrates the proper way to utilize the null terminals on an MC1741SC op amp. It should be noted, however, that different op amps use different techniques for nulling the effects of input offset voltage. Therefore, you must refer to the manufacturer's data sheet for each particular op amp.

The errors caused by the input bias currents and the input offset voltage can be totally eliminated by adjusting R_B and/or utilizing the nulling terminals. Unfortunately, however, the required level of compensation varies with temperature. Thus, although you may completely cancel the nonideal effects at one temperature, the effects will likely return at a different temperature. For many, if not most, comparator applications, this latter drift does not present severe problems. If the application demands greater compensation, then an op amp should be initially selected that offers optimum performance in these areas.

REVIEW QUESTIONS

1. Refer to Fig. 3.7. Which component(s) is (are) used to determine the threshold voltages?
2. Refer to Fig. 3.9. If resistor R_1 was to develop a short circuit, describe the effect on circuit operation. Would there still be a rectangular wave on the output pin?
3. Refer to Fig. 3.9. If resistor R_F is reduced in value, describe the relative effect on the circuit hysteresis.
4. Refer to Fig. 3.7. If resistor R_1 is changed to 68 kilohms, what is the value for the upper threshold voltage? Does this resistor change affect the circuit hysteresis?
5. Refer to Fig. 3.14. If resistor R_1 is reduced in value (and no components are damaged), what is the effect on the negative threshold (assume ideal zeners)?
6. Refer to Fig. 3.14. If diode D_3 was to become open, describe the effect on circuit operation.
7. Refer to Fig. 3.15. Describe the effect on circuit operation if diode D_1 develops a short circuit.
8. Refer to Fig. 3.17. What is the purpose of R_1? Would the circuit still appear to operate correctly if R_1 was shorted?
9. Sketch a simple graph of voltage versus time that illustrates the relationship between the voltages at the following points in Fig. 3.17: v_I, v_O, and the output pin of the op amp. Be sure to indicate relative voltage amplitudes and phase relationships.
10. Refer to Fig. 3.19. What is the effect on circuit operation if R_S is returned to a +20-volt supply instead of the +15-volt source shown in the figure?

CHAPTER 4

OSCILLATORS

4.1 OSCILLATOR FUNDAMENTALS

An oscillator is essentially an amplifier that produces its own input. That is, if we connect an oscillator circuit to a DC power supply, it will generate a signal without having a similar signal available as an input. One of the most fundamental ways to classify oscillator circuits is by the shape of the waveform generated. In this chapter, we shall study oscillator circuits that produce such waveforms as sine wave, rectangular wave, ramp wave, and triangular wave.

In general, in order for a circuit to operate as an oscillator, three basic factors must be provided in the circuit. They are

1. amplification,
2. positive feedback,
3. frequency determining network.

Suppose that many random signal frequencies (e.g., noise voltages) are present at the input of the amplifier shown in Fig. 4.1. All of these frequencies are amplified by the amplifier. They then enter the frequency selective circuit. This portion of the circuit normally introduces a loss or reduction in signal amplitude. Essentially, all frequencies can enter the frequency determining network, but only a single frequency (ideally) is allowed to pass through. In practice, a narrow band of frequencies can pass with minimal attentuation. The narrower the passband of frequencies, the more stable the output frequency of the oscillator.

Once the desired signal emerges from the frequency selective portion of the circuit, it is returned to the input of the amplifier. The amplifier compensates for losses in the frequency selective portion of the circuit. The overall closed-loop gain of the circuit must

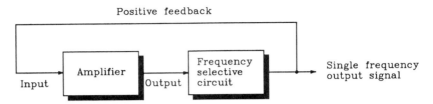

Figure 4.1 An oscillator circuit requires amplification, frequency selection, and positive feedback to operate.

be at least one (unity) in order for the circuit to sustain oscillation. If the overall loop gain is less than one, the oscillations quickly decay (ringing at best). If the loop gain exceeds unity, then the amplitude of the output signal will continue to increase until saturation is reached. If the circuit is intended to produce sinewaves, then the loop gain must be set to unity in order to maintain a constant amplitude, undistorted output signal.

4.2 WEIN-BRIDGE

4.2.1 Operation

Figure 4.2 shows the schematic diagram of a Wein-bridge oscillator circuit built around a 741 op amp. A Wein-bridge oscillator produces sinewaves and uses an RC network as the frequency determining portion of the circuit. The amplification is, of course, provided by the op amp. The op amp is essentially connected as a noninverting amplifier circuit similar to those discussed in Chap. 2. The gain of the op amp portion of the circuit is determined by the ratio of the feedback resistor (R_F) and the effective resistance of the field-effect transistor (FET) in parallel with R_1. The FET's resistance is determined by the amount of bias voltage on the gate. As the voltage on the gate becomes more negative, the channel resistance in the FET is increased.

The gate voltage for the FET is obtained from the output of a half-wave rectifier and filter combination. The input to the rectifier is provided by the output of the oscillator. In short, if the output amplitude were to try to increase, then the output of the rectifier circuit would become more negative. This increased negative voltage would bias the FET more toward cutoff (i.e., higher channel resistance). The increased FET resistance would cause the gain of the op amp circuit to decrease and thus prevent the output amplitude from increasing. A similar, but opposite, effect would occur if the output amplitude tried to decrease.

The output signal is also returned to the (+) input terminal (positive feedback) via the R_1C_1 and R_2C_2 network. This is the frequency selective portion of the oscillator. At the desired frequency of oscillation, the RC network will have a voltage gain of one third and a phase shift of zero (i.e. no phase shift). At all other frequencies, the loss will be even greater and the input/output signals will differ in phase.

Now, if the amplifier portion of the circuit can provide a gain of three and the frequency selection portion of the circuit has a gain (actually a loss) of one third, then

Sec. 4.2   Wein-bridge

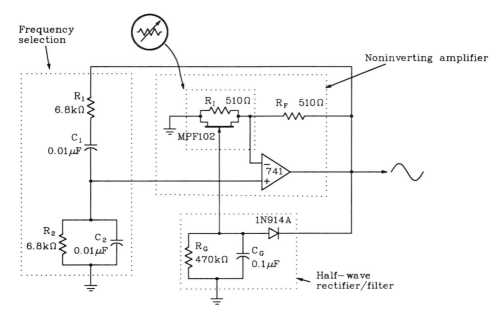

Figure 4.2   A FET-stabilized Wein-bridge oscillator circuit.

the overall closed-loop gain will be one or unity **at the frequency of oscillation.** We now have the conditions necessary for oscillation. Additionally, since the gain of the amplifier is self-adjusting because of Q_1, we also have the conditions necessary for a stable output amplitude.

4.2.2 Numerical Analysis

Let us now analyze the Wein-bridge oscillator circuit shown in Fig. 4.2 in greater detail. The most important characteristic to be evaluated is the frequency of operation. This is solely determined by the R_1C_1 and R_2C_2 networks. Although oscillators can be made with unlike values of resistance and capacitance, the general practice is to use equal values for the resistors and equal values for the capacitors in the R_1C_1 and R_2C_2 networks. This greatly simplifies the design and analysis of the Wein-bridge oscillator. When equal sets of values are used for the bridge, the frequency of oscillation is given by Eq. (4.1).

$$f_o = \frac{1}{2\pi RC} \tag{4.1}$$

In the case of the circuit in Fig. 4.2, the frequency of operation is computed as

$$f_o = \frac{1}{6.28 \times 6.8 \times 10^3 \times 0.01 \times 10^{-6}} = 2.34 \text{ kHz}$$

The voltages at the various points in the circuit are not readily computed since they are highly dependent on the specific FET being used in the circuit. We know from our basic oscillator theory, the amplifier must have a voltage gain of three at the frequency of oscillation. Since the amplifier is configured as a noninverting amplifier, we can compute the voltage gain as we did in Chap. 2.

$$A_V = \frac{R_F}{R_{IN}} + 1,$$

or

$$R_{IN} = \frac{R_F}{A_V - 1}$$
$$= \frac{510}{3 - 1}$$
$$= 255 \ \Omega$$

In the present case, however, R_{IN} is the effective resistance of R_1 in parallel with the FET. The effective resistance of the FET can be computed with our basic parallel resistance formula.

$$R_{IN} = \frac{R_1 R_{FET}}{R_1 + R_{FET}},$$

or

$$R_{FET} = \frac{R_{IN} R_1}{R_1 - R_{IN}}$$
$$= \frac{255 \times 510}{510 - 255}$$
$$= 510 \ \Omega$$

Now we know that the channel resistance of the FET will be 510 ohms during oscillation. How does it get to that value? Well, the output of the op amp will be as large as necessary to produce the exact DC level at the gate of the FET that is needed to cause the 510-ohms channel resistance. Unfortunately, the parameters of the FET vary considerably (see App. 7) and can only be estimated for a particular device. The manufacturer's data sheet (App. 7) gives the value of $V_{GS(OFF)}$ as a maximum of 8 volts. In order to bias the FET in the "resistive" range (i.e., below the knee of the I_D versus V_{DS} curve), the gate voltage will generally be 25 percent of $V_{GS(OFF)}$ or less. In the case of Fig. 4.2, we could anticipate a gate voltage of 2 volts or less. This, of course, restricts our peak output voltage to about 2.7 volts since the output actually produces the FET's gate voltage via the rectifier circuit (D_1). If we try to generate significantly higher voltages, then we can anticipate a distorted output because we will be operating past the knee of the FET curve.

Sec. 4.2   Wein-bridge

4.2.3 Practical Design Techniques

Now let us design a Wein-bridge oscillator circuit that will perform according to the following design goals:

1. frequency of oscillation   10.5 kHz
2. available FET   MPF102

Compute the frequency determining values. We base our design on the accepted practice that $R_1 = R_2 = R$ and $C_1 = C_2 = C$. We shall choose a value for C and compute the associated value for R. The initial selection of C is somewhat arbitrary but will generally produce good results with values at least 100 times greater than the input capacitance of the op amp. For our present example, let us choose an initial capacitance value of 1000 picofarads. We can now compute the required value of R with Eq. (4.1).

$$f_o = \frac{1}{2\pi RC},$$

or

$$R = \frac{1}{2\pi C f_o}$$

$$= \frac{1}{6.28 \times 1000 \times 10^{-12} \times 10.5 \times 10^3}$$

$$= 15.2 \text{ k}\Omega$$

We shall choose a standard value of 15 kilohms. If the computed value of R is below 1 kilohm or greater than 470 kilohms, you might want to select a different value for C and recompute R.

Compute R_F and R_I. R_F and R_I are selected to produce **both** of the following conditions:

1. when the FET is biased off, the gain of the op amp will be less than three.
2. when the FET is in its "Resistive" range, the gain of the op amp can exceed three.

We shall compute R_I as follows:

$$R_I = \frac{V_{GS(off)}}{I_{DSS} \text{ (max)}} \tag{4.2}$$

In our present case, we compute R_I as

$$R_I = \frac{8}{20 \times 10^{-3}} = 400 \text{ }\Omega$$

Let us select a standard value of 390 ohms for R_I. We shall set R_F to be equal to R_I. That is

$$\boxed{R_F = R_I} \quad (4.3)$$

More specifically,

$$R_F = 390 \text{ }\Omega$$

This insures that conditions for oscillation will be met.

Compute rectifier and filter components. The $R_G C_G$ network is a filter for the half-wave rectifier circuit. As with any basic rectifier circuit, the time constant of the filter should be long relative to the period of the input signal. In the case of the circuit shown in Fig. 4.2, the input signal is the basic oscillator frequency (f_O). Thus, the $R_G C_G$ time constant is computed as

$$\boxed{R_G C_G = \frac{100}{f_O}} \quad (4.4)$$

In this particular case, the required RC time constant is computed as

$$R_G C_G = \frac{100}{10.5 \times 10^3} = 9.52 \text{ milliseconds}$$

Let us select a value for R_G and compute the associated value for C_G. The value for R_G is not critical, but it is generally in the range of 10 kilohms to 1 megohm. Let us select 270 kilohms as the value for R_G. Capacitor C_G can now be computed by using Eq. (4.4).

$$R_G C_G = \frac{100}{f_O},$$

or

$$C_G = \frac{100}{R_G f_O}$$

$$= \frac{100}{270 \times 10^3 \times 10.5 \times 10^3}$$

$$= 0.035 \text{ }\mu\text{F}$$

We shall choose a standard value of 0.039 microfarad.

Sec. 4.2 Wein-bridge

The rectifier diode can be any general-purpose diode capable of withstanding the currents and voltages present in this application. Although the actual voltages and currents will be less, the following provides an easy and conservative computation for diode selection:

$$\boxed{V_{PIV} = +V_{SAT} - (-V_{SAT})} \tag{4.5}$$

where V_{PIV} is the minimum reverse voltage breakdown rating for the diode. Also,

$$\boxed{I_F = \frac{-V_{SAT}}{R_G}} \tag{4.6}$$

where I_F is the maximum average forward current rating for the diode. Let us select a 1N914A for this example. The data for this diode is listed in Appendix 6.

Select the op amp. The two most significant considerations regarding op amp selection are unity gain frequency (bandwidth) and slew rate. The minimum required unity gain frequency can be estimated with the following equation:

$$\boxed{f_{UG} = 30 f_o} \tag{4.7}$$

In our particular example, the minimum required unity gain frequency is

$$f_{UG} = 30 \times 10.5 \times 10^3 = 315 \text{ kHz}$$

The minimum required slew rate for our op amp can be estimated from the following equation:

$$\boxed{\text{slew rate(min)} = \pi f_o \left[\frac{V_{GS(OFF)}}{2} + 0.7 \right]} \tag{4.8}$$

The minimum slew rate for our present application is estimated as

$$\text{slew rate(min)} = 3.14 \times 10.5 \times 10^3 \left[\frac{8}{2} + 0.7 \right] = 0.155 \text{ V}/\mu\text{s}$$

The required values for both unity gain frequency and slew rate are well within the values offered by the 741 op amp. We shall choose the 741 for this application.

The final schematic of our Wein-bridge oscillator is shown in Fig. 4.3. Its performance is indicated by the oscilloscope plots presented in Fig. 4.4. Also, indicated in Fig. 4.3 is a potentiometer being used as an amplitude control. As the amount of signal fed to the rectifier circuit decreases, the gain of the op amp increases causing a higher output signal amplitude. However, as stated earlier, if the output amplitude is made too large, then the FET will not be operating in the correct portion of its curve and the signal will have significant distortion.

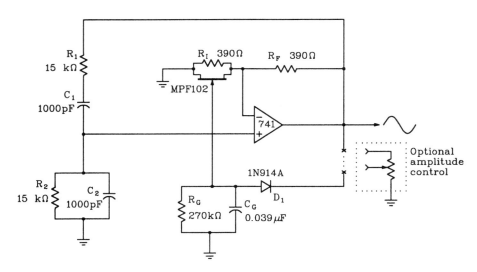

Figure 4.3 A design example of a 10.5 kilohertz Wein-bridge oscillator.

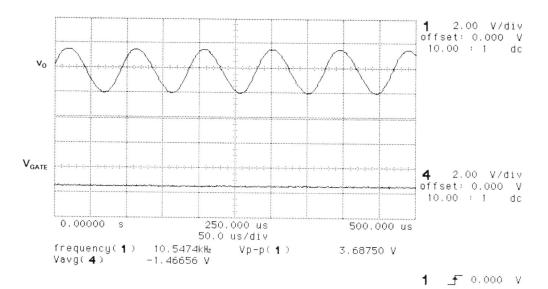

Figure 4.4 Oscilloscope display showing the output of the Wein-bridge oscillator circuit shown in Fig. 4.3. (Test equipment courtesy of Hewlett-Packard Company)

4.3 VOLTAGE-CONTROLLED OSCILLATOR

4.3.1 Operation

A voltage-controlled oscillator (VCO) is an oscillator circuit whose frequency can be controlled or varied by a DC input voltage. This type of circuit is also called a voltage-to-frequency converter (VFC). The output waveform from the VCO may be sine, square, or other waveshape depending on the circuit design. Figure 4.5 shows the schematic of a representative VCO circuit. This circuit produces both triangle and square wave outputs. In both cases, the frequency is determined by the magnitude of the DC input voltage ($+V_{IN}$).

Let's examine the circuit's operation one stage at a time. The left most stage is basically an inverting, summing amplifier with the feedback resistor replaced by a capacitor. The operation of this circuit, called an integrator, is discussed in greater detail in Chap. 7. For now, however, recall that the value of feedback current in an inverting amplifier is determined by the input voltage and the value of the input resistor(s). First, let us assume that diode D_4 is reverse biased and acting as an open. Under these conditions, $+V_{IN}$ and R_1 will determine the value of feedback current for A_1. Since V_{IN} is DC and R_1 does not change, then the value of input current, and therefore, feedback current will be constant. The feedback current must flow through C_1. You may recall from basic electronics theory, that if a capacitor is charged with a constant current source, then the resulting voltage will increase linearly. Since the charging current for C_1 is constant, we can expect a linear ramp of voltage across C_1. Since the left end of C_1 is connected to a

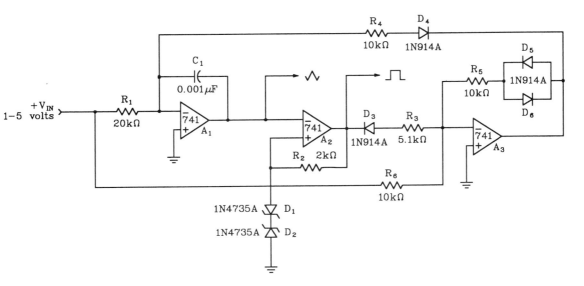

Figure 4.5 A voltage-controlled oscillator providing both triangle and squarewave outputs.

virtual ground point, the other end (output of the op amp) will reflect the linear ramp voltage. Since the input voltage V_{IN} is positive, we know that the output ramp will be increasing in the negative direction.

A_2 is configured as a voltage comparator circuit with the upper and lower thresholds being established by diodes D_1 and D_2. As long as the ramp voltage is above the lower threshold point (established by D_1) then the output of amplifier A_2 will remain at its negative limit ($-V_{SAT}$).

Amplifier A_3 is connected as an inverting summing amplifier. One input comes from A_2 and receives a gain of -2. The other input is provided by $+V_{IN}$ and receives a gain of -1. As long as the output of A_2 is at its $-V_{SAT}$ level, then diode D_3 will be forward biased, and this voltage will be coupled to the input of A_3. Clearly, this high-negative voltage will drive amplifier A_3 into saturation. That is, the output of A_3 will be at the $+V_{SAT}$ level regardless of the value of input voltage ($+V_{IN}$). It is this $+V_{SAT}$ level on the output of A_3 that causes D_4 to remain in a reverse-biased state. This circuit condition remains constant as long as the ramp voltage on the output of A_1 is above the lower threshold voltage of A_2.

Once the decreasing ramp voltage from A_1 falls below the lower threshold voltage of comparator A_2, the output of A_2 changes to its $+V_{SAT}$ level. This reverse biases diode D_3 and causes A_3 to act as a simple inverting amplifier with regard to the input voltage $+V_{IN}$. A voltage level that is equal (but opposite polarity) to $+V_{IN}$ is felt at the right end of R_4. Since R_4 is half as large as R_1 and has the same voltage applied, we can expect the current flow through R_4 to be twice as large as that through R_1 and in the opposite direction.

The current provided by R_4 not only cancels the input current provided via R_1 but supplies an equal (but opposite) current to C_1. That is, C_1 will now continue to charge at the **same** linear rate but in the opposite direction. The ramp voltage at the output of A_1 will rise linearly until it exceeds the upper threshold of A_2. Once the upper threshold has been exceeded, the output of A_2 switches to the $-V_{SAT}$ level. The forces to the output of A_3 go to $+V_{SAT}$ and reverse biases D_4. We are now back to the original circuit state and the cycle repeats.

The frequency of operation is determined by the time it takes C_1 to charge to the threshold levels of A_2. Once the circuit components have been fixed, the only thing that determines frequency is the value of input voltage ($+V_{IN}$). This, of course, gives rise to the name voltage-controlled oscillator.

A triangle wave (or double ramp) signal may be taken from the output of A_1. The output of A_2 provides a square wave output.

4.3.2 Numerical Analysis

Let us now analyze the performance of the circuit in Fig. 4.5 from a numerical point of view. First consider the voltage comparator A_2. Since it has no negative feedback, we know the output will be driven to one of its two extremes ($\pm V_{SAT}$) at all times. For purposes of this analysis, let us use the typical values of ± 13 volts for $\pm V_{SAT}$. The

Sec. 4.3 Voltage-Controlled Oscillator

threshold voltages for the comparator are determined by the zener diodes (D_1 and D_2). Appendix 8 shows that the 1N4735 diodes are designed to regulate at 6.2 volts. Additionally, a typical forward voltage drop is 0.6 volts. Therefore, when the output is at $+V_{SAT}$, the upper threshold (V_{UT}) will be determined by the regulated voltage of D_2 plus the forward voltage drop of D_1. We can express this in equation form as

$$\boxed{V_{UT} = V_{ZD2} + V_{FD1}} \quad (4.9)$$

where V_Z is the rated zener voltage and V_F is the forward voltage drop. In the case of the circuit in Fig. 4.5, the upper threshold is computed as

$$V_{UT} = 6.2 + 0.6 = 6.8 \text{ V}$$

The lower threshold (V_{LT}) is computed in a similar manner.

$$\boxed{V_{LT} = -V_{ZD1} - V_{FD2}} \quad (4.10)$$

In the present case, we have

$$V_{LT} = -6.2 - 0.6 = -6.8 \text{ V}$$

These threshold values are particularly important since they will determine the charging limits of capacitor C_1 which is the heart of the circuit.

Now let us evaluate the numerical performance of A_3 and its associated circuitry. We shall apply the superposition theorem and consider the two inputs to A_3 independently. First, let's consider the $+V_{IN}$ signal input. The voltage gain of this signal is computed in the same manner as we did in Chap. 2 for a simple inverting amplifier. That is

$$A_V = -\frac{R_F}{R_I}$$
$$= -\frac{10 \times 10^3}{10 \times 10^3}$$
$$= -1$$

In this case, the generic R_I is replaced with physical resistor R_6. Notice that we have ignored the effects of D_5 and D_6. You will recall that the output of a closed-loop op amp amplifier will go to whatever level is required to bring the differential input voltage back to near zero. By inserting a forward-biased diode in the feedback loop, the output is forced to rise an additional 0.6 volts (the forward voltage drop of the diode). By paralleling two diodes in opposite polarities (D_5 and D_6), we force the output to be 0.6 volts larger than it would otherwise have been. The actual output voltage, then will be the normal expected output plus a fixed 0.6 volt potential that causes the output to be more positive during positive output times and more negative during negative output times. The reason for D_5 and D_6 will be evident in a moment.

Since the voltage gain for the $+V_{IN}$ signal is -1, the range of output voltages for A_3 as a result of $+V_{IN}$ is

$$v_{O_1} = -1 \times 1 - 0.6 = -1.6 \text{ V}$$

and

$$v_{O_2} = -1 \times 5 - 0.6 = -5.6 \text{ V}$$

Now let's consider the effects of the second A_3 signal input which comes from the output of A_2. The voltage gain for this input is computed in a similar manner.

$$A_V = -\frac{R_F}{R_I} = -\frac{10 \times 10^3}{5.1 \times 10^3} = -1.96 \approx -2$$

The input voltage (output from A_2) is the $\pm V_{SAT}$ levels for A_2. Since diode D_3 will block the positive level, we need only calculate the effects of the $-V_{SAT}$ input. The resulting output voltage from A_3 as a result of this input is computed as shown

$$v_O = -2 \times (-V_{SAT} + 0.6) + 0.6$$
$$= -2 \times (-13 + 0.6) + 0.6$$
$$= +25.4 \text{ V}$$

This computed value exceeds the limits of A_3 since it only has a ± 15-volt power supply. This means that the output of A_3 will be driven to its $+V_{SAT}$ level ($+13$ volts).

Now let us consider the combined effects of the two inputs to A_3. You will recall that the combined effect is found by adding the outputs caused by the two individual inputs. During times when the output of A_2 is positive, it has no effect on the output of A_3 (D_3 is reverse biased), and the output of A_3 will be solely determined by the $+V_{IN}$ signal as computed previously. During the times when A_2 is at its $-V_{SAT}$ level, the output of A_3 will clearly be driven to its $+V_{SAT}$ level. Even in the best case when $+V_{IN}$ is at its most positive ($+5$ V) level, the output of A_3 will be

$$V_O = V_{O_1} + V_{O_2}$$
$$= +25.4 + (-5.5)$$
$$= +19.8 \text{ V}$$

where V_{O_1} and V_{O_2} are the effective output voltages produced by A_2 and $+V_{IN}$, respectively. As you can readily see, this combined value still exceeds the $+V_{SAT}$ level of A_3, so we will expect the output of A_3 to remain at $+13$ volts anytime the output of A_2 is at the $-V_{SAT}$ level. On the other hand, when the output of A_2 is at the $+V_{SAT}$ level, the output of A_3 will be between -1.6 and -5.6 depending on the value of input voltage ($+V_{IN}$).

Sec. 4.3 Voltage-Controlled Oscillator

Finally, let's examine the operation of A_1 more closely. During times that the output of A_3 is at the $+V_{SAT}$ level, diode D_4 will be reverse biased and will isolate or remove that input path for A_1. During these times, A_1 is controlled by the effects of $+V_{IN}$ only. Let us examine the charging rate of C_1 at the two extremes of $+V_{IN}$.

If $+V_{IN}$ is at its lower limit ($+1$ V), then the current through R_1 is computed as

$$I_{R_1} = \frac{+V_{IN}}{R_1}$$

$$= \frac{1}{20 \times 10^3}$$

$$= 50 \ \mu A$$

Similarly, the maximum input current is computed as

$$I_{R_1} = \frac{+V_{IN}}{R_1}$$

$$= \frac{5}{20 \times 10^3}$$

$$= 250 \ \mu A$$

Since D_4 is effectively open (i.e., reverse biased), and since no significant current can flow into or out of the ($-$) input terminal of the op amp, we can infer that all of the input current goes to charge C_1. More specifically, electrons flow from the output of A_1, through C_1 (i.e., charging C_1), and through R_1 to $+V_{IN}$. Further, since this current is constant (unless $+V_{IN}$ changes), capacitor C_1 will charge linearly according to the following expression:

$$\boxed{V_{C_1} = \frac{IT}{C_1}} \qquad (4.11)$$

where I is the input current computed and T is the time C_1 is allowed to charge. This equation can be transposed to produce another very useful form.

$$\boxed{T = \frac{C_1 \times \Delta V_{C_1}}{I}} \qquad (4.12)$$

This allows us to compute the amount of time it takes C_1 to charge to a given voltage change (ΔV_{C_1}) when a given value of charging current is applied.

We already know from earlier discussions that the limits of C_1's charge are set by the upper and lower thresholds of A_2. That is, C_1 will charge linearly between the V_{LT} and V_{UT} values established by A_2. In the present circuit, the **change** in C_1 voltage in going from the V_{LT} to the V_{UT} is

$$\Delta V_{C_1} = V_{UT} - V_{LT} \qquad (4.13)$$

More specifically,

$$\Delta V_{C_1} = +6.8 - (-6.8) = 13.6 \text{ V}$$

If we now compute the time it takes C_1 to make this voltage change, we will know the time for one alternation (negative slope) of the oscillator's output. Let us compute this time for input voltages of $+1$ and $+5$ volts which have been previously shown to produce 50 microamperes and 250 microamperes, respectively. Eq. (4.12) gives us

$$T = \frac{C_1 \times \Delta V_{C_1}}{I}$$

$$T_{+1} = \frac{0.001 \times 10^{-6} \times 13.6}{50 \times 10^{-6}} = 272 \text{ μs},$$

and

$$T_{+5} = \frac{0.001 \times 10^{-6} \times 13.6}{250 \times 10^{-6}} = 54.4 \text{ μs}$$

The remaining alternation (positive ramp) occurs when D_4 is forward biased. This effectively connects the output of A_3 to A_1 via R_4. Recall that during this portion of the cycle, the output of A_3 is 0.6 volts larger than $+V_{IN}$ and is of the opposite polarity. Since diode D_4 drops 0.6 volts when it is forward biased, this means that the voltage applied to the right end of R_4 is exactly the same as the value of $+V_{IN}$, but it is negative instead of positive. That is, this input to A_1 ranges from -1 to -5 volts. The resulting input current through R_4 is computed as

$$I_{R_4} = \frac{V_{R_4}}{R_4}$$

$$= \frac{1}{10 \times 10^3}$$

$$= 100 \text{ μA}$$

for the -1 volt input case. The input current for the -5 volt case is

$$I_{R_4} = \frac{5}{10 \times 10^3} = 500 \text{ μA}$$

Sec. 4.3   Voltage-Controlled Oscillator   191

Let us consider the electron flow in the case of a 1 volt input. Recall that R_1 will have a 50-microampere current flowing in a right-to-left direction. R_4, as computed, will have a current of 100 microamperes flowing from right-to-left. When this latter current gets to the summing point of A_1, it splits. One part, 50 microamperes, goes through R_1 and satisfies the requirements of $+V_{IN}$ and R_1. Kirchoff's Current Law will tell us that the remaining 50 microamperes must flow into C_1 in a left-to-right direction. It is very important to note that the magnitude of this charging current is **identical** to that which flowed on the previous alternation, but it is flowing in the opposite direction. Therefore, C_1 will charge at the **same** rate but in the opposite polarity. Since the charging currents are equal and since the required voltage charge (ΔV_{C1}) is the same, the amount of time for this alternation will be the same as the first. Given this observation, we can now compute the frequency of oscillation for a given input voltage ($+V_{IN}$)

$$f_O = \frac{+V_{IN}}{2R_1C_1(V_{UT} - V_{LT})} \qquad (4.14)$$

This equation will be valid as long as the slew rate of the op amps does not interfere with circuit operation. As the oscillator frequency increases, the slew rate limitations of the op amp tends to reduce the actual frequency from the value computed.

4.3.3 Practical Design Techniques

We shall now design a voltage controlled oscillator that meets the following design criteria:

1. input voltage range   0 to 6 volts DC
2. ramp output voltage   ±4 volts (±3 volts minimum)
3. frequency range   0 to 5.0 kHz

The configuration and gain values for A_3 should stay the same as that shown in Fig. 4.5. Therefore, the following components will be considered as "previously computed": D_3, R_3, R_5, D_5, D_6, R_6, and D_4. If, due to availability, you elect to change any of these resistors, then be sure to keep their ratios such the voltage gain of A_3 is -1 and -2 for the $+V_{IN}$ and A_2 signals, respectively. The diodes may be substituted with any general-purpose diode, but they should all be of the same type in order to have similar voltage drops.

Select the zener diodes.   The voltage rating of D_2 (plus the forward voltage drop of D_1) determine the upper limit of the ramp output voltage. Similarly, D_1 (plus the forward voltage drop of D_2) determine the lower limit of the ramp output voltage. We can express this as an equation for selecting the voltage ratings of D_1 and D_2.

$$V_{D_1} = -V_{RAMP} + 0.6 \quad (4.15)$$

and

$$V_{D_2} = +V_{RAMP} - 0.6 \quad (4.16)$$

For our present design, the required zener ratings are computed as shown

$$V_{D_1} = -4 + 0.6 = -3.3 \text{ V},$$

and

$$V_{D_2} = +4 - 0.6 = +3.3 \text{ V}$$

Appendices 5 and 8 provide a manufacturer's listing of several zener diodes. Either 1N5226 or 1N4728 zeners should work for our application. Let us select the 1N5226 devices for this example.

Compute R_2. Resistor R_2 is a current-limiting resistor that keeps the current through D_1 and D_2 within safe limits. Although the circuit will work well with a wide range of values for R_2, a good choice is to design for a current through the zener diodes given by the following expression:

$$I_Z = 10I_{ZK} \quad (4.17)$$

Resistor R_2 can then be determined from the following equation:

$$R_2 = \frac{V_{SAT} - V_Z - 0.6}{I_Z} \quad (4.18)$$

where V_{SAT} is the highest expected saturation voltage for A_2, V_Z is the **lower** of the two zener voltages (if not equal), and I_Z is the zener current calculated with Eq. (4.17). The data sheet in App. 5 lists a knee current (I_{ZK}) of 0.25 milliamperes for the 1N5226 diodes. The design value of zener current is then computed, Eq. (4.17), as

$$I_Z = 10 \times 0.25 \times 10^{-3} = 2.5 \text{ mA}$$

The value of R_2 can now be computed, Eq. (4.18), as

$$R_2 = \frac{+13 - 3.3 - 0.6}{2.5 \times 10^{-3}} = 3.64 \text{ k}\Omega$$

We shall select a standard value of 3.6 kilohms for R_2. Since this design method inherently uses a zener current which is less than the test current (I_{ZT}), we can expect the regulated voltage to be less than the stated value. Appendix 5 includes a graph that allows us to estimate the error. In the present case, the zeners will have about 2.65 volts instead of the rated 3.3 volts. This will, in turn, cause the ramp output to have an amplitude of 6.5

Sec. 4.3 Voltage-Controlled Oscillator

volts instead of the design goal of 8 volts. If this is an important circuit parameter for a given application, then we should select a zener with a higher voltage rating but continue to operate it below its rated current. We continue with our present selection since the reduced voltage is still within the tolerance stated as part of the original design goals.

Compute R_1 and C_1. Once the zeners have been selected, it is the values of R_1 and C_1 that determine the frequency for a given voltage. The required R_1C_1 product can be found with Eq. (4.19).

$$R_1C_1 = \frac{V_{IN}(MAX)}{2f_{HI}V_{RAMP}} \qquad (4.19)$$

where V_{RAMP} is the amplitude of the ramp output voltage, $V_{IN}(max)$ is the highest input voltage, and f_{HI} is the highest frequency of oscillation. Calculations for our present design example are shown

$$R_1C_1 = \frac{6}{2 \times 5 \times 10^3 \times 8} = 75 \times 10^{-6}$$

At this point, we can either select C_1 and calculate R_1 or vice versa. In either case, we want R_1 to be in the range of 1.0 kilohms to 470 kilohms if practical. Similarly, C_1 should be greater than 470 picofarads and be nonpolarized. Since it is essential that R_4 be exactly one half the value of R_1 and since there are a limited number of resistor pairs that have exactly a 2:1 ratio, it is generally easier to select R_1 and compute C_1.

For purposes of this design, let us select R_1 as 2 kilohms. We can then compute C_1 by transposing the results of Eq. (4.19).

$$R_1C_1 = 75 \times 10^{-6},$$

or

$$C_1 = \frac{75 \times 10^{-6}}{R_1}$$

$$= \frac{75 \times 10^{-6}}{2 \times 10^3}$$

$$= 0.0375 \ \mu F$$

We shall select a standard value of 0.033 microfarad for C_1. If greater accuracy is required, then we could add a second capacitor in parallel with C_1.

Compute R_4. R_4 must be exactly one half the value of R_1. That is

$$R_4 = \frac{R_1}{2} \qquad (4.20)$$

In our particular case, we compute R_4 as

$$R_4 = \frac{2 \times 10^3}{2} = 1000 \; \Omega$$

Select the op amp. The primary op amp characteristic (other than power supply voltage, etc.) in this application is slew rate. If the output of A_1 tries to change faster than the slew rate will allow, then the actual operating frequency will be lower than originally predicted. Similarly, if the switching times for A_2 and A_3 are a substantial percentage of one alternation, then again the actual frequency of oscillation will be below the calculated value. To minimize this effect, we can insure that the slew rate is fast enough to allow the rise and fall times of A_2 and A_3 to be a small part of the time for one alternation. That is

$$\boxed{\text{slew rate} = 40 f_{\text{MAX}} [+V_{\text{SAT}} - (-V_{\text{SAT}})]} \qquad (4.21)$$

This equation insures that switching time is no greater than 20 percent of the time for one alternation. If the factor 40 is changed to 200, this relationship is reduced to 1 percent, but it requires a very high slew-rate op amp to achieve moderate frequencies. If this is an important consideration for your application, then consider the use of an integrated comparator. These devices are readily available with switching times in the tens of nanoseconds range. For our present design, however, let us determine the required slew rate for a 20 percent rise time factor.

$$\begin{aligned}\text{slew rate} &= 40 \times 5 \times 10^3 [+13 - (-13)] \\ &= 5.2 \; \text{V}/\mu\text{s}\end{aligned}$$

This requirement exceeds the 0.5 volts per microsecond slew-rate rating of the standard 741, but falls within the capabilities of the MC741SC. We shall utilize this latter device in our design.

Figure 4.6 shows the schematic diagram of our design example. The oscilloscope displays in Figure 4.7 show the performance of the circuit. Additionally, Table 4.1 contrasts the design goals with the actual measured performance of the circuit.

TABLE 4.1

	DESIGN GOAL	MEASURED RESULTS
Input voltage range	0–6 V	0–6 V
Frequency range	0–5 kHz	0–5.01 kHz
Ramp voltage	±4 V ±3 V (min)	−4, +3.75 V

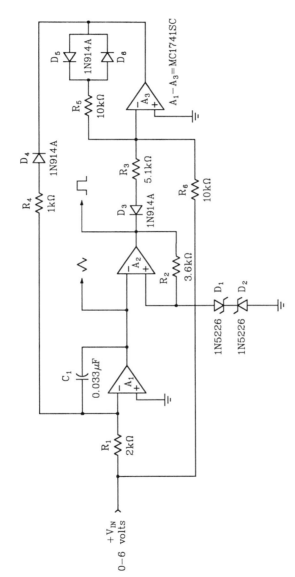

Figure 4.6 A design example of a 200-hertz 5-kilohertz voltage-controlled oscillator.

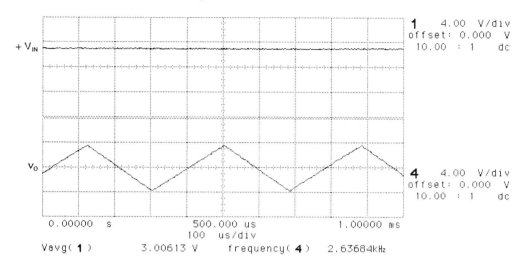

(a)

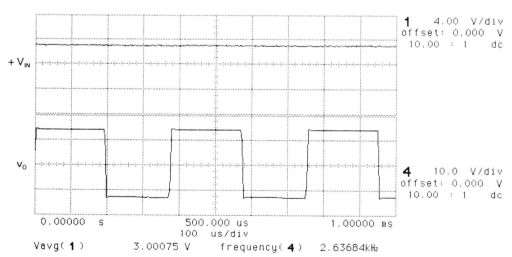

(b)

Figure 4.7 Oscilloscope displays showing the performance of the voltage-controlled oscillator circuit shown in Fig. 4.6. (Test equipment courtesy of Hewlett-Packard Company)

Sec. 4.4 Variable-Duty Cycle

more positive than the negative reference voltage on the (+) input. Once the voltage on C_1 falls below the reference voltage on the (+) input, the circuit quickly switches back to its original state and the cycle repeats.

The time it takes C_1 to charge during the positive output alternation is determined by the values of C_1, R_3, R_4, and D_5. Since R_3 is adjustable, it can be used to control the period of the positive alternation without affecting the negative alternation. Similarly, the values of C_1, R_1, R_2, and D_6 determine the charge time for C_1 during the negative output alternation. Resistor R_1 can be used to control this time period without affecting the positive alternation.

Resistors R_5 and R_6 are current limiting resistors for the two sets of back-to-back zener regulators.

4.4.2 Numerical Analysis

Let us now extend our analysis of the variable-duty cycle oscillator shown in Fig. 4.8 to include a numerical understanding of its operation. We shall first consider the two sets of back-to-back zener regulators. The $\pm V_{SAT}$ output of the op amp will be regulated by D_3 and D_4 to provide a reference voltage for the (+) input. The value of this reference is computed with Eqs. (4.22) and (4.23).

$$\boxed{+V_{REF} = V_{D_3} + 0.6} \qquad (4.22)$$

$$\boxed{-V_{REF} = -V_{D_4} - 0.6} \qquad (4.23)$$

For many analytical purposes, the rated voltages of the zeners may be used. If you require greater accuracy, then you can compute the current through the zeners and refer to the manufacturer's data sheet to determine the actual voltage. For the present example, let us compute the reference voltages in both ways. First, the approximate reference voltage can be found by applying Eqs. (4.22) and (4.23) using the rated voltages for the zeners. In our present case we have

$$+V_{REF} = 3.3 + 0.6 = 3.9 \text{ V},$$

and

$$-V_{REF} = -3.3 - 0.6 = -3.9 \text{ V}$$

If we compute the actual zener current, we can make a closer approximation. Since the circuit is utilizing similar zeners, we can calculate either one. Suppose we work out the current during the positive output alternation.

$$\boxed{I_{D_3} = \frac{+V_{SAT} - V_{D_3} - 0.6}{R_5}} \qquad (4.24)$$

Substituting values give us

$$I_{D_3} = \frac{13 - 3.3 - 0.6}{4.7 \times 10^3} = 1.94 \text{ mA}$$

Of course, even this is not an exact value since we know that the 3.3 zener drop is actually less, but our overall result will be very close to the actual value. If we now refer to the manufacturer's data sheet in App. 5, we can estimate the actual zener voltage for a 1N5226 with about 2 milliamperes of current. The graph in Fig. 2 of App. 5 indicates that our zener will have a voltage of about 2.6 volts. If we use this value and recompute the reference voltages with Eqs. (4.22) and (4.23), we will get more accurate values.

$$+V_{\text{REF}} = 2.6 + 0.6 = 3.2 \text{ V},$$

and

$$-V_{\text{REF}} = -2.6 - 0.6 = -3.2 \text{ V}$$

Let us now perform a similar calculation for D_5 and D_6. We shall use the nominal values for the zener voltages and apply Eqs. (4.25) and (4.26).

$$\boxed{+V_O = V_{D_5} + 0.6} \quad (4.25)$$

$$\boxed{-V_O = -V_{D_6} - 0.6} \quad (4.26)$$

Substituting values gives us the following estimates:

$$+V_O = 6.2 + 0.6 = 6.8 \text{ V},$$

and

$$-V_O = -6.2 - 0.6 = -6.8 \text{ V}$$

It is important to note that the actual measured zener voltages will vary somewhat from their nominal value. Nevertheless, we shall use our computed values for the remainder of the analysis.

We are now in a position to compute the operating frequency, duty cycle, pulse width, and so on. Equation 4.27 is used to determine the duration of the positive output alternation.

$$\boxed{t^+ = (R_3 + R_4)C_1 ln \frac{V_{D_6} - V_{D_3}}{V_{D_6} - V_{D_4}}} \quad (4.27)$$

Similarly, Eq. (4.28) is used to determine the time for the negative output alternation.

$$\boxed{t^- = (R_1 + R_2)C_1 ln \frac{V_{D_5} - V_{D_4}}{V_{D_5} - V_{D_3}}} \quad (4.28)$$

For illustrative purposes, let us compute the minimum and maximum time for both alternations by repeatedly applying Eqs. (4.27) and (4.28).

Sec. 4.4 Variable-Duty Cycle

$$t^+(\text{min}) = (0 + 1 \times 10^3) \times 0.1 \times 10^{-6} \ln \frac{6.2 - (-3.3)}{6.2 - 3.3} = 119 \ \mu s,$$

and

$$t^+(\text{max}) = (100 \times 10^3 + 1 \times 10^3) \times 0.1 \times 10^{-6} \ln \frac{6.2 - (-3.3)}{6.2 - 3.3}$$

$$= 12 \text{ milliseconds}$$

We can already see that the circuit gives us a 101:1 range of control on the positive alternation. A similar calculation for the negative alternation gives us

$$t^-(\text{min}) = (0 + 1 \times 10^3) \times 0.1 \times 10^{-6} \ln \frac{6.2 - (-3.3)}{6.2 - 3.3} = 119 \ \mu s,$$

and

$$t^-(\text{max}) = (100 \times 10^3 + 1 \times 10^3) \times 0.1 \times 10^{-6} \ln \frac{6.2 - (-3.3)}{6.2 - 3.3}$$

$$= 12 \text{ milliseconds}$$

Since Eqs. (4.27) and (4.28) ignore the forward voltage drops of D_1, D_2 and the effective resistance of the zener regulator circuit, the actual times for t^+ and t^- will be somewhat longer than our calculations predict.

Since the components are matched, the results are the same for each alternation. In practice, the two alternations do not have to have equal ranges.

Now let us extend our analysis to determine the minimum and maximum frequency of operation. These two extremes are given by Eqs. (4.29) and (4.30).

$$f_{\text{MIN}} = \frac{1}{t^+(\text{max}) + t^-(\text{max})} \quad (4.29)$$

$$f_{\text{MAX}} = \frac{1}{t^+(\text{min}) + t^-(\text{min})} \quad (4.30)$$

Substituting values for our particular case gives the following results for a frequency range.

$$f_{\text{MIN}} = \frac{1}{12 \times 10^{-3} + 12 \times 10^{-3}} = 41.7 \text{ Hz},$$

and

$$f_{\text{MAX}} = \frac{1}{119 \times 10^{-6} + 119 \times 10^{-6}} = 4.2 \text{ kHz}$$

This, of course, equates to a 101:1 frequency range.

Finally, let us determine the range of duty cycles. Recall from your basic electronics theory that duty cycle is defined as the ratio of pulse width to total period.

$$\% \text{ duty} = \frac{PW}{\text{period}} \times 100 \qquad (4.31)$$

For this calculation, we will consider pulse width to be the positive alternation of the output signal. The range of duty cycles then is computed.

$$\% \text{ duty(min)} = \frac{119 \times 10^{-6}}{119 \times 10^{-6} + 12 \times 10^{-3}} \times 100 = 0.98\%,$$

and

$$\% \text{ duty(max)} = \frac{12 \times 10^{-3}}{12 \times 10^{-3} + 119 \times 10^{-6}} \times 100 = 99\%$$

As you might suspect, this is also a 101:1 range of control.

4.4.3 Practical Design Techniques

For purposes of our design example, let us design a circuit similar to the one in Fig. 4.8 that displays the following behavior:

1. positive output time   1 to 10 milliseconds
2. negative output time   2 to 20 milliseconds
3. output amplitude ± 7 volts (± 6 minimum)

Select the output zeners. The amplitude of the output voltage specification dictates the zener diodes that will be used. If the required output amplitude is less than 6 to 7 volts, then it is best to design for a higher voltage and subsequently reduce it with an output voltage divider. For proper circuit operation, it is essential that the output swing be larger than the reference swing felt on the (+) input. Equation (4.25) can be used to determine the required voltage rating for D_5.

$$+V_o = V_{D_5} + 0.6,$$

or

$$V_{D_5} = +V_o - 0.6$$
$$= +7 - 0.6$$
$$= 6.4 \text{ V}$$

Similarly, Eq. (4.26) tells us the nominal voltage rating for D_6.

$$-V_o = -V_{D_6} - 0.6,$$

Sec. 4.4   Variable-Duty Cycle   203

or

$$-V_{D_6} = -V_O + 0.6$$
$$= -7 + 0.6$$
$$= -6.4 \text{ V}$$

Referring to Apps. 5 and 8, we see that there will be difficulty getting a 6.4 volt zener. However, if we design for a zener current that is less than the test current, then the actual zener voltage will be less than the rated value. With this in mind, let us select the next higher standard value. More specifically, let us plan to use 1N5235 zeners for diodes D_5 and D_6.

Compute the value for R_6. R_6 can be found with our basic zener equation, Eq. (4.24). In our present case, we will be finding the value of R_6 with $+V_{SAT}$, V_{D5}, and I_{D5} known. Since the output of the op amp must supply currents to two zener circuits and the C_1 timing circuit, let's plan to limit the zener currents to no more than 20 percent of the short-circuit output current. That is

$$\boxed{I_Z = 0.2 \, I_{OS}} \qquad (4.32)$$

In our present case

$$I_Z = 0.2 \times 20 \times 10^{-3} = 4 \text{ mA}$$

If we use the typical $+V_{SAT}$ value of $+13$ volts, we can calculate a value for R_6 by applying Eq. (4.24).

$$R_6 = \frac{+V_{SAT} - V_{D5} - 0.6}{I_Z}$$
$$= \frac{13 - 6.4 - 0.6}{4 \times 10^{-3}}$$
$$= 1.5 \text{ k}\Omega$$

Select the reference zener.   Although the selection of diodes D_3 and D_4 is not critical, the following equation provides a good rule of thumb

$$\boxed{V_{ZREF} = \frac{V_{ZOUT}}{2}} \qquad (4.33)$$

Let us utilize this practice and determine the required voltage for the reference diodes.

$$V_{ZREF} = \frac{6.4}{2} = 3.2 \text{ V}$$

Let us choose to use 1N5226 zeners for our application. As App. 5 shows, these diodes have a rated voltage of 3.3 volts, but with less than 20 milliamperes of zener current, the actual voltage will be somewhat lower.

Compute the value for R_5. Resistor R_5 is computed in the same way as resistor R_6. We use the same guideline Eq. (4.32), that sets the zener current to 20 percent of I_{OS}. This has been previously computed to be 4 milliamperes. The value for R_5 can be found by applying the principle represented in Eq. (4.24).

$$R_5 = \frac{+13 - 3.3 - 0.6}{4 \times 10^{-3}} = 2.28 \text{ k}\Omega$$

We shall choose the standard value of 2.2 kilohms.

Compute C_1 and the timing resistors. The first step in determining values for C_1 and $R_1 - R_4$, is to determine the required RC time constant for the **shorter** period in the design requirement. For this, we utilize a transposed version of Eq. (4.27) or (4.28). If the positive alternation is the shorter, then use a transposed version of Eq. (4.27). Equation (4.28) should be utilized if the negative alternation is shorter. For our particular case, the 1 millisecond positive output time is clearly the shorter. Therefore, we shall apply Eq. (4.27) to determine the required RC product.

$$R_4 C_1 = \frac{t^+}{\ln\left(\dfrac{V_{D_6} - V_{D_3}}{V_{D_6} - V_{D_4}}\right)}$$

$$= \frac{1 \times 10^{-3}}{\ln\left(\dfrac{6.4 - (-3.3)}{6.4 - 3.3}\right)}$$

$$= 877 \times 10^{-6}$$

To insure that we come up with practical values, it is generally best to select R_4 at this point and compute C_1. Additionally, since we are working with the shortest time period, we should select a fairly small value for R_4 as long as we don't go below 1000 ohms. For this example, let us use a 1.8 kilohm resistor for R_4. We can utilize the results of our previous calculation and determine the value of C_1.

$$R_4 C_1 = 877 \times 10^{-6},$$

or

$$C_1 = \frac{877 \times 10^{-6}}{R_4}$$

Sec. 4.4 Variable-Duty Cycle

$$= \frac{877 \times 10^{-6}}{1.8 \times 10^3}$$

$$= 0.487 \ \mu F$$

We shall select a standard value of 0.47 microfarad for C_1.

We can utilize Eq. (4.27) to compute the required value for R_3. Its value will establish the maximum time for the positive alternation.

$$t^+ = (R_3 + R_4)C_1 ln\frac{V_{D_6} - V_{D_3}}{V_{D_6} - V_{D_4}},$$

or

$$R_3 = \frac{t^+}{C_1 ln\left(\frac{V_{D_6} - V_{D_3}}{V_{D_6} - V_{D_4}}\right)} - R_4$$

$$= \frac{10 \times 10^{-3}}{0.47 \times 10^{-6} ln\left(\frac{6.4 - (-3.3)}{6.4 - 3.3}\right)} - 1.8 \times 10^3$$

$$= 16.85 \ k\Omega$$

The nearest standard potentiometer value is 20 kilohms. We shall use a 20 kilohm variable resistor for R_3.

Resistors R_1 and R_2 are computed in the same way as R_3 and R_4 except that Eq. (4.28) is utilized along with the times associated with the negative alternation. These calculations are

$$t^- = (R_1 + R_2)C_1 ln\frac{V_{D_5} - V_{D_4}}{V_{D_5} - V_{D_3}},$$

or

$$R_2 = \frac{t^-(min)}{C_1 ln\left(\frac{V_{D_5} - V_{D_4}}{V_{D_5} - V_{D_3}}\right)}$$

$$= \frac{2 \times 10^{-3}}{0.47 \times 10^{-6} ln\left(\frac{6.4 - (-3.3)}{6.4 - 3.3}\right)}$$

$$= 3.73 \ k\Omega$$

We shall select a standard value of 3.6 kilohms for R_2. Finally, we compute the value of R_1 by applying Eq. (4.28).

$$t^- = (R_1 + R_2)C_1 \ln \frac{V_{D5} - V_{D4}}{V_{D5} - V_{D3}},$$

or

$$R_1 = \frac{t^-(\max)}{C_1 \ln\left(\dfrac{V_{D5} - V_{D4}}{V_{D5} - V_{D3}}\right)} - R_2$$

$$= \frac{20 \times 10^{-3}}{0.47 \times 10^{-6} \ln\left(\dfrac{6.4 - (-3.3)}{6.4 - 3.3}\right)} - 3.6 \times 10^3$$

$$= 33.7 \text{ k}\Omega$$

The nearest standard value for R_1 is 25 kilohms. It might be a better choice, however, to go to the next higher value so we can be sure that the maximum pulse width in our original design goal can be achieved. With this in mind, let us select a 50 kilohm variable resistor for R_1.

Select D_1 and D_2. Diodes D_1 and D_2 are simple isolation diodes and have no critical characteristics as long as the V_{PIV} rating of the diode exceeds about 30 volts and the I_F rating is greater than the I_{OS} rating of the op amp. Let's use 1N914A diodes for our example design.

Select the op amp. The primary op amp parameter that must be considered in this application is the slew rate. If the slew rate of the op amp causes the rise and fall times of the output waveform to be a significant part of either alternation, then the alternation will be longer than originally predicted.

For purposes of our present example, let us accept a rise and fall time of 10 percent of the shortest alternation period. In our case, this means that the rise and fall times can be no longer than 10 percent of 1 millisecond or 100 microseconds. Having established the longest acceptable switching time, we can apply Eq. (4.34) to determine the required slew rate.

$$\boxed{\text{slew rate(min)} = 10 \frac{+V_{SAT} - (-V_{SAT})}{t_{MIN}}} \qquad (4.34)$$

where t_{MIN} is the shortest alternation for the circuit. In our particular design, the shortest alternation occurs on the positive half cycle and is 1.0 milliseconds. Let's use typical values for $\pm V_{SAT}$ and compute our minimum slew rate.

$$\text{slew rate(min)} = 10 \frac{13 - (-13)}{1 \times 10^{-3}} = 0.26 \text{ V}/\mu\text{s}$$

Sec. 4.4 Variable-Duty Cycle

This is below the 0.5 volts per microseconds slew rate of the 741 op amp so let us select this device for our design.

The completed design is shown in Fig. 4.9. The actual performance of the circuit is indicated by the oscilloscope waveforms presented in Fig. 4.10. Finally, Table 4.2 contrasts the original design goals with the measured performance of the circuit.

These first-try values will satisfy the requirements of many applications. If greater accuracy is needed in a particular parameter, then simple tweaking in the laboratory will bring the circuit into compliance.

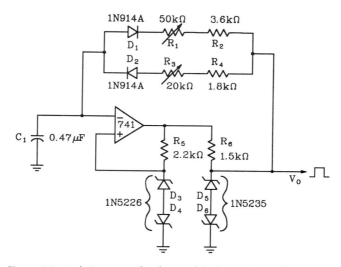

Figure 4.9 A design example of a variable duty-cycle oscillator circuit.

TABLE 4.2

		DESIGN GOAL	MEASURED VALUE
Positive output time	Minimum	1.0 ms	1.1 ms
	Maximum	10 ms	12.8 ms
Negative output time	Minimum	2.0 ms	2.2 ms
	Maximum	20 ms	26.5 ms
Output amplitude		±7 V (±6 V min)	−7.6, +7.2 V

208 Oscillators Chap. 4

Minimum Pulse Widths

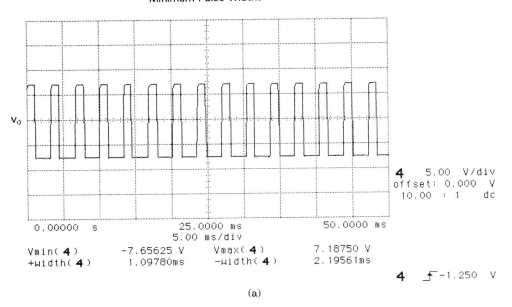

(a)

Maximum Pulse Widths

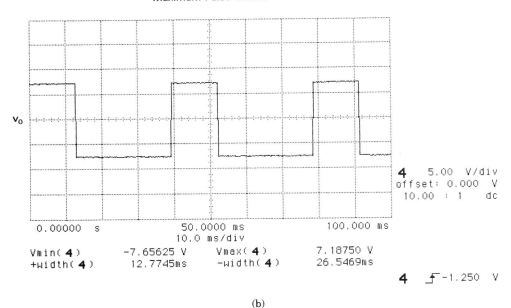

(b)

Figure 4.10 Oscilloscope displays of the output of the oscillator circuit shown in Fig. 4.9. (Test equipment courtesy of Hewlett-Packard Company)

Sec. 4.5 Triangle-Wave Oscillator

4.5 TRIANGLE-WAVE OSCILLATOR

4.5.1 Operation

Figure 4.11 shows the schematic of an oscillator circuit that generates a dual ramp (triangle) output. The heart of the circuit is amplifer A_1 which uses a capacitor as the feedback element. This operates as an integrator. It is similar in operation to amplifier A_1 in Fig. 4.5 discussed in an earlier section.

Let us assume that the output of A_2 is at its $-V_{SAT}$ level. Under these conditions, electrons will flow from the negative potential at the output of A_2, through R_1, and through C_1 as a charging current. The value of this current is determined by the voltage at the output of A_2 and the value of R_1. Since neither of these are changing at the moment, we assume the charging current is constant.

Whenever a capacitor is charged from a constant current source, the voltage across it accumulates linearly. Therefore, the voltage across C_1 will be increasing linearly with the right side becoming more positive. Since the left end of C_1 is connected to a virtual ground point, the right end of C_1 has a positive-going ramp with reference to ground. This is, of course, our output signal.

When the positive-going ramp exceeds the upper threshold voltage of A_2, which you should recognize as a noninverting voltage comparator, the output of A_2 will quickly switch to its $+V_{SAT}$ level.

The electron flow through C_1 now reverses and flows from C_1 through R_1 toward the positive potential at the A_2 output. Again, the value of current is constant and determined by R_1 and the voltage at the output of A_2. The voltage across C_1 will decay linearly until it passes through zero. It will then begin to charge at the same rate in the opposite polarity. This produces the negative slope on our output.

The output of A_1 continues to become more negative until it falls below the lower threshold voltage of A_2. At this time, the output of A_2 switches to its $-V_{SAT}$ level and the cycle repeats.

4.5.2 Numerical Analysis

Let us now numerically analyze the performance of the circuit shown in Fig. 4.11. First, we compute the upper ($+V_{TH}$) and lower ($-V_{TH}$) threshold voltages for the noninverting comparator A_2. The value of either threshold can be determined with Eq. 4.35.

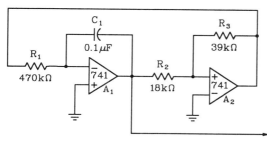

Figure 4.11 A triangle-wave oscillator circuit.

$$\boxed{\pm V_{TH} = \frac{\pm V_{SAT}R_2}{R_3}} \qquad (4.35)$$

In the case of the circuit in Fig. 4.11, the threshold voltages are computed as

$$\pm V_{TH} = \frac{\pm 13 \times 18 \times 10^3}{39 \times 10^3} = \pm 6 \text{ V}$$

We are now ready to compute the frequency of operation. Since both positive and negative saturation voltages, as well as both thresholds of A_2 are equal in this circuit, the time for either alternation can be computed with Eq. (4.36).

$$\boxed{t^{\pm} = R_1C_1 \ln \frac{V_{SAT} + V_{TH}}{V_{SAT} - V_{TH}}} \qquad (4.36)$$

In our present case, these times are computed as follows:

$$t^{\pm} = 470 \times 10^3 \times 0.1 \times 10^{-6} \ln \frac{13 + 6}{13 - 6} = 46.93 \text{ milliseconds}$$

The total period for one cycle is, of course, twice the time computed with Eq. (4.36). The frequency is simply the inverse of the total period. That is

$$\boxed{f = \frac{1}{t^+ + t^-}} \qquad (4.37)$$

In our particular circuit, the frequency of oscillation is found as follows:

$$f = \frac{1}{46.93 \times 10^{-3} + 46.93 \times 10^{-3}} = 10.65 \text{ Hz}$$

4.5.3 Practical Design Techniques

Now let's design a dual-ramp oscillator similar to the one shown in Fig. 4.11. For this design, we shall strive to achieve the following design goals:

1. frequency of oscillation   1.5 kilohertz
2. ramp amplitude ± 3 volts

Calculate the values for R_2 and R_3. Resistors R_2 and R_3 establish the threshold voltages for comparator A_2. These voltages in turn determine the output amplitude of the ramp voltage. The ratio of R_2 to R_3 can be found by applying Eq. (4.35).

Sec. 4.5 Triangle-Wave Oscillator 211

$$\pm V_{TH} = \frac{V_{SAT} R_2}{R_3},$$

or

$$\frac{R_2}{R_3} = \frac{V_{TH}}{V_{SAT}}$$

$$= \frac{3}{13}$$

$$= 0.231$$

We can now select R_2 and compute R_3. Both resistors should be in the range of 1 kilohm to 680 kilohms unless there is a compelling reason to exceed these suggested extremes. Let us select R_2 to be 56 kilohms. R_3 can now be computed by using the results of our previous calculation.

$$\frac{R_2}{R_3} = 0.231,$$

or

$$R_3 = \frac{R_2}{0.231}$$

$$= \frac{56 \times 10^3}{0.231}$$

$$= 242.4 \text{ k}\Omega$$

We shall use the nearest standard value of 240 kilohms.

Compute R_1 and C_1. Once the thresholds have been established on the comparator circuit, it is R_1 and C_1 that determine the frequency of oscillation. The required $R_1 C_1$ time constant can be found by applying Eqs. (4.36) and (4.37). First, let us use Eq. 4.37 to determine the total period for one cycle.

$$f = \frac{1}{t^+ + t^-},$$

or

$$t^+ + t^- = \frac{1}{f}$$

$$= \frac{1}{1.5 \times 10^3}$$

$$= 666.7 \times 10^{-6}$$

Since we know the two alternations are equal, we can determine the time for either alternation by dividing the total time by 2. That is,

$$t^+ = t^- = \frac{666.7 \times 10^{-6}}{2} = 333.35 \times 10^{-6}$$

We can now apply Eq. (4.36) to determine the R_1C_1 time constant.

$$t^\pm = R_1 C_1 \ln\left(\frac{V_{SAT} + V_{TH}}{V_{SAT} - V_{TH}}\right),$$

or

$$R_1 C_1 = \frac{t^\pm}{\ln\left(\dfrac{V_{SAT} + V_{TH}}{V_{SAT} - V_{TH}}\right)}$$

$$= \frac{333.35 \times 10^{-6}}{\ln\left(\dfrac{13 + 3}{13 - 3}\right)}$$

$$= 709.25 \times 10^{-6}$$

We now select either R_1 or C_1 and compute the other. In either case, we want R_1 to be in the range of 1.0 kilohms to 470 kilohms if practical. Similarly, C_1 should be greater than 470 picofarads and be nonpolarized. For our present example, let us select C_1 as 0.0047 microfarad. We now compute R_1 by dividing C_1 into the R_1C_1 time constant.

$$R_1 = \frac{R_1 C_1}{C_1}$$

$$= \frac{709.25 \times 10^{-6}}{0.0047 \times 10^{-6}}$$

$$= 150.9 \text{ k}\Omega$$

Let's use a standard value of 150 kilohms for R_1.

Select the op amps. Other than obvious things like supply voltage ratings, the most critical op amp parameter is slew rate. In order for our calculations regarding frequency of operation to be valid, the rise and fall time in the output of A_2 must be a small part of the time for either alternation. The greater the switching times, the greater the error in calculations. If we will accept rise and fall times of 10 percent of one alternation of the triangle wave, then we can apply Eq. (4.34).

Sec. 4.5 Triangle-Wave Oscillator

$$\text{slew rate(min)} = 10 \frac{+V_{SAT} - (-V_{SAT})}{t_\pm}$$

$$= 10 \frac{13 - (-13)}{333.35 \times 10^{-6}}$$

$$= 0.78 \text{ V}/\mu\text{s}$$

This exceeds the 0.5 volts per microseconds rating of the standard 741 but falls well within the capability of the MC1741SC. We shall select this device for our op amps.

Figure 4.12 shows the schematic diagram of the completed design. The oscilloscope plots in Fig. 4.13 indicate the actual performance of the circuit. Finally, Table 4.3 contrasts the original design goals with the measured performance of the circuit. An interesting modification to the circuit involves paralleling R_1 with a smaller resistor in series with a diode. This will cause one alternation to be significantly shorter and will generate a

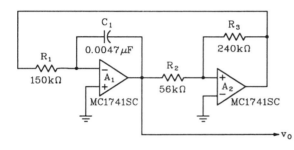

Figure 4.12 A triangle-wave oscillator designed for a 1.5 kilohertz operating frequency.

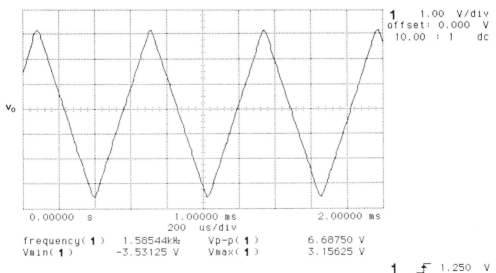

Figure 4.13 Oscilloscope displays showing the actual performance of the triangle-wave oscillator shown in Fig. 4.12. (Test equipment courtesy of Hewlett-Packard Company)

TABLE 4.3

	DESIGN GOAL	MEASURED VALUE
Frequency	1.5 kHz	1.59 kHz
Ramp amplitude	±3 V	+3.2 V, −3.3 V

forward or reverse sawtooth waveform (determined by the polarity of the added diode). The slew rate of the op amp ultimately limits the minimum ramp time.

4.6 TROUBLESHOOTING TIPS FOR OSCILLATOR CIRCUITS

The problems with op amp oscillator circuits generally fall into one of three categories

1. completely inoperative (i.e., no output signal),
2. distorted output waveform,
3. incorrect frequency of oscillation.

In many cases, items 2 and 3 occur simultaneously. As always, your best troubleshooting tool is your basic electronics theory and your complete understanding of correct circuit operation.

Oscillator completely inoperative. The first thing to check if there is no output signal is the power supply voltages. A quick check **directly** on the $\pm V_{CC}$ pins of the op amp will reveal or eliminate this potential problem. When checking for missing voltages, be certain to measure directly on the pin of the op amp. If you measure at some other point, you may fail to detect a poor solder joint, a broken printed circuit trace, and so on.

If the supplies are proper, but the oscillator has no output (and assuming it is correctly designed to oscillate), then measure the DC level of the output pin of the op amp. Some oscillators (e.g., Weinbridge) use AC coupling for some of the feedback. If the output is near zero volts DC (i.e., not driven to either saturation level) then suspect an open in the AC feedback path. If the output is driven to V_{SAT} then the circuit has a DC problem.

If the circuit is found to have a DC problem and the output is at one of the saturation levels, then note the polarity of the output voltage and then measure the two input pins. Ask yourself if the polarity of the input pins would *cause* the output to be at the present saturation limit. if the answer is no, then the op amp is probably defective.

If the input polarity would indeed cause the present output polarity, then mentally examine the circuit paths to determine what signal is supposed to cause the input polarity to change. That is, in order for the output to change (e.g., oscillate), the input pins must have a change. Further, since this is an oscillator, the changing input signal originates at

Sec. 4.7 Nonideal Considerations   215

the output. So, if we are missing the changing input signal, then trace the path between input and output and determine where the signal is lost.

A useful technique in some cases is to force the oscillator (or input) to a given state and monitor the effects elsewhere in the circuit. If the op amp you are using is short-circuit protected, then you can directly short the output to ground momentarily while observing the input pins. Ask yourself if the results agree with the behavior of a properly connected circuit.

Distorted output waveform. If the oscillator being analyzed is a new design, then one common cause for output distortion is an improperly selected op amp. More specifically, if the slew rate of the op amp is not sufficiently high relative to the demands of the oscillator, then the output will be distorted.

If the oscillator circuit uses active components as part of the basic oscillator loop (e.g., transistors), then a shift in the DC levels in the circuit can cause the active device to move out of its normal range of operation and introduce distortion.

If the waveform distortion is caused by clipping at one of the V_{SAT} levels (unless this is normal behavior for the circuit), then look for defects that would affect the DC operating point of the circuit. The first thing to do in this case is to verify proper power supply voltages. Some oscillator configurations can continue to oscillate with dramatic changes in power supply voltages. The symmetry and purity of the output signal, however, may suffer.

Incorrect frequency of operation. In certain oscillator designs, nearly every component in the circuit affects the frequency of operation. Troubleshooting a circuit of this type can be streamlined by noting, but not concentrating on, the frequency error. Rather, verify all other aspects of the oscillator's operation (e.g., DC level, waveshape, duty cycle). If one of these other characteristics are found to be abnormal, then focus your attention to this latter problem. The off-frequency problem is probably only a symptom and will be corrected when the other, more easily detected, problem is corrected. If all other characteristics appear to be normal, then suspect the components whose sole purpose is for frequency determination and that a change in value would not alter the DC levels in the circuit. There will be very few components that can qualify for this category.

4.7 NONIDEAL CONSIDERATIONS

We have already discussed one of the most significant nonideal op amp characteristics—the slew rate. If the op amp's slew rate is not high enough, then the output will be distorted (at best), and the frequency of operation will generally be lower than expected.

Another limitation that can cause problems is the limited current capability of the op amp output. For best performance, stability, and so forth, it is generally wise to avoid heavy loading of an oscillator output. This is especially true if the loads vary. This

REVIEW QUESTIONS

1. In order for a sinewave oscillator to maintain a constant amplitude (undistorted) output waveform, the overall closed-loop gain must be _____.
2. What is the general name for an oscillator circuit whose frequency is determined by the magnitude of input voltage?
3. Refer to Fig. 4.2. This circuit uses __(positive, negative)__ feedback that is __(AC, DC)__ coupled.
4. Refer to Fig. 4.2. Will the circuit oscillate if the 1N914A diode were to become open? Explain your answer.
5. Refer to Fig. 4.3. What is the primary purpose of resistor R_2?
6. Refer to Fig. 4.5. If capacitor C_1 is made larger, what happens to the frequency at the output of A_1? Does the amplitude of the signal at the output of A_2 change?
7. Refer to Fig. 4.5. Explain the effect on circuit operation if resistor R_4 was changed to 20 kilohms.
8. Refer to Fig. 4.8. If capacitor C_1 is changed to 0.33 microfarad, R_1 is set to 100 kilohms and R_3 is set to zero, compute the following.
   a. positive pulse width
   b. duty cycle
   c. frequency
9. Refer to Fig. 4.12. Amplifier A_2 operates as a _____ circuit.
10. Refer to Fig. 4.3. Compute the frequency of operation if R_1, R_2, C_1, and C_2 are changed to the following values.

    $R_1 = R_2 = 4.7 \text{ k}\Omega$

    $C_1 = C_2 = 1500 \text{ pF}$

CHAPTER 5

ACTIVE FILTERS

5.1 FILTER FUNDAMENTALS

A filter, be it an oil filter, a lint filter, a furnace filter, or an active filter, accepts a wide spectrum of inputs but only passes certain of these inputs through to the output. In some cases, the "good stuff" may be passed through the filter while the filter catches the "bad stuff." An oil filter in your car is an example of this type of filter. Other applications require a filter to catch the "good stuff" and let the "bad stuff" pass through. A gold prospector's sieve is an example of this type of filter action. In both of the preceding examples, the filter discriminated between "good" and "bad" on the basis of physical size (i.e., size of the dirt particle). In the filters that are discussed in this chapter, the "good" signals and "bad" signals will be classified on the basis of their frequency. The input will be a broad range of signal frequencies. The filter will allow a certain range of these frequencies to pass and reject other ranges.

Electronic filters designed to discriminate as a function of frequency can be broadly grouped into five classes.

1. Low pass.   Allows all frequencies below a specified frequency to pass through the filter circuit.
2. High pass.   Allows frequencies above a specified frequency to pass through the filter circuit.
3. Bandpass.   Allows a range or band of frequencies to pass through the filter circuit while rejecting frequencies higher or lower than the desired band.
4. Band reject. Rejects all frequencies within a certain band but passes frequencies higher or lower than the specified band. Also called a band-stop filter.

5. Notch. A notch filter is essentially a band-stop filter with a very narrow range of frequencies which are rejected.

Figure 5.1 shows the general frequency response curves for each of the basic filter types. The exact nature of a given curve will vary with the type of circuit implementation. Most notably, the slope of the curve between the "pass" and "reject" regions of the filter varies greatly with different types of filter designs.

There are seemingly endless ways to achieve the various filter functions listed. Each method of implementation has its individual advantages and disadvantages for a particular application. In this chapter, we will select a representative filter design for each basic filter type. In each case, we will discuss its operation, numerically evaluate its performance, and finally, design one to satisfy a given design goal. Band reject and notch filters will be treated as one general class.

An important consideration regarding active filters is how sharply the frequency response drops off for frequencies outside of the passband of the filter. In general, the steeper the slope of the curve, the more ideal the behavior of the filter. If the slope becomes too steep, however, the filter becomes unstable and is prone to oscillations. It is common to express the steepness of the slope in terms of dB per decade where a decade represents a factor of 10 increase or decrease in frequency. For example, suppose a low-pass filter had a 20 dB per decade slope beyond the cutoff frequency. This means that if the input frequency is increased by a factor of 10, then the output will decrease by 20

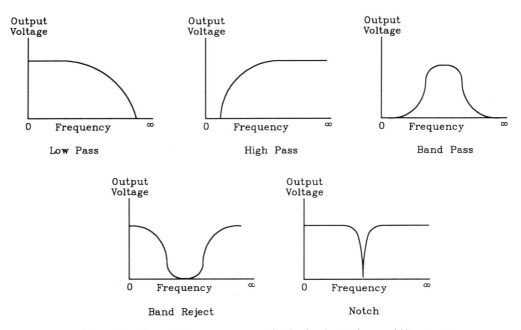

Figure 5.1 Theoretical response curves for the five basic classes of filter circuits.

Sec. 5.2 Low-Pass Filter

dB. If the input frequency is again increased by a factor of 10, then the output will decrease another 20 dB or 40 dB from the first measurement. Typical filter circuits have slopes ranging from 6 to 60 dB per decade or more.

In the case of bandpass, band stop, and notch filters, we often describe the steepness of the slopes in another way. The ratio of the center frequency (f_C) to the bandwidth (bw) gives us an indication of the sharpness of the cutoff region. The ratio f_C/bw is called the Q of the circuit. The higher the Q, the sharper the cutoff slopes of the filter.

The term Q is also used with reference to low-pass and high-pass filters, but it must be interpreted differently. The output of some filters peaks just before the edge of the passband. The Q of the filter indicates the degree of peaking. A Q of one has only a slight peaking effect. Q values of less than one reduce this peaking while values of Q greater than one cause a more pronounced peaking. There is usually a trade-off between peaking (generally undesired) and steepness (generally desired) of the slope. The high- and low-pass filter designs in this chapter use a Q of 0.707 which produces a very flat response.

5.2 LOW-PASS FILTER

Figure 5.2 shows one of the most common implementations of the low-pass filter circuit. This particular configuration is called a Butterworth filter and is characterized by having a very flat response in the passband portion of its response curve.

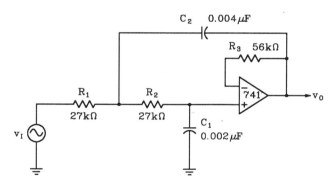

Figure 5.2 A low-pass Butterworth filter circuit.

Ideally, a low-pass filter will pass frequencies from DC up through a specified frequency, called the cutoff frequency, with no attenuation or loss. Beyond the cutoff frequency, the filter ideally offers infinite attenuation to the signal. In practice, however, the transition from passband to stopband is a gradual one. The cutoff frequency is defined as the frequency which passes with a 70.7 percent response. This, of course, is the familiar half-power point referenced in basic electronics theory.

220 Active Filters   Chap. 5

5.2.1 Operation

Let's try to understand the operation of the low-pass filter circuit shown in Fig. 5.2 from an intuitive or logical standpoint before evaluating it numerically. First, mentally open-circuit the capacitors. This modified circuit is shown in Fig. 5.3.

This is essentially how the circuit will look at low frequencies when the capacitive reactance of the capacitors is high. We can see that this amplifier is connected as a simple voltage follower circuit. Resistor R_3 is included in the feedback loop to compensate for the effects of bias currents flowing through R_1 and R_2. For low frequencies, then, we shall expect to have a voltage gain of about unity.

Now let's mentally short-circuit the capacitors in Fig. 5.2 to get an idea of how the circuit looks to high frequencies where the capacitive reactance is quite low. This equivalent circuit is shown in Fig. 5.4.

First, notice that the (+) input of the amplifier is essentially grounded. This should eliminate any chance of signals passing beyond this point. The junction of R_1 and R_2 is effectively connected to the output of the op amp. This, you will recall, is a very low impedance point. Thus, for high frequencies, the junction of R_1 and R_2 also has a low impedance to ground.

So, as our preliminary analysis indicates, the low frequencies should receive a voltage gain of about one, and the high frequencies should be severely attenuated. We are ready to confirm this numerically.

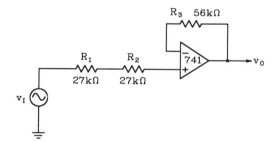

Figure 5.3 A low-frequency equivalent circuit for the low-pass filter shown in Fig. 5.2.

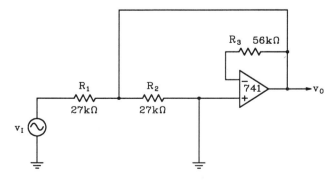

Figure 5.4 A high-frequency equivalent circuit for the low-pass filter shown in Fig. 5.2.

5.2.2 Numerical Analysis

The three primary considerations in active filters are

1. cutoff frequency
2. Q
3. input impedance

Cutoff frequency. In the case of the circuit in Fig. 5.2, the cutoff frequency is the frequency that causes the output amplitude to be 70.7 percent of the input. We can compute this frequency with Eq. (5.1).

$$f_C = \frac{1}{2\pi R\sqrt{C_1 C_2}} \quad (5.1)$$

where $R = R_1 = R_2$. For the circuit in Fig. 5.2, the cutoff frequency is computed as follows:

$$f_C = \frac{1}{6.28 \times 27 \times 10^3 \sqrt{0.002 \times 10^{-6} \times 0.004 \times 10^{-6}}} = 2.09 \text{ kHz}$$

Filter Q. The Q of the circuit in Fig. 5.2 is computed with Eq. (5.2).

$$Q = \frac{1}{2}\sqrt{\frac{C_2}{C_1}} \quad (5.2)$$

In our present case, we have

$$Q = \frac{1}{2}\sqrt{\frac{0.004 \times 10^{-6}}{0.002 \times 10^{-6}}} = 0.707$$

The value of 0.707 produces a maximally flat curve in the passband. That is, the response curve has minimal peaking at the edge of the passband. This is a common choice for Q.

Input impedance. The input impedance is an important consideration since it determines the amount of loading presented by the filter to the circuit driving the filter. The exact value of input impedance will vary dramatically with frequency. At very low frequencies, the input impedance approaches that of the standard voltage follower amplifier. As the input frequency increases, the input impedance decreases. The ultimate limit for the dropping input impedance is the value of R_1. Expressing this as an equation gives us

$$Z_{IN}(\min) = R_1 \quad (5.3)$$

In the case of the circuit in Fig. 5.2, we can be assured that the input impedance will never be lower than 27 kilohms.

5.2.3 Practical Design Techniques

Let's now design a low-pass filter similar to the circuit in Fig. 5.2. The design goal for our filter is

1. cutoff frequency   1.5 kilohertz
2. Q   0.707
3. input impedance   > 10 kilohms

Compute the ratio of C_2/C_1. From Eq. (5.2), we can see that the ratio of C_2/C_1 determines the Q of the circuit. Therefore, since we know Q (from the design criteria), we can compute the capacitor ratio by transposing Eq. (5.2).

$$Q = \frac{1}{2}\sqrt{\frac{C_2}{C_1}},$$

or

$$\frac{C_2}{C_1} = 4Q^2$$
$$= 4(0.707)^2$$
$$= 2$$

This tells us that C_2 will have to be twice as large as C_1. In general, the value of C_2 is determined with Eq. (5.4).

$$\boxed{C_2 = \left(\frac{C_2}{C_1}\right) \times C_1} \tag{5.4}$$

where the quantity (C_2/C_1) is determined from Eq. (5.2).

We can now select C_1 to be any convenient value and then double it to get C_2. For this design, let us choose C_1 as a 3300-picofarad capacitor. We can then make C_2 a 6600-picofarad ideally, or perhaps a 6800-picofarad since this is a standard size.

Compute R_1 and R_2. R_1 and R_2 should be within the general range of 1.0 kilohm to 220 kilohms. And, of course, R_1 must be larger than the minimum required input impedance (10 kilohms in this case). If the following calculation produces a value for R_1 and R_2 that does not comply with these restrictions, then a different value must be selected for C and then recalculate the resistor values. We compute the resistance value by applying Eq. (5.1).

Sec. 5.2   Low-Pass Filter

$$f_C = \frac{1}{2\pi R\sqrt{C_2 C_1}},$$

or

$$R = \frac{1}{2\pi f_C \sqrt{C_2 C_1}}$$

$$= \frac{1}{6.28 \times 1.5 \times 10^3 \sqrt{6800 \times 10^{-12} \times 3300 \times 10^{-12}}}$$

$$= 22.4 \text{ k}\Omega$$

We shall use a standard value of 22 kilohms for R_1 and R_2.

Determine the value of R_3. The value for resistor R_3 is calculated in the same way we did for a simple voltage follower. That is, we want equal DC resistances between each op amp input and ground. For the circuit in Fig. 5.2, we can compute R_3 with Eq. (5.5).

$$\boxed{R_3 = R_1 + R_2} \qquad (5.5)$$

Substituting values for our present circuit gives us

$$R_3 = 22000 + 22000 = 44 \text{ k}\Omega$$

We shall select the nearest standard value of 47 kilohms for R_3.

Select the op amp. There are three op amp parameters that we should evaluate before specifying a particular op amp for our low-pass filter

1. bandwidth
2. slew rate
3. op amp corner frequency

Since our op amp is operated as a voltage follower, the required bandwidth of the amplifier is essentially the same as the cutoff frequency. That is

$$\boxed{f_{UG}(\min) = f_C} \qquad (5.6)$$

In the case of our present circuit, our op amp must have a bandwidth of greater than 1.5 kilohertz. In many cases, including this one, the bandwidth will not be a limiting factor since the op amp is operated at unity gain.

The minimum required slew rate for the op amp can be estimated with Eq. (5.7).

$$\boxed{\text{slew rate}(\min) = \pi f_C v_o(\max)} \qquad (5.7)$$

where f_C is the filter cutoff frequency and v_O is the highest expected peak-to-peak output swing. If the application clearly has externally imposed limits on the maximum output amplitude, then use these limits. If the maximum output amplitude is not specifically known, as in the present case, then you should design for worst case and assume that the signal will swing between the $\pm V_{SAT}$ levels. In the present circuit, the required op amp slew rate (using ± 13 volts as the saturation limits) is computed as

$$\text{slew rate(min)} = 3.14 \times 1.5 \times 10^3 \times 26 = 0.122 \text{ V}/\mu\text{s}$$

Finally, the minimum op amp corner frequency may be estimated with Eq. (5.8).

$$\boxed{f_{BREAK}(\min) = \frac{40 f_C}{A_{OL}}} \tag{5.8}$$

where A_{OL} is the low frequency, open-loop gain of the op amp and f_C is the filter cutoff frequency. The corner frequency of the op amp is that frequency where the open-loop gain has dropped to 70.7 percent of its low frequency or DC value. If we choose a 741, it must have a corner frequency greater than

$$f_{BREAK}(\min) = \frac{40 \times 1.5 \times 10^3}{50 \times 10^3} = 1.2 \text{ Hz}$$

Let us consider a 741 op amp for this application. Appendix 1 lists the data for the 741. The minimum bandwidth and slew-rate requirements for our application are exceeded by the ratings of the 741. By referring to the graph of open-loop frequency response in App. 1, we can estimate the corner frequency of the 741 as about 5 hertz. Again, this exceeds our requirements so let us choose to use a 741 for our design.

The schematic diagram of our completed low-pass filter design is shown in Fig. 5.5. This circuit configuration provides a theoretical roll-off slope of 40 dB per decade. The oscilloscope plots in Fig. 5.6 indicate the actual behavior of the circuit. The filter shifts the phase of different frequencies by differing amounts as evidenced in Fig. 5.6. This is important with certain applications. Finally, Table 5.1 contrasts the original design goal with the measured performance of the circuit.

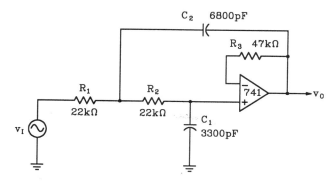

Figure 5.5 A low-pass filter designed for a cutoff frequency of 1.5 kilohertz.

Sec. 5.2 Low-Pass Filter

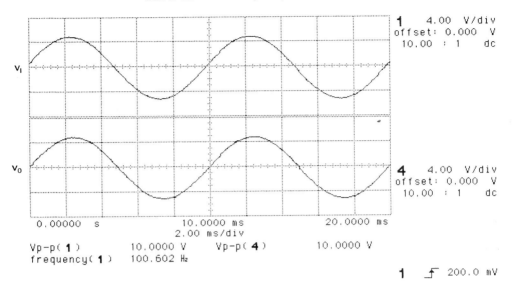

(a)

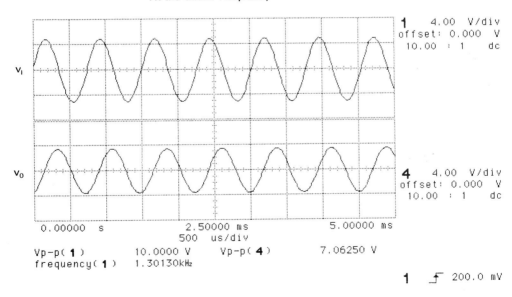

(b)

Figure 5.6 Actual circuit performance of low-pass filter shown in Fig. 5.5. (Test equipment courtesy of Hewlett-Packard Company) (continued)

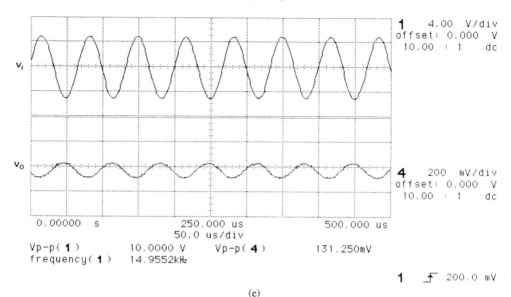

(c)

Figure 5.6c

TABLE 5.1

	DESIGN GOAL	MEASURED VALUE
Cutoff frequency	1.5 kHz	1.3 kHz
Input impedance (min)	> 10 kΩ	> 22 kΩ

5.3 HIGH-PASS FILTER

Figure 5.7 shows the schematic diagram of a high-pass filter circuit that provides a theoretical roll-off slope of 40 dB per decade. The circuit configuration is obtained by changing positions with all of the resistors and capacitors (except R_3) in the low-pass equivalent (Fig. 5.2). As a high-pass filter, we will expect it to severely attenuate signals below a certain frequency and pass the higher frequencies with minimal attenuation.

5.3.1 Operation

An intuitive feel for the operation of the circuit in Fig. 5.7 can be gained by picturing the equivalent circuit at very low and very high frequencies. At very low frequencies,

Sec. 5.3 High-Pass Filter

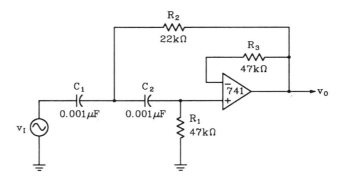

Figure 5.7 A 40 dB per decade high-pass filter circuit.

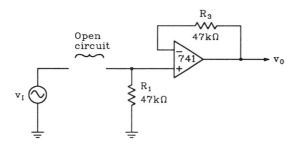

Figure 5.8 A low-frequency equivalent circuit for the high-pass filter shown in Fig. 5.7.

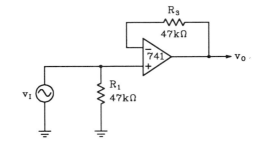

Figure 5.9 A high-frequency equivalent circuit for the high-pass filter shown in Fig. 5.7.

the capacitors will have a high reactance and will begin to appear as open circuits. Figure 5.8 shows the low-frequency equivalent circuit for the high-pass filter shown in Fig. 5.7. As you can readily see, the amplifier acts as a unity gain circuit, but **it has no input signal.** At low frequencies, we will expect little or no output signal.

At high frequencies, the capacitors will have a low reactance and will begin to appear as short circuits. The high-frequency equivalent circuit is shown in Fig. 5.9. Here we see the capacitors have been replaced with direct connections. Also, resistor R_2 has been removed. This is because it is connected between two points which have the same signal amplitude and phase (i.e., input and output of a voltage follower). Since it has the same potential on both ends, it will have no current flow and is essentially open. The resulting equivalent circuit indicates that for high frequencies, our high-pass filter will act as a simple voltage follower.

5.3.2 Numerical Analysis

Let us now extend our analysis of the high-pass filter shown in Fig. 5.7 to include a numerical analysis. There are three primary characteristics that we will want to determine

1. cutoff frequency
2. Q
3. input impedance

Cutoff frequency. The cutoff frequency for a high-pass filter is the frequency which causes the output voltage to be 70.7 percent of the amplitude of signals in the passband (i.e., the higher range of frequencies in this case). We can determine the cutoff frequency for the circuit in Fig. 5.7 by applying Eq. (5.9).

$$f_C = \frac{1}{2\pi C \sqrt{R_1 R_2}} \quad (5.9)$$

For the values in our present circuit, the cutoff frequency is computed as

$$f_C = \frac{1}{6.28 \times 0.001 \times 10^{-6} \sqrt{47 \times 10^3 \times 22 \times 10^3}} = 4.95 \text{ kHz}$$

Filter Q. The Q of the circuit shown in Fig. 5.7 is computed with Eq. (5.10).

$$Q = \frac{1}{2}\sqrt{\frac{R_1}{R_2}} \quad (5.10)$$

For the values given in Fig. 5.7, the Q is computed as

$$Q = \frac{1}{2}\sqrt{\frac{47 \times 10^3}{22 \times 10^3}} = 0.731$$

If the resistor values had an exact ratio of 2:1, then the Q would equal 0.707 and the passband response would be maximally flat.

Input impedance. The input impedance of the circuit shown in Fig. 5.7 varies inversely with the input frequency. The limit, however, is established by R_1 in parallel with the input impedance of the voltage follower. Therefore, for practical purposes, the limit is established by the value of R_1. In our present case, the minimum input impedance will be 47 kilohms.

Sec. 5.3   High-Pass Filter

5.3.3 Practical Design Techniques

Let us now design a high-pass filter similar to the circuit shown in Fig. 5.7. We shall use the following as our design goals

1. cutoff frequency   300 hertz
2. Q   0.707
3. input impedance   > 2000 ohms
4. highest input frequency   5000 hertz

Determine the R_1/R_2 ratio. As indicated by Eq. (5.10), the ratio of R_1 to R_2 determines the Q of the circuit. Let us apply Eq. (5.10) to determine the required ratio for our present design.

$$Q = \frac{1}{2}\sqrt{\frac{R_1}{R_2}},$$

or

$$\frac{R_1}{R_2} = 4Q^2$$

$$= 4(0.707)^2$$

$$= 2$$

Now we know that resistor R_1 will be twice as large as R_2. In general, R_1 is computed with Eq. (5.11).

$$\boxed{R_1 = \left(\frac{R_1}{R_2}\right) \times R_2} \qquad (5.11)$$

where the quantity (R_1/R_2) is computed with Eq. (5.10).

We can pick any convenient set of values for these resistors provided they fall within the suggested range of 1.0 to 470 kilohms, and as long as R_1 is larger than the minimum input impedance specified in the design criteria. For our present example, let us choose R_2 as 10 kilohms. R_1 is simply twice this value or 20 kilohms.

Compute the value of C_1 and C_2. Capacitors C_1 and C_2 are equal in value and can be computed by applying Eq. (5.9).

$$f_C = \frac{1}{2\pi C\sqrt{R_1 R_2}},$$

or

$$C = \frac{1}{2\pi f_C \sqrt{R_1 R_2}}$$

$$= \frac{1}{6.28 \times 300\sqrt{20 \times 10^3 \times 10 \times 10^3}}$$

$$= 0.0375 \; \mu F$$

We shall select a standard value of 0.033 microfarad for both capacitors.

Compute the value of R_3. Resistor R_3 is included to reduce the effects of the op amp bias current that flows through R_1. R_3 is set equal to R_1, or simply

$$\boxed{R_3 = R_1} \tag{5.12}$$

In our case, we shall use a 20 kilohm resistor for R_3.

Select the op amp. The following op amp parameters are the most essential when designing a high-pass filter circuit

1. bandwidth
2. slew rate

Additionally, if it is necessary to use high-resistance values, then every effort should be made to use an op amp with low-bias currents.

Bandwidth. When we construct a high-pass filter with an op amp, we inherently build a bandpass filter. That is, our filter circuit, by design, will attenuate all frequencies below the cutoff frequency. Ideally, all frequencies above the cutoff frequency should be passed with minimal attenuation. In practice, however, the gain of our op amp falls off at high frequencies. Thus, the very high frequencies are attenuated by the reduced gain of the op amp.

When selecting an op amp for a particular application, we must insure that the amplifier gain is still adequate at the highest expected frequency of operation. Since the op amp is configured for unity gain, we simply need to be sure that the op amp has a unity-gain bandwidth (f_{UG}) that is higher than the highest input frequency. In the present case, the highest input frequency is cited as 5000 hertz. Therefore, our choice of op amps must have a unity-gain frequency greater than 5000 hertz. This should be an easy task.

Slew rate. The slew-rate limitation of the op amp restricts the highest frequency that we can properly amplify at a given amplitude. Since the maximum input amplitude was not specified in the design goals, we shall assume that the output may be expected to produce a full swing between $\pm V_{SAT}$. Eq. (5.7) can be used to estimate the required op amp slew rate for our present application

$$\text{slew rate(min)} = \pi f_C v_o(\text{max})$$
$$= 3.14 \times 5000 \times 26$$
$$= 0.408 \; V/\mu s$$

Sec. 5.3 High-Pass Filter

Both bandwidth and slew-rate requirements are within the capabilities of a standard 741 op amp. Let's plan to use this device in our design.

This completes the design of our high-pass filter circuit. The final schematic is shown in Fig. 5.10. The oscilloscope plots shown in Fig. 5.11 indicate the performance of the circuit. Note the varying phase shifts for different frequencies. Additionally, Table 5.2 compares the original design goals with the actual measured circuit performance. The measured cutoff frequency is somewhat higher than the original design goal. The reason

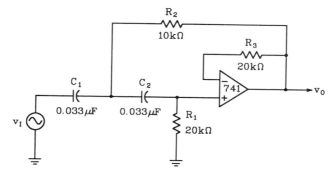

Figure 5.10 A high-pass filter designed for a flat response and a cutoff frequency of 300 hertz.

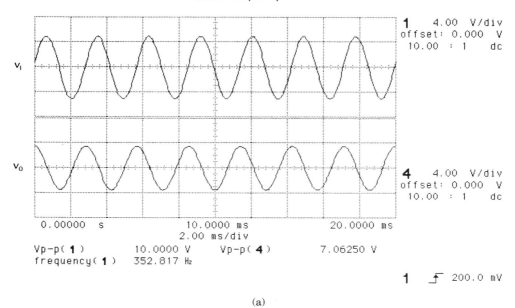

(a)

Figure 5.11 Actual circuit performance of the high-pass filter shown in Fig. 5.10. (Test equipment courtesy of Hewlett-Packard Company) (continued)

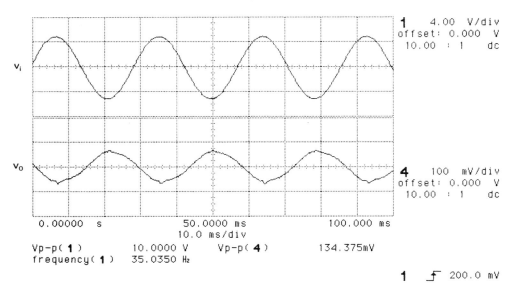

(b)

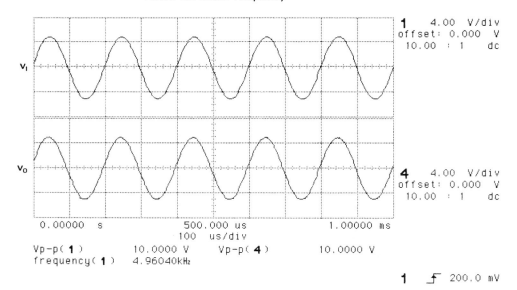

(c)

Figure 5.11b and c

Sec. 5.4   Bandpass Filter

TABLE 5.2

	DESIGN GOAL	MEASURED VALUE
Cutoff frequency	300 Hz	352.8 Hz
Input impedance	> 2000 Ω	> 10 kΩ

is twofold. First, we chose to use standard values of 0.033 microfarad for the capacitors when the correct value was 0.0375 microfarad. Second, the capacitors used to build the circuit were actually 0.032 microfarad (0.022 µF‖0.01 µF) due to availability. If the exact cutoff frequency is needed, then the resistors could be made variable.

5.4 BANDPASS FILTER

Figure 5.12 shows the schematic diagram of a bandpass filter. This circuit provides maximum gain (or minimum loss) to a specific frequency called the **resonant or center frequency** (even though it may not actually be in the center). Additionally, it allows a range of frequencies on either side of the resonant frequency to pass with little or no attenuation but severely reduces frequencies outside of this band. The edges of the passband are identified by the frequencies where the response is 70.7 percent of the response for the resonant frequency.

The range of frequencies which make up the passband is called the **bandwidth** of the filter. This can be stated as

$$\text{bw} = f_H - f_L \tag{5.13}$$

where f_H and f_L are the frequencies that mark the edges of the passband. The Q of the circuit is a way to describe the ratio of the resonant frequency (f_R) to the bandwidth (bw). That is

$$Q = \frac{f_R}{\text{bw}} \tag{5.14}$$

If the Q of the circuit is 10 or less, we call the filter a wide-band filter. Narrow-band filters have values of Q over 10. In general, higher Qs produce sharper, more well-defined responses. If the application requires a Q of 20 or less, then a single op amp filter circuit can be used. For higher Qs, a cascaded filter should be used to reduce potential oscillation problems.

5.4.1 Operation

To help us gain an intuitive understanding of the circuit's operation, let us draw equivalent circuits for the filter shown in Fig. 5.12 at very high and very low frequencies. To obtain

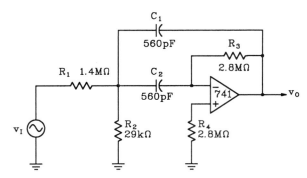

Figure 5.12 A bandpass filter used for a numerical analysis example.

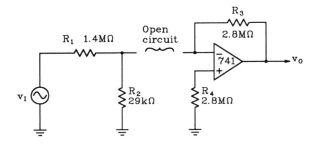

Figure 5.13 A low-frequency equivalent circuit for the bandpass filter shown in Fig. 5.12.

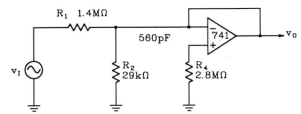

Figure 5.14 A high-frequency equivalent circuit for the bandpass filter shown in Fig. 5.12.

the low-frequency equivalent where the capacitors have a high reactance, we simply open-circuit the capacitors. Figure 5.13 shows the low-frequency equivalent. It is obvious from this equivalent circuit that the low frequencies will never reach the amplifier's input, and therefore, cannot pass through the filter.

At high frequencies, the reactance of the capacitors will be low, and the capacitors will begin to act like short circuits. Figure 5.14 shows the high-frequency equivalent circuit which was obtained by short-circuiting all of the capacitors. From this equivalent circuit, we can see that the high frequencies will be attenuated by the voltage divider action of R_1 and R_2. Additionally, the amplifier has zero resistance in the feedback loop which causes our voltage gain to be zero for the op amp (i.e., no output).

At some intermediate frequency (determined by the component values) the gain of the amplifier will offset the loss in the voltage divider (R_1 and R_2) and the signals will be allowed to pass through. The circuit is frequently designed to have unity gain at the resonant frequency, but may be set up to provide some amplification.

Sec. 5.4   Bandpass Filter

5.4.2 Numerical Analysis

The numerical analysis of the filter shown in Fig. 5.12 can range from very "messy" to straightforward, depending on the ratios of the components. That is, each component except R_4 affects both frequency and Q of the filter. The following analytical method assumes that the filter was designed according to standard practices. The following checks will provide you with a reasonable degree of assurance that the filter design is compatible with the analytical procedure to be described

1. Is R_3 approximately twice the size of R_1?
2. Are C_1 and C_2 equal in value?
3. Is R_1 at least 10 times the size of R_2?

If the answer is yes to *all* of these questions (which is the typical case), then the filter can be analyzed as described below. We shall compute the following characteristics:

1. resonant frequency
2. Q
3. bandwidth
4. voltage gain

Filter Q.   The Q of the filter shown in Fig. 5.12 can be computed with the following equation:

$$Q = \sqrt{\frac{R_1 R_2}{2R_2}} \qquad (5.15)$$

Substituting values gives us

$$Q = \sqrt{\frac{1.4 \times 10^6 + 29 \times 10^3}{2 \times 29 \times 10^3}} = 4.96$$

Since the Q is less than 10, we will classify this circuit as a wide-band filter.

Resonant frequency.   The resonant frequency of the filter shown in Fig. 5.12 can be computed with the following equation:

$$f_R = \frac{2Q}{2\pi R_3 C} \qquad (5.16)$$

where C is the value of either C_1 or C_2. Calculations for the present circuit are

$$f_R = \frac{2 \times 4.96}{6.28 \times 2.8 \times 10^6 \times 560 \times 10^{-12}} = 1.01 \text{ kHz}$$

It is this frequency which should receive the most amplification (or least attenuation) from the filter circuit.

Bandwidth. The bandwidth of the filter can be calculated by applying a transposed version of Eq. (5.14).

$$Q = \frac{f_R}{\text{bw}},$$

or

$$\text{bw} = \frac{f_R}{Q}$$

$$= \frac{1.01 \times 10^3}{4.96}$$

$$= 204 \text{ Hz}$$

Thus, the range of frequencies that is amplified at least 70.7 percent as much as the resonant frequency is 204 hertz wide.

Voltage gain at the resonant frequency. The voltage gain at the resonant frequency can be estimated with Eq. (5.17).

$$\boxed{A_V = \frac{Q}{2\pi R_1 F_R C}} \quad (5.17)$$

In the case of the circuit in Fig. 5.12, the voltage gain is computed as

$$A_V = \frac{4.96}{6.28 \times 1.4 \times 10^6 \times 1.01 \times 10^3 \times 560 \times 10^{-12}} = 0.997$$

5.4.3 Practical Design Techniques

We are now ready to design a bandpass filter to satisfy a given design requirement. Let's design a filter similar to the circuit in Fig. 5.12, that will perform according to the following design goals:

1. resonant frequency   8 kilohertz
2. Q   10

Sec. 5.4 Bandpass Filter

3. voltage gain at f_R unity
4. bandwidth   800 hertz

The following design is based on the assumption that the circuit provides unity gain at the resonant frequency. Although this is common in practice, it will also be shown how to design a filter to have a voltage gain of greater than unity at the resonant frequency.

Select C_1 and C_2. Although the selection of these capacitors is somewhat arbitrary, our choice of values for C_1 and C_2 will ultimately determine the values for the resistors. If our subsequent calculations result in an impractical resistance value, then we will have to select a different value for C_1 and C_2 and recompute the resistance values. Due to the wide range of resistance values which are typically required in a given design, it is not uncommon to have resistance values ranging from 1000 or less to well into the megohm ranges. Nevertheless, it should remain a goal to keep the resistance values above 1 kilohm and below 1 megohm if practical. The lower limit is established by the output drive of the op amp and the effects on input impedance. The upper limit is established by the op amp bias currents and circuit sensitivity. That is, if the resistance values are very large, then the voltage drops due to op amp bias currents become more significant. Additionally, if the resistances in the circuit are excessively high, then the circuit is far more prone to interference from outside noise, nearby circuit noise, or even unwanted coupling from one part of the filter to another. For our initial selection, let us choose to use 0.001 microfarad capacitors for C_1 and C_2.

Compute the value of R_1. The value of resistor R_1 is computed with Eq. (5.18).

$$R_1 = \frac{Q}{2\pi A_v f_R C} \tag{5.18}$$

In the case of the present design, we compute the value for R_1 as

$$R_1 = \frac{10}{6.28 \times 1 \times 8 \times 10^3 \times 0.001 \times 10^{-6}} = 199 \text{ k}\Omega$$

We shall plan to use a standard value of 200 kilohms. It should be noted, however, that the component values in an active filter are generally more critical than for many other types of circuits. Therefore, if close adherence to the original design goals is required, then either use variable resistors for trimming, fixed resistors in a series and/or parallel combination, or use precision resistors.

Compute the value for R_2. Resistor R_2 is calculated with Eq. (5.19).

$$R_2 = \frac{R_1 A_v}{2(2\pi f_R C R_1 A_v)^2 - A_v} \tag{5.19}$$

For purposes of our present design, we compute R_2 as follows:

$$R_2 = \frac{199 \times 10^3 \times 1}{2(6.28 \times 8 \times 10^3 \times 0.001 \times 10^{-6} \times 199 \times 10^3 \times 1)^2 - 1} = 1.0 \text{ k}\Omega$$

Compute the value for R_3. Resistor R_3 is computed by simply doubling the value of R_1. That is

$$\boxed{R_3 = 2R_1} \qquad (5.20)$$

For our design, we compute R_3 as

$$R_3 = 2 \times 200 \times 10^3 = 400 \text{ k}\Omega$$

The nearest standard value is 390 kilohms. As previously stated, if the application requires greater compliance with the original design goals, then either use a variable resistor or a combination of fixed resistors to achieve the exact value required. In our case, we shall use two 200 kilohm resistors in series for R_3.

Determine the value for R_4. Resistor R_4 has no direct effect on the frequency response of the filter circuit. Rather, it is included to help compensate for the effects of the op amp bias current which flows through R_3. You will recall that we try to keep the resistances between ground and the $(+)$ and $(-)$ input pins of the op amp equal. Therefore, we shall set R_4 equal to R_3. In equation form, we have

$$\boxed{R_4 = R_3} \qquad (5.21)$$

In this case, it is probably not necessary to use a variable resistor or fixed resistor combination to obtain an exact resistance. We simply use the nearest standard value of 390 kilohms for R_4.

Select the op amp. We shall pay particular attention to the following op amp parameters when selecting an op amp for our active filter

1. bandwidth
2. slew rate

If the resistance values turn out to be quite high, then an op amp with particularly low bias current would be important. If the capacitance values must be below, about 270 picofarads, then select an op amp with minimum internal capacitances.

Bandwidth. The required bandwidth of our op amp is determined by the highest frequency that must pass the circuit. This frequency is, of course, the upper cutoff frequency (f_H), and can be approximated with Eq. (5.22).

Sec. 5.4 Bandpass Filter

$$\boxed{f_H = f_R + \frac{bw}{2}} \qquad (5.22)$$

In the present case, f_H is estimated as

$$f_H = 8000 + \frac{800}{2} = 8400 \text{ Hz}$$

The required bandwidth for the op amp is computed in the same manner described in Chap. 2. The required bandwidth can be computed as

$$f_{UG} = A_V f_H$$
$$= 1 \times 8400$$
$$= 8.4 \text{ kHz}$$

This is well within the capabilities of the standard 741 op amp.

Slew rate. The minimum slew rate for the op amp can be computed with Eq. (5.7).

$$\text{slew rate(min)} = \pi f_{HI} v_O(\text{max})$$
$$= 3.14 \times 8400 \times 26$$
$$= 0.686 \text{ V}/\mu\text{s}$$

This exceeds the capability of the standard 741 which has a 0.5 volts per microsecond slew rate. We shall use an MC1741SC for our design. It satisfies both the bandwidth and slew rate requirements of our design.

The schematic of our final design is shown in Fig. 5.15. Its performance is indicated by the oscilloscope plots in Fig. 5.16. Be sure to note the varying phase shifts at different frequencies. Finally, the design goals are contrasted with the actual measured performance of the circuit in Table 5.3.

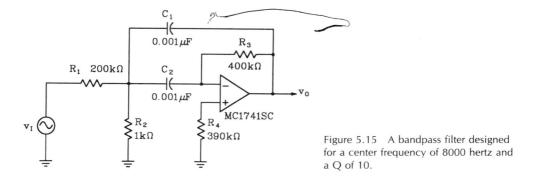

Figure 5.15 A bandpass filter designed for a center frequency of 8000 hertz and a Q of 10.

At Center Frequency

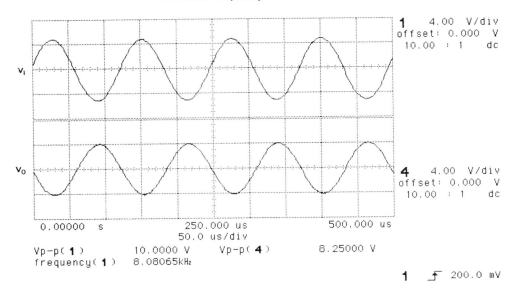

(a)

At the Lower Cutoff Frequency

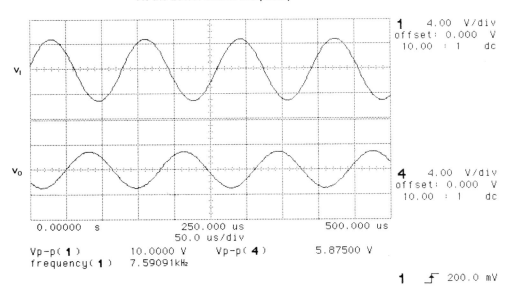

(b)

Figure 5.16 Oscilloscope displays showing the performance of the bandpass filter shown in Fig. 5.15. (Test equipment courtesy of Hewlett-Packard Company) (continued)

Sec. 5.5 Band-Reject Filter

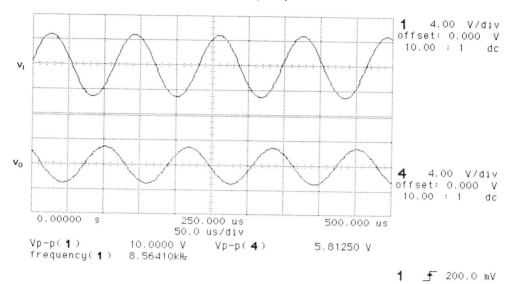

(c)

Figure 5.16c

TABLE 5.3

	DESIGN GOAL	MEASURED VALUE
Resonant frequency	8000 Hz	8080 Hz
Q	10	8.3
Bandwidth	800 Hz	973 Hz
Voltage gain at f_R	1.0	0.825

5.5 BAND-REJECT FILTER

A band reject filter is a circuit that allows frequencies to pass which are either lower than the lower cutoff frequency or higher than the upper cutoff frequency. That is, only those frequencies that fall between the two cutoff frequencies are rejected or at least severely attenuated.

5.5.1 Operation

Figure 5.17 shows an active filter that is based on the common twin "T" configuration. The twin T gets its name from the two RC T networks on the input. For purposes of

Active Filters Chap. 5

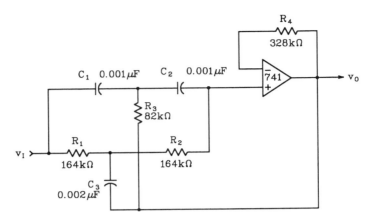

Figure 5.17 A band-reject filter circuit used for a numerical analysis example.

analysis, let us consider the lower ends of R_3 and C_3 to be grounded. This is a reasonable approximation since the output impedance of an op amp is generally quite low. The T circuit consisting of C_1, C_2, and R_3 is, by itself, a high-pass filter. That is, the low frequencies are prevented from reaching the input of the op amp because of the high reactance of C_1 and C_2. The high frequencies, on the other hand, find an easy path to the op amp since the reactance of C_1 and C_2 is low at higher frequencies.

The second T network is made up of R_1, R_2, and C_3 and forms a low-pass filter. Here the low frequencies find C_3's high reactance to be essentially an open so they pass on to the op amp input. High frequencies, on the other hand, are essentially shorted to ground by the low reactance of C_3. It would seem, both low and high frequencies have a way to get to the (+) input of the op amp, and therefore, to be passed through to the output. If, however, the cutoff frequencies of the two T networks do not overlap, then there is a frequency (f_R) that results in a net voltage of zero at the (+) terminal of the op amp. To understand this effect, we must also consider the phase shifts given to a signal as it passes through the two networks. At the center, or resonant, frequency (f_R), the signal is shifted in the negative direction while passing through one T network. It receives the same amount of positive phase shift while passing through the other T network. These two shifted signals pass through equal impedances (R_2 and X_{C2}) to the (+) input. Thus, at any instant in time (at the center frequency), the effective voltage on the (+) input is zero. The more the input frequency deviates from the center frequency, the less the cancellation effect. Thus, as we initially expected, this circuit rejects a band of frequencies and passes those frequencies which are higher or lower than the cutoff frequencies of the filter.

The op amp offers a high impedance to the T networks thus reducing the loading effects, and therefore, increasing the Q of the circuit. Additionally, by connecting the "ground" point of C_3 and R_3 to the output of the op amp, we have another increase in Q as a result of the feedback signal. At or very near the center frequency, very little signal makes it to the (+) input of the op amp. Therefore, very little signal appears at the output of the op amp. Under these conditions the output of the op amp merely provides

Sec. 5.5 Band-Reject Filter

a ground (i.e., low impedance return to ground) for the T networks. For the other frequencies though, the feedback essentially raises the impedance offered by C_3 and R_3 at a particular frequency. Therefore, they don't attenuate the off-resonance signals as much. This has the effect of narrowing the bandwidth or, we could say, increasing the Q.

Resistor R_4 is to compensate for the voltage drops caused by the op amp bias current flowing through R_1 and R_2. It is generally equal in value to the sum of R_1 and R_2.

5.5.2 Numerical Analysis

The component values for the twin-T circuit normally have the following ratios:

1. $R_1 = R_2$
2. $R_1 = 2R_3$
3. $C_1 = C_2$
4. $C_3 = 2C_1$
5. $0 \leq R_4 \leq (R_1 + R_2)$

Under these conditions, let us compute the following circuit characteristics:

1. center frequency
2. input impedance

Center frequency. The center frequency for the twin-T filter is that frequency which causes the reactance of C_3 to equal the resistance of R_3. At this same frequency, $X_{C1} = X_{C2} = R_1 = R_2$. The equation for the center frequency then is simply a transposed version of the basic capacitive reactance equation:

$$X_C = \frac{1}{2\pi f C},$$

or

$$f = \frac{1}{2\pi X_C C}$$

Since, at the center frequency, $X_{C1} = R_1$, we can substitute R_1 for X_C in the preceding equation to yield our equation for the center frequency of the twin-T filter.

$$\boxed{f_o = \frac{1}{2\pi R_1 C_1}} \qquad (5.23)$$

In the case of the circuit shown in Fig. 5.17, we can compute the center frequency as follows:

$$f_R = \frac{1}{6.28 \times 164 \times 10^3 \times 0.001 \times 10^{-6}} = 971 \text{ Hz}$$

On either side of the center frequency, we can expect the signals to pass with a voltage gain of nearly unity.

Input impedance. As with many filter circuits, the input impedance of the circuit shown in Fig. 5.17 varies with frequency. The lowest impedance occurs at the higher frequencies and is approximately equal to R_1, R_2, and R_3 in parallel. That is

$$\boxed{Z_{IN}(min) \approx R_1 \| R_2 \| R_3 = \frac{1}{\frac{1}{R_1} + \frac{1}{R_2} + \frac{1}{R_3}}} \qquad (5.24)$$

In the case of Fig. 5.17, we can estimate the minimum input impedance as

$$Z_{IN}(min) = \frac{1}{\frac{1}{164 \times 10^3} + \frac{1}{164 \times 10^3} + \frac{1}{82 \times 10^3}} = 41 \text{ k}\Omega$$

5.5.3 Practical Design Techniques

Now let us design a twin T, band-reject filter to satisfy a specific design requirement. We shall design a filter that will deliver the following performance:

1. center frequency 5500 hertz
2. minimum input impedance 10 kilohms
3. highest input frequency 18 kilohertz

Select a preliminary value for R_3. The minimum value for R_3 is determined by the specification for the minimum input impedance. More specifically, the minimum value for R_3 is determined according to Eq. (5.25).

$$\boxed{R_3(min) = 2Z_{IN}(min)} \qquad (5.25)$$

In our present case, the minimum value for R_3 is determined as follows:

$$R_3(min) = 2 \times 10 \times 10^3 = 20 \text{ k}\Omega$$

The upper limit for the value of R_3 is established by two factors:

1. effects of op amp bias currents
2. minimum practical values for C_1 to C_3

Sec. 5.5 Band-Reject Filter

The effects of op amp bias currents can be made tolerable if we keep the value of R_3 below about 270 kilohms. Although the bias currents do not actually flow through R_3, it is the value of R_3 that will determine the values for R_1 and R_2. The minimum practical value for the capacitors is affected by several things including the op amp used and the required degree of stability. A workable goal, however, is to design the circuit such that all capacitor values are greater than 100 picofarads.

Let us initially choose to use a value of 22 kilohms for R_3. If the computed values for the capacitors are found to be too small, or there are no reasonably close standard values, then we will have to select a different value for R_3 and recompute.

Determine the value for C_1. First we compute the ideal value for C_1, then we select the nearest standard value. We can compute the required value of capacitance by applying Eq. (5.26).

$$\boxed{C_1 = \frac{1}{4\pi f_R R_3}}$$

For the present design, we compute the value for C_1 as follows:

$$C_1 = \frac{1}{4 \times 3.14 \times 5500 \times 22 \times 10^3} = 658 \text{ pF}$$

Let us choose the nearest standard value of 680 picofarads for C_1.

Compute the exact value for R_3. Now that a standard value for C_1 is selected, we can determine the exact value required for R_3. It should be noted that the performance of the twin T filter design relies heavily on accurate selection and matching of component values. Therefore, "ballpark" values are usually inappropriate. The exact value needed for R_3 can now be computed by applying a transposed version of Eq. (5.26).

$$C_1 = \frac{1}{4\pi f_R R_3},$$

or

$$R_3 = \frac{1}{4\pi f_R C_1}$$

$$= \frac{1}{4 \times 3.14 \times 5500 \times 680 \times 10^{-12}}$$

$$= 21.3 \text{ k}\Omega$$

We can obtain this value by combining fixed resistances (e.g., 27 kΩ in parallel with 100 kΩ), or by using a fixed resistance and a variable resistor in series (e.g., 18 kilohm fixed resistor in series with a 5 kilohm variable resistor). In either case, however, every effort should be made to obtain the correct value.

Compute C_2 and C_3. Capacitor C_2 is always the same value as C_1. That is

$$\boxed{C_2 = C_1} \tag{5.27}$$

For our purposes, we compute C_2 as

$$C_2 = C_1 = 680 \text{ pF}$$

Capacitor C_3 is twice the size of C_1 or simply

$$\boxed{C_3 = 2C_1} \tag{5.28}$$

In our particular case

$$C_3 = 2C_1 = 2 \times 680 \text{ pF} = 1360 \text{ pF}$$

The nearest standard value is 1500 picofarads. While this is close enough for many applications, it would undermine the performance of our filter. The simplest way to obtain more precise values in this case is to simply parallel two 680-picofarad capacitors. This will give us the value we need.

Compute R_1 and R_2. Resistors R_1 and R_2 are equal in size and are twice the value of R_3. We can express this as an equation.

$$\boxed{R_1 = R_2 = 2R_3} \tag{5.29}$$

In the present design case, we compute these resistors as

$$R_1 = R_2 = 2 \times 21.3 \times 10^3 = 42.6 \text{ k}\Omega$$

The nearest standard value would be 43 kilohms. We might be able to get by with this, but in general, we must try to obtain the exact values required. To this end, let us choose to either use a combination of fixed resistors (e.g., 39 kΩ in series with a 3.6 kΩ) or a fixed resistor and a variable resistor in combination (e.g., a 39 k fixed resistor in series with a 5 k variable resistor).

Compute R_4. Resistor R_4 helps to compensate for the voltage drop across R_1 and R_2 that is produced by the op amp bias currents. To minimize the effects of the bias currents, we set R_4 equal to the series combination of R_1 and R_2. That is

$$\boxed{R_4 = R_1 + R_2} \tag{5.30}$$

For purposes of our present design, R_4 is computed as

$$R_4 = (42.6 \times 10^3) + (42.6 \times 10^3) = 85.2 \text{ k}\Omega$$

Sec. 5.5 Band-Reject Filter

For many applications, this is not a critical value. In our case, let us use the nearest standard value of 82 kilohms.

Select the op amp. As with previous filter designs, there are two op amp parameters that we want to focus on in order to select an appropriate op amp.

1. bandwidth
2. slew rate

As mentioned previously, if the resistance values turn out to be quite high, then an op amp with particularly low-bias current would be important. And, if the capacitance values must be below about 270 picofarads, then select an op amp with low-internal capacitances.

Bandwidth. The required bandwidth of our op amp is determined by the highest frequency that must pass the circuit. The design specifications for our present case specify the highest input frequency as 18 kilohertz. The required bandwidth for the op amp is computed in the same manner described in Chap. 2. The required bandwidth can be computed as

$$f_{UG} = A_v f$$
$$= 1 \times 18000$$
$$= 18 \text{ kHz}$$

This is well within the capabilities of the standard 741 op amp.

Slew rate. The minimum slew rate for the op amp can be computed with Eq. (5.7).

$$\text{slew rate(min)} = \pi f_{HI} v_O(\text{max})$$
$$= 3.14 \times 18000 \times 26$$
$$= 1.469 \text{ V}/\mu s$$

This exceeds the capability of the standard 741 which has a 0.5 volts per microsecond slew rate. We shall use an MC1741SC for our design. It satisfies both the bandwidth and slew-rate requirements of our design.

This completes the design of our 5500 hertz band-reject filter. The final schematic is shown in Fig. 5.18. The oscilloscope displays in Fig. 5.19 indicate the performance of the circuit at resonance, at the two cutoff frequencies, and at two far removed frequencies in the passband of the filter. Table 5.4 contrasts the design goals for the filter with the actual measured performance of the final design. It should be noted that 5 percent tolerance components were used to construct the circuit. More precise values would yield performance correspondingly closer to the design goals. The Q of the circuit can go as high as 40 or 50 with careful selection of components.

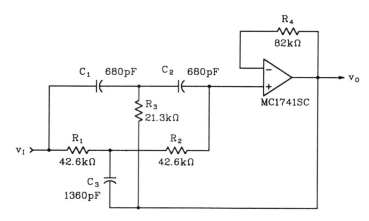

Figure 5.18 A twin-T band-reject filter designed for a center frequency of 5500 hertz.

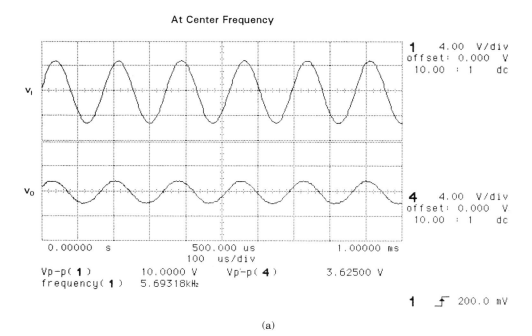

(a)

Figure 5.19 Oscilloscope displays showing the performance of the twin-T filter shown in Fig. 5.18. (Test equipment courtesy of Hewlett-Packard Company) (continued)

Sec. 5.5 Band-Reject Filter

At the Upper Cutoff Frequency

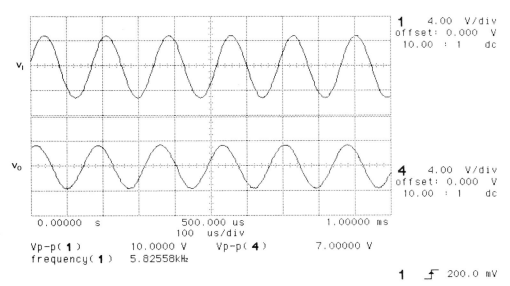

(b)

At the Lower Cutoff Frequency

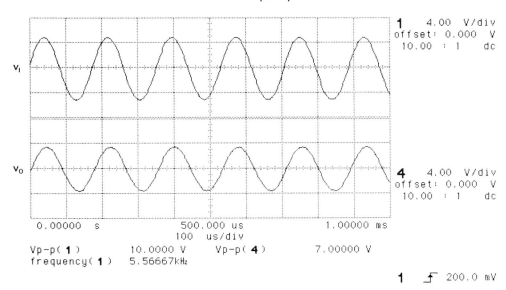

(c)

Figure 5.19b and c

250   Active Filters   Chap. 5

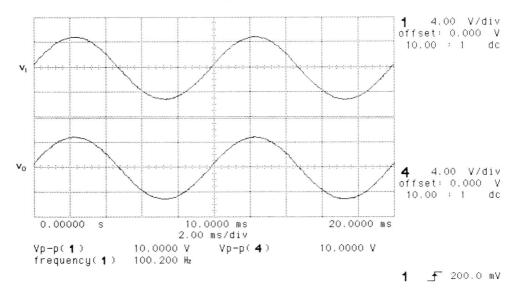

(d)

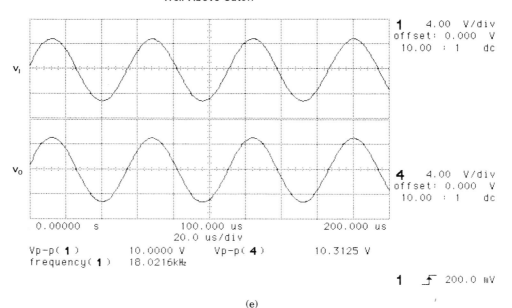

(e)

Figure 5.19d and e

Sec. 5.6 Troubleshooting Tips for Active Filters

TABLE 5.4

	DESIGN GOAL	MEASURED PERFORMANCE
Resonant frequency	5500 Hz	5693 Hz
Minimum Z_{IN}	10 kΩ	10.65 kΩ
Q	Not specified	22

5.6 TROUBLESHOOTING TIPS FOR ACTIVE FILTERS

We can generally classify potential troubles in active filters into two broad classes

1. DC problems
2. AC problems

Our first measurements will quickly tell which type of problem exists. We can then focus our efforts on areas that could cause this type of problem.

DC problems. Problems that cause the output of the op amp to be at some abnormal DC level are generally located in the same manner as described in previous chapters. Basically, you need to insure that the proper V^+ and V^- are present **directly** on the appropriate pins of the op amp. If these voltages are correct, compare the polarity of the differential input voltage (v_d) with the polarity on the output pin. If the polarities contradict normal op amp behavior, then suspect the op amp. If the polarities are correct for a normal op amp, then measure the DC level on the input to the op amp circuit. A prior stage may be sending an abnormal DC level into this stage which makes it appear to be defective.

If all of these appear normal, then verify the integrity of the feedback circuit. If it is open, the output of the op amp will be at one of the two extremes.

AC problems. If the DC voltages are correct in the filter circuit, but the filter does not correctly discriminate against certain frequencies, then suspect the frequency determining components. If the resonant, or cutoff frequencies, have simply shifted slightly, you could suspect a change in component values. On the other hand, if the AC operation of the circuit has been altered dramatically, then suspect an open component. Normal DC values with abnormal AC values often point to an open capacitor or to a resistor that is isolated from DC (e.g., R_3 in Fig. 5.18).

As with all op amp troubleshooting tasks, it is essential that you understand the proper operation of the circuit and continuously contrast the actual performance with the known expected performance.

REVIEW QUESTIONS

1. Refer to Fig. 5.2. What is the effect on circuit operation if capacitor C_2 was to become open?
2. If the DC level on the output of the op amp in Fig. 5.2 was normal but there were no AC signals present, would capacitor C_1 open be a possible cause? Explain your answer.
3. Refer to Fig. 5.7. If capacitor C_2 was to open, would the DC level on the output of the op amp be affected? Explain.
4. If resistor R_3 was to short in Fig. 5.7, what would be the effect on circuit operation?
5. A circuit which passes all frequencies below the cutoff frequency is called a _____ filter.
6. A circuit which rejects a very narrow band of frequencies is called a _____ filter.
7. Refer to Fig. 5.12. The ratio of resistor R_4 and resistor R_3 establishes the gain of the amplifier. (True or False)
8. Refer to Fig. 5.15. What is the effect on circuit operation if capacitor C_2 becomes open?
9. Refer to Fig. 5.15. What is the effect on the DC level on the output of the op amp if resistor R_2 becomes shorted?
10. Refer to Fig. 5.18. Which of the following defects could cause the circuit to respond like a high-pass filter?
    a. C_1 open
    b. R_1 open
    c. R_4 shorted
    d. C_2 open
    e. R_3 shorted

CHAPTER 6

POWER SUPPLY CIRCUITS

6.1 VOLTAGE REGULATION FUNDAMENTALS

Nearly all electronic systems require one or more sources of stable DC voltage. Yet, many systems get their input power from the standard 120 VAC power line. Even battery powered units may require stable DC voltages at levels other than those provided directly by the battery. Figure 6.1 shows the basic role played by a voltage regulator circuit and where it fits in the power distribution scheme.

If the system receives its power from the 120 VAC power line, then the first step is usually voltage reduction via a step-down transformer. The output(s) of the transformer is then rectified to produce pulsating DC. The rectified waveform is then filtered with large capacitors to produce a relatively smooth, but *unregulated* source of DC voltage. An unregulated voltage source is one which varies with changes in load current or changes in applied voltage. All of the functions just described are represented by the first block in Fig. 6.1.

The voltage at the output of the rectifier/filter is smooth DC but it is not regulated. Thus, the value of voltage will change with changes in input voltage or with changes in load current. To eliminate these changes and produce a solid source of DC voltage, we route the filtered DC to a voltage regulator circuit (the second block in Fig. 6.1).

There are many types of voltage regulator circuits, but the purpose remains the same. That is, to maintain a constant output voltage even though the input voltage and

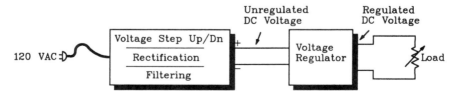

Figure 6.1　A voltage regulator circuit provides constant voltage to a load.

253

the load current may both be changing. The regulated output voltage is always less than the unregulated input voltage. We will examine three basic classes of voltage regulator circuits

1. series
2. shunt
3. switching

6.1.1 Series Regulation

Figure 6.2 illustrates the basic concept of series voltage regulation. The voltage regulator circuit is designed to act as a variable resistance in series with the load. The regulator senses changes in load voltage (whether caused by changes in input voltage or by changes in load current) and adjusts its resistance such that the voltage across the load remains constant. This is one of the most common voltage regulation techniques. The regulator can also be designed to protect against short circuits on the regulated output. In practice, the "variable resistor" shown as the regulating element in Fig. 6.2 is actually a transistor or an integrated voltage regulator circuit.

6.1.2 Shunt Regulation

The concept of a shunt-voltage regulator is illustrated in Fig. 6.3. Here the regulating element (shown as a variable resistor) is connected in parallel, or shunt, with the load. The regulator circuit senses changes in load voltage and adjusts the effective resistance of the regulating element to compensate. If, for example, the load current were to drop, then the output voltage would tend to rise (i.e., less drop across R_s). The regulator circuit detects this change, however, and decreases the resistance of the shunt regulator element.

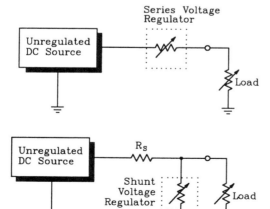

Figure 6.2 A series voltage regulator acts as a variable resistor in series with the load.

Figure 6.3 A shunt-voltage regulator acts as a variable resistor in parallel with the load.

Sec. 6.1 Voltage Regulation Fundamentals

This causes the regulator branch to draw more current and causes the current through R_S to remain constant, and prevents the output voltage from rising.

The shunt regulator is generally used for low-current applications since it consumes a significant amount of power. A simple zener diode regulator is an example of a shunt regulator. By adding an op amp, however, the degree of regulation can be improved.

6.1.3 Switching Regulation

The basic operation of a switching voltage regulator circuit is shown in Fig. 6.4. Here, the regulating element (usually a transistor) is operated either *full on* (closed switch) or *full off* (open switch). The switching usually occurs at tens or hundreds of kilohertz.

During the time the switch is closed, the unregulated source supplies current to the load via L_1. The inductance of L_1 smoothes the current changes that might be caused by the switching circuit. During this time, energy is stored in the magnetic field which builds up around the coil. When the switch opens, the magnetic field begins to collapse and the stored energy is returned to the circuit. The collapsing field now acts as a voltage source and keeps the load current flowing steadily through the alternate path of D_1.

Many switching regulator circuits adjust the duty cycle of the switching action to compensate for changing load or input voltage conditions. That is, if the on time of the switching action is lengthened (relative to the off time), then the average (DC) output voltage will be higher. As with the other regulator circuits, the switching regulator must sense changes in the output voltage in order to compensate (i.e., regulate).

6.1.4 Line and Load Regulation

In order to express the regulator's ability to compensate for changes in the line voltage or the load current, we compute two percentages. The first percentage, called *line regulation*, provides an indication of the regulator's ability to compensate for changes in the input voltage. It is a simple ratio of the change in output voltage to the change in line voltage. That is,

$$\% \text{ line regulation} = \frac{V_{REG}(\max) - V_{REG}(\min)}{V_{IN}(\max) - V_{IN}(\min)} \times 100$$

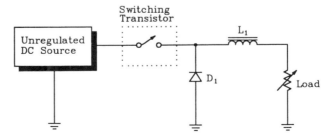

Figure 6.4 A switching voltage regulator offers high efficiency of operation.

The second percentage, called *load regulation,* provides an indication of the regulator's ability to compensate for changes in load current. It is computed as

$$\% \text{ load regulation} = \frac{V_{REG}(\text{no load}) - V_{REG}(\text{full load})}{V_{REG}(\text{full load})} \times 100$$

6.1.5 Voltage References

All of the regulator circuits described in this chapter require a stable reference voltage. The actual load voltage is continuously compared against this reference to determine what changes are required by the regulator circuit. In essence, the voltage reference is in itself a voltage regulator circuit.

Although a zener diode is a low cost, practical reference source, the actual zener voltage changes significantly with changes in current through the zener. Therefore, if we want a more stable source, we must go beyond the simple zener regulator. Figure 6.5 shows a circuit that combines a zener diode and an op amp to produce a simple, but stable reference voltage. We shall utilize this circuit in all of the regulator circuits described in this chapter.

The MC3401 op amp is somewhat different than the other op amps discussed so far in the text. It is designed for operation from a single power supply. That is, only one power source is required for normal operation. A more complete discussion of single-supply op amps is presented in Chap. 11. For now, suffice it to say that the input terminals are essentially PN junctions connected to ground. This means that the voltage on either input will remain at 0.6 volts or less. You may think of the input as responding to current changes the same as the emitter-base circuit of a transistor.

Since the voltage across R_1 is constant (approximately 0.6 volts), then its current is constant. This is essentially equal to the zener diode current since the op amp bias current is insignificant. Since the zener current is constant, the zener voltage will be constant.

If the output voltage attempts to change, this change is felt on the $(-)$ pin via D_1. Since the voltage on the $(-)$ input is essentially limited by an internal junction, the changes fed back have only minimal effect on the voltage on the $(-)$ pin, but rather cause changes in bias currents. In any case, the result is that the output of the op amp

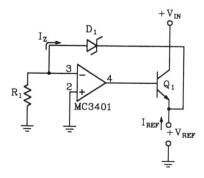

Figure 6.5 A simple, but stable, voltage reference can be built around a single-supply op amp and zener diode.

Sec. 6.1 Voltage Regulation Fundamentals

changes in a polarity that tends to compensate for the changing output voltage. All of this closed-loop action occurs almost instantly so that the actual load voltage never really changes significantly. Although the circuit compensates for changes in load current and for changes in input voltage, it may still drift due to changes in temperature. This latter effect can be essentially eliminated by selecting a zener diode with a temperature coefficient that is opposite ($+2$ mV/°C) from that in the op amp. For our purposes, we shall ignore the effects of temperature changes.

The output or reference voltage for the circuit shown in Fig. 6.5 can be approximated with Kirchoff's Voltage Law. That is

$$V_{REF} = V_{D_1} + 0.6 \qquad (6.1)$$

Transistor Q_1 is a simple current booster (as discussed in Chap. 2). The output current of the op amp is limited to about 5 milliamperes, but required zener currents may be substantially higher than this. Assuming the junction breakdown voltages are adequate, there are only three critical parameters for the selection of Q_1

1. current gain (β or h_{FE})
2. power dissipation (P_D)
3. collector current

The minimum required current gain for Q_1 can be determined from the basic transistor equation for current gain

$$\beta_{MIN} = \frac{I_{REF} + I_Z}{5 \times 10^{-3}} \qquad (6.2)$$

where 5 milliamps is the maximum recommended output current for the MC3401 op amp.

The power dissipation for Q_1 can be determined from the basic power equation

$$P_D = (I_{REF} + I_Z)(V_{IN} - V_{REF}) \qquad (6.3)$$

Resistor R_1 simply establishes the desired zener current. Ohm's Law gives us an approximate value

$$R_1 = \frac{0.6}{I_Z} \qquad (6.4)$$

Let us now design a voltage reference to be used as a stable source for the regulator circuits presented in this chapter. We use the following design goals:

1. unregulated input voltage $+10$ to 15 volts DC
2. regulated output voltage $+4$ volts

3. percent of voltage regulation   0.1%
4. maximum reference current   1 milliampere

Choose D_1.   The required voltage for D_1 can be determined by applying Eq. (6.1). In our case

$$V_{D_1} = V_{REF} - 0.6$$
$$= 4 - 0.6$$
$$= 3.4 \text{ V}$$

This is **not** necessarily the *value* of the zener diode. Rather, it is the required voltage across it. We shall refer to a manufacturer's data sheet (App. 5) and select a diode that is close to the required voltage and then adjust the zener current to obtain the exact value needed. For the present case, let us choose to use a 1N5227 zener. It is rated for 3.6 volts when a 20-milliampere current is passed through it. We shall adjust the value of R_1 to cause more or less current through the zener, and therefore, obtain a higher or lower voltage drop across the zener. Recall that the zener voltage varies nonlinearly with zener current, and that the exact zener voltage at a certain current varies between similar devices. Although the exact value for R_1 will have to be obtained experimentally, the circuit is exceptionally stable once it is constructed.

Compute R_1.   We can now calculate a starting value for R_1 with Eq. (6.4).

$$R_1 = \frac{0.6}{I_{ZT}} = \frac{0.6}{20 \times 10^{-3}} = 30 \text{ } \Omega$$

We shall use a standard value of 27 ohms for R_1. Once the circuit has been constructed, we have to adjust R_1 slightly to obtain the exact output voltage.

Select Q_1.   Since the highest DC input voltage was listed as +15 volts, our collector-to-emitter and collector-to-base breakdown voltages should be greater than 15 volts. The minimum current gain is computed with Eq. (6.2).

$$\beta_{MIN} = \frac{I_{REF} + I_Z}{5 \times 10^{-3}} = \frac{1 \times 10^{-3} + 20 \times 10^{-3}}{5 \times 10^{-3}} = 4.2$$

This, of course is not a challenging goal. In fact, if the actual zener current were less than 5 milliamps, then we could omit Q_1 from the design.

The power dissipation for Q_1 can be estimated with Eq. (6.3) as

$$P_D = (I_{REF} + I_Z)(V_{IN} - V_{REF})$$
$$= (1 \times 10^{-3} + 20 \times 10^{-3})(15 - 4)$$
$$= 231 \text{ mW}$$

Sec. 6.1 Voltage Regulation Fundamentals

Finally, Q_1 must be able to handle the combined currents of I_Z and I_{REF} as collector current. In our particular case

$$I_C = I_Z + I_{REF} = 20 \times 10^{-3} + 1 \times 10^{-3} = 21 \text{ mA}$$

Let us choose a common 2N2222A as the current booster for our design. The data sheet in App. 3 indicates that it will exceed our requirements. By following the process presented in App. 10, we can determine that no heat sink will be necessary, but the transistor will operate fairly hot. It might be desirable to add a small heat sink.

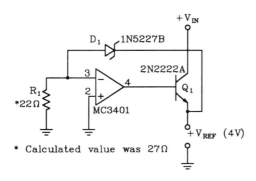

Figure 6.6 A simple, stable voltage reference which is used throughout Chap. 6.

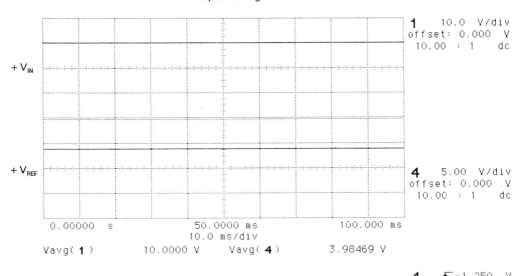

(a)

Figure 6.7 Oscilloscope displays showing the stability of the voltage reference shown in Fig. 6.6. (Test equipment courtesy of Hewlett-Packard Company) (continued)

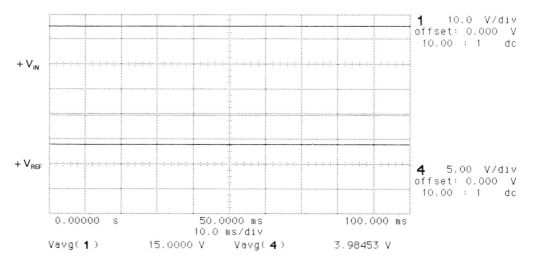

(b)

Figure 6.7b

TABLE 6.1

	DESIGN GOAL	MEASURED VALUE
Input voltage (DC)	+10–15 V	10–15 V
Output voltage (V_{REF})	+4 V	+3.985 V
Percent regulation	0.1%	0.004%
Reference current	0–1 mA	0–1 mA

Figure 6.6 shows the final schematic of our 4-volt reference circuit. The input and output voltage levels are shown on the oscilloscope displays in Fig. 6.7 for minimum and maximum input voltage conditions. Finally, Table 6.1 contrasts the actual circuit performance with the original design goals.

As Table 6.1 and the oscilloscope displays in Fig. 6.7 indicate, the actual circuit performance exceeds our design requirements. Be sure to note in Fig. 6.6 that the actual value of R_1 was adjusted to 22 ohms to trim the output voltage to the required value.

6.2 SERIES-VOLTAGE REGULATORS

6.2.1 Operation

Figure 6.8 shows the schematic diagram of a basic series regulator circuit. The input to the circuit is filtered, but unregulated DC voltage. The output, of course, is regulated DC voltage that remains constant in spite of changes in the load current or changes in the input voltage.

Transistor Q_1 in Fig. 6.8 is known as the series *pass* transistor. Kirchoff's Voltage Law would tell us that the voltage across the load plus the voltage across the series-pass transistor must always be equal to the applied voltage. Thus, if we can control the amount of voltage dropped across the pass transistor then we inherently have control over the load voltage.

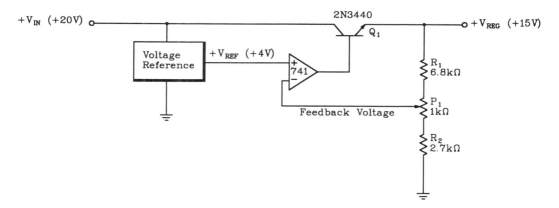

Figure 6.8 A series-voltage regulator with an adjustable output.

The output voltage of the regulator circuit is sampled with the voltage divider made up of resistors R_1, R_2, and potentiometer P_1. The portion of the output that appears on the wiper arm of P_1 is called the *feedback voltage*. Potentiometer P_1 is used to adjust the amount of feedback voltage and thus is used to adjust the output voltage level.

The voltage reference circuit indicated in Fig. 6.8 was discussed in an earlier section. Its purpose is to provide a constant voltage level that can be used as a stable reference. The schematic of a representative voltage reference circuit was presented in Fig. 6.5.

The op amp in Fig. 6.8 is called the *error amplifier*. It continuously compares the magnitude of the reference voltage with the level of the feedback signal (which represents the output voltage). Any difference between these two voltages (both magnitude and polarity) is amplified and applied to the base of the pass transistor. The polarity is such that the output voltage is returned to its correct value. As an example, let us assume that the load current suddenly decreased. This would tend to make the output voltage rise.

However, as soon as the output voltage starts to increase (i.e., becomes more positive), the feedback voltage on the wiper arm of P_1 also becomes more positive. This increasing positive on the inverting pin of the op amp causes the output of the op amp to become less positive (i.e., moves in the negative direction). Recall that the reference voltage remains constant so any changes in the feedback voltage are immediately reflected in the output of the op amp. This reduced positive voltage on the base of Q_1 reduces the amount of forward bias, and therefore, increases the effective resistance of the pass transistor causing an increased voltage drop across it. Since we are now dropping more voltage across the pass transistor, we will have less dropped across the load (Kirchoff's Voltage Law). Thus, the initial tendency for the load voltage to rise has been offset by an increased voltage drop across the pass transistor. This process happens nearly instantaneously so that the load voltage never really sees a significant increase. Of course, the better the degree of regulation, the smaller the changes in load voltage.

To further clarify the operation of the error amplifier, let's examine the circuit from a different viewpoint. First, consider the wiper arm to be at some fixed point. We can now view the resistor network as a simple, two-resistor voltage divider. A redrawn circuit for the error amplifier is shown in Fig. 6.9. Here R_1' is equivalent to R_1 and that portion of P_1 which is above the wiper arm. Similarly, R_2' is equivalent to R_2 and that portion of P_1 which is below the wiper arm. It is readily apparent that the resulting circuit is a simple noninverting amplifier circuit with a current boost transistor. This circuit was discussed in detail in Chap. 2.

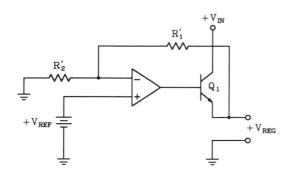

Figure 6.9 A simplified circuit of the error amplifier portion of Fig. 6.8. It is actually a simple noninverting voltage amplifier with a current boost transistor.

Potentiometer P_1 in Fig. 6.8 is used to adjust the output voltage to a particular level. If we move the wiper arm up, we will increase the feedback voltage (i.e., more positive), decrease the bias on Q_1, increase the voltage drop across Q_1, and ultimately bring the output voltage down to a new lower level. Similarly, if we move the wiper arm down, we will reduce the amount of feedback voltage (i.e., less positive), increase the bias on Q_1, decrease the voltage drop across Q_1, and cause the load voltage to increase to a higher regulated level.

It is important to note that the series-regulator circuit shown in Fig. 6.8 is not immune to short-circuits. That is, if the output of the regulator were accidentally shorted

Sec. 6.2 Series-Voltage Regulators

to ground, the pass transistor would undoubtedly be destroyed by the resulting high-current flow. Many, if not most, regulated supplies are designed to be current limited. Section 6.5 discusses this option in greater detail.

6.2.2 Numerical Analysis

Let us now analyze the series-voltage regulator circuit presented in Fig. 6.8. The output voltage of the regulator can be computed with Eq. (6.5).

$$+V_{REG} = +V_{REF}\left(\frac{R'_1}{R'_2} + 1\right) \quad (6.5)$$

where R'_1 and R'_2 are the equivalent values shown in Fig. 6.9 and as discussed. If we assume that the wiper arm of P_1 is moved to the upper most extreme, then we can apply Eq. (6.5) to compute the minimum output voltage as shown

$$+V_{REG} = 4\left(\frac{6.8 \times 10^3}{3.7 \times 10^3} + 1\right)$$
$$= 4 \times 2.84$$
$$= 11.36 \text{ V}$$

Similarly, we can move the wiper arm to the lowest position and compute the highest output voltage with Eq. (6.5) as follows:

$$+V_{REG} = 4\left(\frac{7.8 \times 10^3}{2.7 \times 10^3} + 1\right)$$
$$= 4 \times 3.89$$
$$= 15.56 \text{ V}$$

So the range of regulated output voltages that can be obtained by adjusting P_1 is 11.36 volts through 15.56 volts.

The maximum allowable output current is determined by one of the following:

1. maximum collector current rating of Q_1
2. maximum power dissipation rating of Q_1
3. current limitation of $+V_{IN}$

Whichever of these limitations is reached first will determine the maximum allowable current that can be drawn from the regulator output.

The manufacturer's data sheet for a 2N3440 lists the maximum collector current as 1.0 amps. The data sheet also lists the maximum power dissipation as 1.0 watts (at 25 °C), gives a thermal resistance from junction-to-case as 17.5 °C per watt, and lists the thermal resistance from junction-to-air as 175 °C per watt. No current limit is shown in Fig. 6.8 for $+V_{IN}$. The current limit imposed by the power rating can be computed as follows:

$$I_O(\text{max}) = \frac{P_D}{+V_{IN}(\text{max}) - V_{REG}(\text{min})} \quad (6.6)$$

where P_D is the maximum power as determined with Eq. (A10.3) (see App. 10). In the case of the circuit shown in Fig. 6.8, the current limit imposed by the power rating of the transistor (for $T_A = 40$ °C) is computed as

$$P_D = \frac{T_J(\text{max}) - T_A}{\Theta_{JA}} = \frac{200 - 40}{175} = 0.91 \text{ W},$$

therefore

$$I_O(\text{max}) = \frac{.91}{20 - 11.36} = 105 \text{ mA}$$

Since this current is lower than the 1.0 amp maximum collector current rating, it will be the limiting factor. Thus, the regulator circuit shown in Fig. 6.8 has a maximum output current of about 105 milliamperes.

6.2.3 Practical Design Techniques

Let us now design a series-voltage regulator similar to the one shown in Fig. 6.8. We shall use the following as design goals:

1. input voltage   +12–18 volts
2. output voltage   +6–9 volts
3. output current   0–0.5 amps
4. line regulation   <1%
5. load regulation   <1%
6. error amplifier   741

Select the pass transistor. The characteristics of the pass transistor are determined by the input voltage, the output voltage and current requirement, and the output drive capability of the op amp. First, the collector current rating of the transistor must be greater than the value of load current. In our case, this means that our transistor must have a maximum DC current rating of greater than 500 milliamperes.

The transistor power dissipation can be found by applying Eq. (6.6).

Sec. 6.2 Series-Voltage Regulators

$$I_O(\text{max}) = \frac{P_D}{+V_{IN}(\text{max}) - V_{REG}(\text{min})},$$

or

$$P_D = I_O(\text{max})[+V_{IN}(\text{max}) - V_{REG}(\text{min})]$$
$$= 0.5(18 - 6)$$
$$= 6 \text{ W}$$

The minimum current gain (h_{FE} or β) for the transistor can be found by applying our basic transistor formula for current gain

$$\beta = \frac{I_C}{I_B}$$

In the case of a circuit like that shown in Fig. 6.8, base current is provided by the output of the op amp. We shall establish the maximum current to be provided by the op amp as one fourth of the short-circuit output current rating of the op amp. In the case of a 741, the short-circuit output current rating is listed in the data sheet as 20 milliamps. Therefore, we shall limit the output current (and therefore the base current of Q_1) to one fourth of 20 milliamperes or 5 milliamperes. The minimum current gain for Q_1 can now be determined as shown

$$\beta(\text{min}) = \frac{I_C}{I_B}$$
$$= \frac{0.5}{5 \times 10^{-3}}$$
$$= 100$$

The minimum collector-to-emitter voltage breakdown rating for Q_1 is found by determining the maximum voltage across Q_1. That is

$$\boxed{V_{CEO} = V_{IN}(\text{max}) - V_{REG}(\text{min})} \quad (6.7)$$

In our particular case, the collector-to-emitter voltage rating is computed as

$$V_{CEO} = 18 - 6 = 12 \text{ V}$$

There are many transistors that will satisfy the requirements for Q_1. Let us select an MJE1103 (refer to App. 2) for this application. The calculations presented in App. 10 indicate that the transistor will require a heat sink for safe operation.

Determine the required op amp voltage gain. The equivalent circuit shown in Fig. 6.9 is useful for determining the required voltage gain of our op amp. We must consider the circuit under both minimum and maximum output voltage conditions. The minimum and maximum voltage gains are determined from the basic amplifier gain formula ($A_V = V_O/V_I$. That is

$$A_V(\text{min}) = \frac{+V_{REG}(\text{min})}{+V_{REF}} \qquad (6.8)$$

Similarly

$$A_V(\text{max}) = \frac{+V_{REG}(\text{max})}{+V_{REF}} \qquad (6.9)$$

In our present application, the required voltage gains for the op amp are determined as follows:

$$A_V(\text{min}) = \frac{6}{4} = 1.5,$$

and

$$A_V(\text{max}) = \frac{9}{4} = 2.25$$

Select the value for P_1. Selection of P_1 is largely arbitrary, but some guidelines may be established. The minimum value for P_1 should be at least 20 times the minimum equivalent load resistance. That is

$$P_1(\text{min}) = 20\left(\frac{V_{REG}(\text{min})}{I_O(\text{max})}\right) \qquad (6.10)$$

In our particular case, the minimum recommended value for P_1 is

$$P_1(\text{min}) = 20\left(\frac{6}{0.5}\right) = 240 \text{ }\Omega$$

The maximum value is also somewhat arbitrary, but there would not generally be any reason for going beyond a few tens of thousands of ohms. Let us decide to use a 5 kilohm potentiometer for P_1 in this particular application.

Compute R_1 and R_2. The values for R_1 and R_2 can be determined by applying the basic equation for voltage gain in a noninverting amplifier (refer to Fig. 6.9). Recall that minimum voltage gain occurs when the wiper arm of P_1 is at its upper extreme. Under these conditions, the equation for R_1 can be determined as follows:

Sec. 6.2 Series-Voltage Regulators

$$A_V(\min) = \frac{R_1}{R_2 + P_1} + 1,$$

or

$$\boxed{R_1 = [A_V(\min) - 1](R_2 + P_1)} \qquad (6.11)$$

Similarly, an equation for R_2 can derived from the basic gain equation as shown

$$A_V(\max) = \frac{R_1 + P_1}{R_2} + 1,$$

or

$$R_2 = \frac{R_1 + P_1}{A_V(\max) - 1}$$

Substituting Eq. (6.11) for R_1 in this equation and performing some algebraic transposing gives us our final equation for the value of R_2.

$$\boxed{R_2 = \frac{P_1 A_V(\min)}{A_V(\max) - A_V(\min)}} \qquad (6.12)$$

We can now compute the required values for R_1 and R_2. First, we apply Eq. (6.12) to find R_2.

$$R_2 = \frac{5 \times 10^3 \times 1.5}{2.25 - 1.5}$$

$$= 10 \text{ k}\Omega$$

We are now in a position to apply Eq. (6.11) to find the value of R_1.

$$R_1 = (1.5 - 1)(10000 + 5000)$$

$$= 7.5 \text{ k}\Omega$$

Since the computed values for R_1 and R_2 are both standard values, we do not have to make any decisions regarding standard values. For most applications, we would simply choose the nearest standard value.

This completes the design of our series regulator circuit. The final schematic is shown in Fig. 6.10. The performance of the circuit is indicated by the oscilloscope displays in Fig. 6.11. The waveforms show the effects of minimum and maximum load current and minimum and maximum line voltage. Figure 6.11(a) shows the output under no-load conditions. Figure 6.11(b) illustrates the effect of adding a 500-milliampere load.

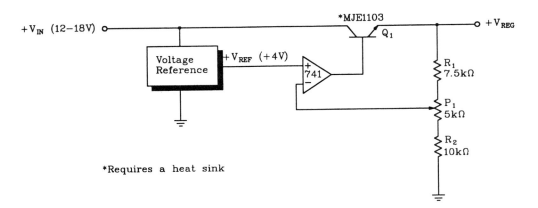

Figure 6.10 A series-voltage regulator circuit designed to deliver 6 to 9 volts at 0 to 500 milliamperes.

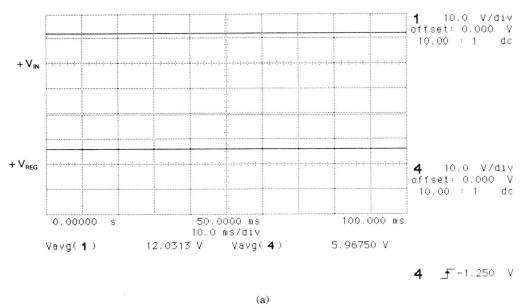

(a)

Figure 6.11 Oscilloscope displays showing the performance of the series-voltage regulator shown in Fig. 6.10. (Test equipment courtesy of Hewlett-Packard Company)   (continued)

Sec. 6.2 Series-Voltage Regulators

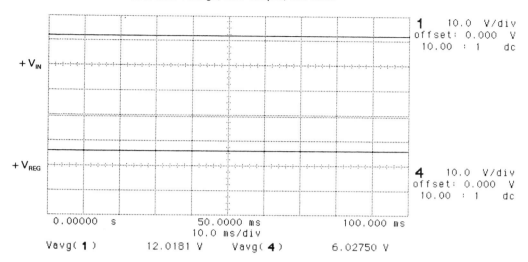

(b)

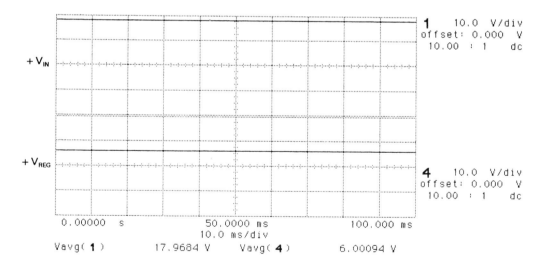

(c)

Figure 6.11b and c

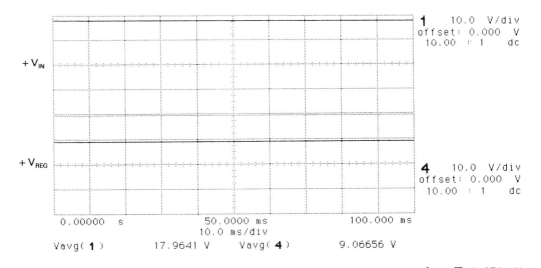

(d)

Figure 6.11d

TABLE 6.2

	DESIGN GOAL	MEASURED VALUE
Input voltage	+12–+18 V	+12–+18 V
Output voltage	+6–+9 V	+5.97–+9.07 V
Output current	0–500 mA	0–500 mA
Load regulation	<1%	0.99%
Line regulation	<1%	0.56%

In both cases, the output voltage was adjusted to minimum (+6 volts). Figure 6.11(c) shows the results of maximum input voltage under no-load conditions. Finally, Fig. 6.11(d) illustrates the circuit performance under conditions of maximum input voltage and a 500-milliampere load. The measured performance of the circuit is summarized and contrasted with the original design goals in Table 6.2.

Sec. 6.3 Shunt-Voltage Regulation

6.3 SHUNT-VOLTAGE REGULATION

6.3.1 Operation

Figure 6.12 shows the schematic diagram of a basic shunt-voltage regulator circuit. To understand its operation, let us assume that the output voltage starts to increase (perhaps as a result of a decreased load current). When the load voltage starts to rise, the voltage across R_2 also increases. This is the feedback voltage for the regulator circuit and is essentially a sample of the output voltage. When the voltage across R_2 increases (i.e., becomes more positive), the output of the op amp becomes less positive since the voltage across R_2 is applied to the inverting input. This falling voltage on the output of the op amp is the base voltage for Q_1. Since Q_1 is connected as an emitter follower, the emitter voltage, and therefore, the regulated output voltage will decrease. Actually, the decrease merely offsets the original increase so the output remains essentially constant. If the output voltage were to attempt to decrease, a similar closed-loop action would compensate for the change and maintain a constant output voltage.

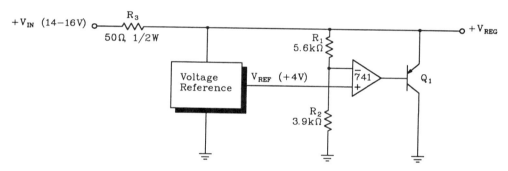

Figure 6.12 A shunt voltage regulator circuit.

Another way to view the regulator action is to consider that the current in Q_1 will increase in response to an increase in the regulated output voltage. This increased transistor current causes an increased voltage drop across R_3 thus returning the output voltage to its initial level. Since the current through Q_1 increases and decreases to compensate for load voltage changes, the highest transistor current will occur during times when the load current is minimum.

The circuit would still work if the voltage reference circuit were powered directly from the unregulated input voltage. But, since we have a convenient source of regulated voltage, we can increase the overall performance of the circuit by allowing the reference circuit to use the regulated output as its input voltage.

Resistor R_3 ultimately determines the maximum current that can be drawn from the regulator. If too much current is drawn, then Q_1 is cutoff and current is limited by R_3. Under these conditions, the output voltage is not regulated and will decrease with increasing load currents. The circuit does have a distinct advantage in that it is inherently current limited. That is, if a short-circuit to ground were to occur on the regulated voltage line, the current would be limited by resistor R_3. No other regulator components would experience an overload condition. If this resistor has an adequate power rating, then no damage will result from shorted outputs.

We can change the level of the regulated output voltage by altering the values of R_1 and/or R_2. In fact, we can include a potentiometer in the feedback circuit and make an adjustable shunt-regulator circuit.

6.3.2 Numerical Analysis

Let us now extend our understanding of the shunt regulator circuit shown in Fig. 6.12 to include a numerical analysis of the important characteristics. Two of the most important characteristics of the regulator circuit are

1. output voltage
2. current capability

We can redraw the circuit somewhat to more clearly see how the op amp is connected. Figure 6.13 clearly shows that the op amp is essentially connected as a simple noninverting amplifier with a current-boost transistor.

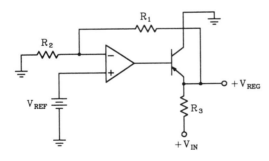

Figure 6.13 The error amplifier portion of the regulator circuit shown in Fig. 6.12 is essentially a simple noninverting voltage amplifier.

The voltage gain of this circuit is simply

$$A_V = \frac{R_1}{R_2} + 1$$

The output voltage of the circuit, then is computed by applying the basic gain equation.

$$+V_{REG} = +V_{REF} \times A_V = +V_{REF}\left(\frac{R_1}{R_2} + 1\right) \quad (6.13)$$

Sec. 6.3 Shunt-Voltage Regulation

In the case of the circuit shown in Fig. 6.12, we can compute the regulated output voltage as

$$+V_{REG} = +4 \times \left(\frac{5.6 \times 10^3}{3.9 \times 10^3} + 1\right) = 9.7 \text{ V}$$

We estimate the current capability of the circuit by considering the case when $+V_{IN}$ is at its lowest level. Under these conditions, the maximum load current can be estimated with Ohm's Law.

$$\boxed{I_O(\text{max}) = \frac{V_{IN}(\text{min}) - V_{REG}}{R_3}} \quad (6.14)$$

For the present case, we compute the highest allowable load current as

$$I_O(\text{max}) = \frac{14 - 9.7}{50} = 86 \text{ mA}$$

Although there could be a higher load current under higher input voltage conditions, the computed value of 86 milliamperes is the highest load current that we can supply under all input voltage conditions and still expect the circuit to remain in a regulated condition. Ohm's Law can be used to determine the minimum value load resistor that can be used with the circuit. That is,

$$R = \frac{E}{I} = \frac{9.7}{86 \times 10^{-3}} = 112.8 \text{ }\Omega$$

Transistor Q_1 must be able to safely conduct the difference between the highest possible input current and the minimum possible load current. For many applications, we assume that the load *could* be disconnected, and therefore, consider the minimum load current to be zero. If the regulator were an integral part of a system that made it impossible for the load to be disconnected (e.g., all part of the same printed circuit board), then the minimum load current may be greater than zero. For purposes of this analysis, we shall assume a worst-case situation which means that Q_1 must be able to handle a value of current given by Eq. (6.15).

$$\boxed{I_O(\text{max}) = \frac{V_{IN}(\text{max}) - V_{REG}}{R_3}} \quad (6.15)$$

For the circuit shown in Fig. 6.12 the maximum transistor current can be calculated as

$$I_{Q_1}(\text{max}) = \frac{16 - 9.7}{50} = 126 \text{ mA}$$

We can estimate the worst-case power dissipation in Q_1 by applying our basic power formula.

$$P_D(\text{max}) = I_{Q_1}(\text{max}) \times V_{REG} \qquad (6.16)$$

For our present circuit, the transistor dissipation is estimated as

$$P_D(\text{max}) = 126 \times 10^{-3} \times 9.7 = 1.22 \text{ W}$$

For many applications, the heat dissipated in the transistor will require a heat sink to keep the transistor within its safe operating range.

The shunt-regulator circuit is inherently current limited. That is, if we try to draw more current than it is designed to deliver, the output voltage will drop. Even if we short the output directly to ground, the current will be limited by resistor R_3. If resistor R_3 has sufficiently high-power rating, then the duration of the short can be any length. If R_3 has a lower power rating, then the regulator is still short-circuit proof, but the duration of the overload must be less. If the output of the circuit in Fig. 6.12 were shorted directly to ground, then the maximum current flow can be computed with Ohm's Law as

$$I_{R_3} = \frac{E_{R_3}}{R_3}$$

$$= \frac{16}{50}$$

$$= 320 \text{ mA}$$

The resulting power dissipation in R_3 is computed with the basic power formula as

$$P_{R_3} = \frac{E_3^2}{R_3}$$

$$= \frac{16^2}{50}$$

$$= 5.12 \text{ W}$$

Since this clearly exceeds the 1/2 watt rating of the resistor as listed in Fig. 6.12, we could expect the resistor to overheat and burn open. However, it could withstand momentary short circuits without having a higher power rating. A good way for momentary short circuits to occur is for your probe to slip off of a test point while troubleshooting a circuit.

6.3.3 Practical Design Techniques

Let's now design a shunt-regulator circuit similar to the one shown in Fig. 6.12. We shall design it to meet the following design specifications:

1. unregulated input voltage   +18 to +22 volts DC
2. regulated output voltage   +12 to +15 volts DC

Sec. 6.3 Shunt-Voltage Regulation

3. load current   0 to 150 milliamps
4. line regulation   <2%
5. load regulation   <2%
6. op amp   741

Determine the error amp voltage gain. Since the design calls for a variable output voltage, we need to compute a range of error amp gains. As with the previous design, we shall elect to use the +4-volt reference circuit designed earlier in the chapter. The required voltage gains can be computed with Eqs. (6.8) and (6.9).

$$A_V(\text{min}) = \frac{+V_{REG}(\text{min})}{+V_{REF}} = \frac{12}{4} = 3,$$

and

$$A_V(\text{max}) = \frac{+V_{REG}(\text{max})}{+V_{REF}} = \frac{15}{4} = 3.75$$

Let's plan to use a potentiometer between R_1 and R_2 of Fig. 6.12 to adjust the gain of the error amp. This is similar to the method used with the series-voltage regulator discussed in a prior section.

Select the potentiometer. Selection of P_1 is not critical, and the guidelines discussed for the series regulator may be followed. That is, the minimum value for P_1 should be at least 20 times the minimum equivalent load resistance. This is computed with Eq. (6.10).

$$P_1(\text{min}) = 20\left(\frac{V_{REG}(\text{min})}{I_O(\text{max})}\right) = 20\left(\frac{12}{.15}\right) = 1.6 \text{ k}\Omega$$

The maximum value is generally no more than a few tens of thousands of ohms. Let us decide to use a 10 kilohms potentiometer for P_1 in this particular application.

Compute R_2. Equation 6.12 provides the tool necessary to compute the value for R_2. We calculate it as follows:

$$R_2 = \frac{P_1 A_V(\text{min})}{A_V(\text{max}) - A_V(\text{min})}$$

$$= \frac{10 \times 10^3 \times 3}{3.75 - 3}$$

$$= 40 \text{ k}\Omega$$

For this application, we select the nearest standard value of 39 kilohms for R_2.

Compute R_1. Resistor R_1 can be calculated with Eq. (6.11). For our present application, R_1 is computed as shown

$$R_1 = (A_V(\min) - 1)(R_2 + P_1)$$
$$= (3 - 1)(39000 + 10000)$$
$$= 98 \text{ k}\Omega$$

Again, we use the nearest standard value of 100 kilohms since the application is not critical.

Determine the value of R_3. Resistor R_3 establishes the maximum possible load current from the shunt-regulator circuit. We can determine its value for a given application by applying Ohm's Law at a time when the load current and output voltage are at maximum and the input voltage is at minimum. That is

$$\boxed{R_3 = \frac{V_{IN}(\min) - V_{REG}(\max)}{I_O(\max)}} \qquad (6.17)$$

In our particular case, the required value for R_3 is found as follows.

$$R_3 = \frac{18 - 15}{150 \times 10^{-3}} = 20 \text{ }\Omega$$

The power rating for R_3 is determined by using the basic power formula under worst-case conditions. That is

$$P_{R_3} = \frac{E_3^2}{R_3} = \frac{(V_{IN}(\max) - V_{REG}(\min))^2}{R_3}$$
$$= \frac{(22 - 12)^2}{20} = 5 \text{ W}$$

Select transistor Q_1. There are several transistor characteristics which must be considered when selecting a particular device for Q_1

1. maximum collector current
2. current gain (h_{FE} or β)
3. breakdown voltages
4. power dissipation

The maximum collector current that the transistor will be expected to carry can be computed with Eq. (6.15).

$$I_{Q_1}(\max) = \frac{V_{IN}(\max) - V_{REG}(\min)}{R_3}$$
$$= \frac{22 - 12}{20}$$
$$= 500 \text{ mA}$$

Sec. 6.3 Shunt-Voltage Regulation

We shall estimate the highest current that should be supplied by the op amp as 25 percent of the short-circuit output current of the op amp. The short-circuit output current for a 741 is listed in the manufacturer's data sheet as 20 milliamperes. Therefore, we shall plan to restrict the output current of the op amp to 25 percent of 20 milliamperes or 5 milliamperes. We can now apply the basic current gain equation for transistors to determine the required transistor current gain.

$$\beta = \frac{I_C}{I_B} = \frac{500 \times 10^{-3}}{5 \times 10^{-3}} = 100$$

This is a fairly high-current gain for a power transistor and may necessitate the use of a Darlington pair for Q_1.

The power dissipation in Q_1 can be found by applying Eq. (6.16) under conditions of maximum output voltage.

$$P_D(\max) = I_{Q_1}(\max) \times V_{REG}(\max)$$
$$= 500 \times 10^{-3} \times 15$$
$$= 7.5 \text{ W}$$

This high-power dissipation is one major disadvantage of shunt-regulator circuits.

The collector-to-emitter breakdown voltage must be higher than $V_{REG}(\max)$. In the present case, the transistor breakdown rating for V_{CEO} must be greater than 15 volts.

There are many transistors that will satisfy the requirements of our design. For purposes of illustration, let us select an MJE2090. As the manufacturer's data sheet (App. 2) indicates, the MJE2090 is a Darlington power transistor which satisfies all of our requirements. The calculations presented in App. 10 dictate the use of a heat sink for the transistor.

The complete schematic of our shunt-regulator circuit is shown in Fig. 6.14. Its performance is demonstrated by the oscilloscope displays in Fig. 6.15. Figures 6.15(a) and 6.15(b) show the effect of adjusting P_1 between its limits. Figures 6.15(c) and 6.15(d) illustrate the circuit's response to a change in load current from zero, Fig. 6.15(c), to

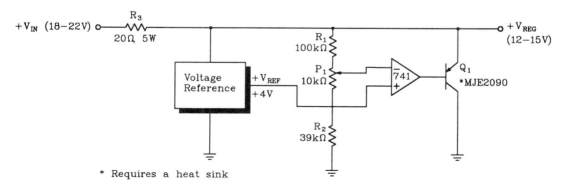

* Requires a heat sink

Figure 6.14 A shunt regulator circuit designed to provide a variable output voltage and to supply a load current of 0 to 150 milliamperes.

278 Power Supply Circuits   Chap. 6

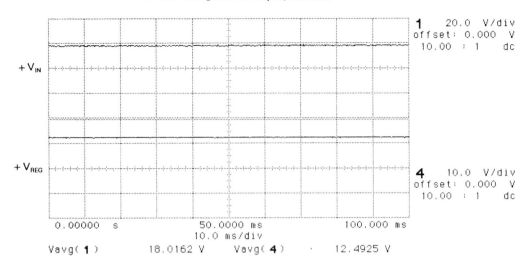

(a)

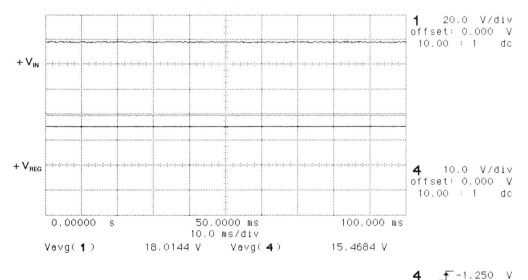

(b)

Figure 6.15 Oscilloscope displays showing the performance of the regulator circuit shown in Fig. 6.14. (Test equipment courtesy of Hewlett-Packard Company)   (continued)

Sec. 6.3 Shunt-Voltage Regulation

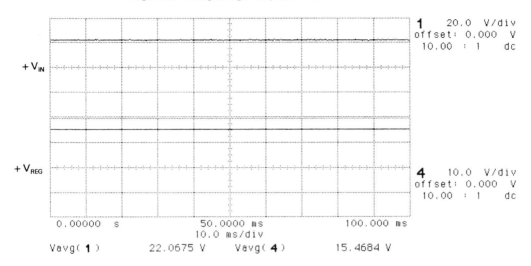

(c)

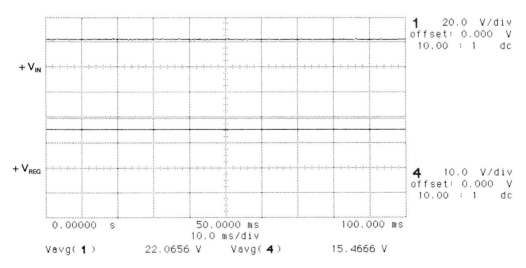

(d)

Figure 6.15c and d

TABLE 6.3

	DESIGN GOAL	MEASURED VALUE
Input voltage	18–22 V	18–22 V
Output voltage	12–15 V	12.5–15.5 V
Output current	0–150 mA	0–150 mA
Line regulation	<2%	0.8%
Load regulation	<2%	0.012%

150 milliamperes, Fig. 6.15(d). Finally, the original design goals are contrasted with the actual measured performance in Table 6.3.

6.4 SWITCHING VOLTAGE REGULATORS

Our discussion on switching regulators will be limited to the theory of operation. Although a switching regulator can be designed around an op amp, most are built using specialized regulator ICs. This not only simplifies the design, it generally improves the overall performance of the regulator circuit. Nevertheless, an understanding of the operation of switching regulators is very important to an engineer or technician working with equipment being designed today, and no discussion of regulated power supplies would be complete without switching regulators.

6.4.1 Principles of Operation

Let us begin by examining the simplest of equivalent circuits. Figure 6.16 shows a simple switching circuit. Assume that the switch is operated at periodic intervals with equal open and closed times. When the switch is closed, the capacitor is charged by current flow through the coil. As current flows through the coil, a magnetic field builds out around it (i.e., energy is stored in the coil). When the switch is opened, the magnetic field around the coil begins to collapse which makes the coil act as a power source (i.e., the stored energy is being returned to the circuit). You will recall from basic electronics theory that inductors tend to oppose changes in current. When the switch opens and the fields begins to collapse, the resulting coil voltage causes circuit current to continue uninterrupted. The path for this electron current is right-to-left through the inductor, down through diode D, up through C (and the load) to the coil. This current will continue (although decaying) until the magnetic field around the coil has completely collapsed.

Now, if the switch were to close again before the coil current had time to decay significantly, and if it continued to open and close at a rapid rate, then there would be some average current through the coil. Similarly, this average current value will produce

Sec. 6.4 Switching Voltage Regulators

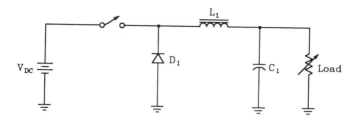

Figure 6.16 A simple circuit to help explain the principles of switching voltage regulators.

some average value of voltage across the capacitor, and therefore, across the parallel load resistor.

Suppose now that the ratio of closed time to open time on the switch is shortened. That is, the switch is left opened longer than it is closed. Can you see that the inductor's field will collapse more completely, and that the *average* current through the coil (and therefore, load voltage) will decrease? On the other hand, if we lengthened the closed time of the switch relative to the open time, then the average load voltage would increase.

The basic behavior of capacitors is to oppose changes in voltage. That is, we cannot change the voltage across a capacitor instantaneously. The voltage can only change as fast as the capacitor can charge or discharge. The capacitor in Fig. 6.16 is a filter and smoothes the otherwise pulsating load voltage. So, even though the switch is interrupting the DC supply at regular intervals, the voltage is steady across the load because of the combined effects of the coil and capacitor. By varying the ratio of on time to off time (i.e., the duty cycle), we can vary the DC voltage across the load. If we sample the actual load voltage and use its value to control the duty cycle of the switch (a transistor in practice), then we will have constructed a switching regulator circuit.

Figure 6.17 shows a more accurate representation of a switching voltage regulator. Here the DC input voltage is provided by a standard transformer-coupled, bridge-rectifier circuit followed by a brute filter (C_1). The interrupting device is an *n*-channel, power MOSFET (Q_1). Gate drive for the MOSFET comes from a pulse width modulator circuit. This could be built around an op amp comparator/oscillator circuit, but is generally an integral part of an integrated circuit designed specifically for use in switching regulators.

The pulse width modulator has two inputs. One is the sample output voltage derived from a voltage divider (R_1 and R_2). The second input to the pulse width modulator circuit is provided by a stable voltage reference. The pulse width modulator compares the reference voltage with the sampled output voltage (just like the series and shunt-linear regulators discussed in previous sections do), and alters the pulse width (effectively duty cycle) of the signal going to the MOSFET. As the duty cycle of the MOSFET is altered, the average (i.e., DC) output voltage is adjusted and maintained at a constant value. If the load voltage tried to change, perhaps in response to a changing current demand, then this change would be fed back through the voltage divider to the pulse width modulator circuit. The pulse width going to the MOSFET would quickly be adjusted to bring the load voltage back to the correct value.

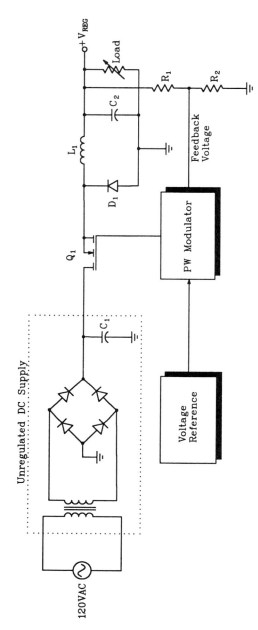

Figure 6.17 A switching regulator controls the switching operation of the pass transistor to regulate load voltage.

6.4.2 Switching versus Linear-Voltage Regulators

So why go to all the trouble of interrupting the DC voltage and then turning right around and smoothing it back into DC again? Well, switching regulators do have some outstanding advantages over linear regulators. One of the primary advantages of switching regulators as compared to their linear equivalents is power dissipation.

In a linear regulator, the series pass transistor or the shunt regulator transistor dissipates a significant amount of power. Typical efficiencies for linear regulators are 30 to 40 percent. This means, for example, that a linear supply designed to deliver 12 volts at 3 amps DC would actually draw at least 90 watts from the power line. The internal power loss results in heat which in turn leads to cooling requirements like fans and heat sinks. Most of the power loss in a linear regulator is in the regulator transistors. Recall that their power dissipation is computed as

$$P_D = I_C V_{CE}$$

The switching regulator, on the other hand, typically achieves efficiencies on the order of 75 percent. This improvement is primarily caused by a dramatic reduction in power dissipation in the regulator transistor. Although the power dissipation is computed in the same basic way, the results are quite different since the transistor is operated in either saturation or cutoff at all times. Thus, although the power at any given time is expressed as

$$P_D = I_C V_{CE}$$

either I_C is very low (at cutoff $I_C \approx 0$) or V_{CE} is very low (during saturation V_{CE} is *ideally* zero). Thus, the only time the switching transistor dissipates significant power is **during** the actual switching time (a few microseconds).

The reduced power dissipation results in other advantages. Since the cooling requirement is less for a given output power, both size and cost of the associated circuitry and support components are less. It is reasonable to expect size reductions on the order of five or more.

Another advantage of switching regulators is that the output voltage can be stepped up, stepped down and/or changed in polarity in the process of being regulated. This can simplify some designs.

Switching regulators are not without their disadvantages, however. First, they require more complex circuitry for control. This disadvantage is becoming less as more specialized regulator ICs are being provided to the power supply designer.

Another major disadvantage of switching regulators is electrical noise generation. Anytime a circuit changes states quickly, high-frequency signals are generated. You may recall from basic electronics theory that a square wave is made up of an infinite number of odd harmonics. So, if we have a 100-kilohertz square wave, we will be generating signal frequencies of 300 kilohertz, 500 kilohertz, 700 kilohertz, and so on. The Federal Communications Commission (FCC) in the United States and other agencies in other countries (such as VDE in Europe) restrict the amount of electromagnetic emissions that may leave an electronic device. For example, suppose you have designed a new computer that fits in the palm of your hand. The FCC will prevent you from marketing your new computer unless it can pass the FCC-defined emissions tests. One reason that many new computer designs fail to pass these tests is because of the electrical noise generated by switching power supplies. Now, this doesn't mean you can never use a switching regulator in a computer. Quite the contrary, most computers do use switching regulators. But additional components will have to be included to filter the high-frequency noise that is generated. This noise can easily extended into the 30 to 150 megahertz band.

Finally, although switching regulators are good, they cannot respond as quickly to sudden changes in line voltage or load current. That is, they do not regulate as well as their linear counterparts if the line and load changes are rapid.

6.4.3 Classes of Switching Regulators

We can categorize switching regulators into four general groups based on the method used to control the switching transistor

1. fixed off-time, variable on-time
2. fixed on-time, variable off-time
3. fixed frequency, variable duty cycle
4. burst regulators

First, it should be noted that all of the switching regulator types listed work by switching the regulator transistor from full off to full on. Additionally, they all regulate by altering the ratio of on time to off time of the transistor. The various methods refer to the actual circuitry and waveform driving the switching transistor.

The first two regulator types listed are similar in that one alternation of the transistor drive signal is fixed. Regulation is achieved by adjusting the time for the remaining alternation. Since one alternation is fixed and one is variable, the frequency of operation inherently varies. These are sometimes called variable frequency regulators.

The third class of switching voltage regulators uses a constant frequency, but alters the duty cycle of the signal applied to the switching transistor. That is, if the on time is increased then the off time is decreased proportionately. Thus, the output voltage can be controlled without altering the basic frequency of operation. This is one of the most common classes of switching regulators.

Finally, the burst regulator operates by gating a fixed-frequency, fixed-pulse width oscillator on and off. The duty cycle of the switching waveform is such that the output voltage would be too high if the switching were continuous. The circuit senses this excessive output voltage and interrupts or stops the switching completely. With the switching transistor turned off continuously, the output voltage will quickly decay. As soon as it decays to the correct voltage, the switching is resumed. Thus, the regulation is actually achieved by periodically interrupting the switching waveform going to the switching transistor.

6.5 OVER-CURRENT PROTECTION

Regulated power supplies are often designed to be short-circuit protected. That is, if the output of the supply is accidentally shorted to ground or at least tries to draw excessive current, the supply will not be damaged. There are several classes of over-current protection

1. load interruption
2. constant current limiting
3. foldback current limiting

6.5.1 Load Interruption

The simplest form of over-current protection is shown in Fig. 6.18. The protective device is generally a fuse (as shown in Fig. 6.18), a fusable resistor, or a circuit breaker. In any case, once a certain value of current has been reached the protective device opens and completely isolates the load from the output of the supply. As long as the protective device is designed to operate at a lower current value than the absolute maximum safe current from the supply, the power supply will be protected from damage. Since the protective element has resistance, it can adversely affect the overall regulation of the circuit.

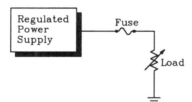

Figure 6.18 Load interruption is the simplest form of over-current protection.

6.5.2 Constant-current Limiting

Fig. 6.19 shows a common example of a constant-current limiting circuit. This is identical to the series regulators discussed earlier in the chapter with the addition of R_1 and Q_2. These are the current-limiting components. Under ordinary conditions, the voltage drop across R_1 is less than the turn-on voltage for the base-to-emitter junction of Q_2 (about 0.6 volts). This means that Q_2 is off and the circuit operates identically to the standard unprotected series regulator.

Now suppose the load current increases. This will cause an increased voltage drop across R_1. As soon as the R_1 voltage drop reaches the threshold of Q_2's base junction, transistor Q_2 will start to conduct. The conduction of Q_2 essentially bypasses the emitter-base junction of Q_1 which prevents any further increase in current flow through Q_1. We can also better understand the operation of Q_2 if we view it in terms of voltage drops.

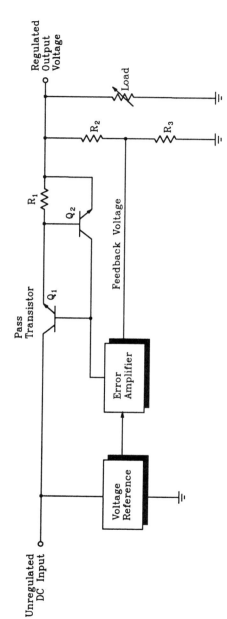

Figure 6.19 A constant-current limiting technique is often used to protect series regulator circuits from over-current conditions.

Sec. 6.5   Over-Current Protection

At the instant Q_2 begins to turn on, there must be approximately 0.6 volts across R_1 and another 0.6 to 0.7 volts across the emitter-base junction of Q_1. Kirchoff's Voltage Law can be used to show us that there must, therefore, be about 1.2 to 1.3 volts between the emitter and collector of Q_2 when it starts to conduct since the emitter-collector circuit of Q_2 is in parallel with the voltage drops of R_1 and the emitter-base circuit of Q_1. Any further attempt to increase current beyond this point, will cause a decrease in the emitter-collector voltage of Q_2. Since this voltage is in parallel with the series combination of Q_1's base-emitter junction and R_1, these voltages tend to decrease, also. However, if the base-emitter voltage of Q_1 actually decreased, then the emitter current of Q_1 would decrease causing the voltage drop across R_1 to decrease resulting in *less* conduction in Q_2 (the opposite of what is really occurring). So, in essence, the current reaches a certain maximum limit and is then forced to remain constant. Any effort to increase the current beyond this point merely lowers the output voltage.

The value of current required to activate Q_2 is determined with Ohm's Law. We simply find the amount of current through R_1 that it takes to get a 0.6 volt drop. That is, short-circuit current (I_{SC}) is computed as follows:

$$\boxed{I_{SC} = \frac{0.6}{R_1}} \qquad (6.18)$$

6.5.3 Foldback Current Limiting

Figure 6.20 shows a simplifed schematic diagram of a voltage regulator circuit which uses foldback current limiting. Note that resistors R_4 and R_5 have been added to the constant current limiting circuit presented in Fig. 6.19. Under normal conditions, transistor Q_2 is off and the circuit works like the unprotected regulator circuit discussed in an earlier section. Voltage divider action causes a voltage drop across R_4 with the upper end being the most positive.

As load current increases, the voltage drop across R_1 increases as it did in the constant current circuit. However, the voltage across R_1 must not only exceed the turn-on voltage of the base-emitter junction of Q_2 in order to turn Q_2 on, it must also overcome the voltage across R_4. Once this point occurs, however, Q_2 begins to conduct and reduces the conduction of Q_1. This, of course, causes both the output voltage and the base voltage of Q_2 to decrease. However, since the base voltage of Q_2 is obtained through a voltage divider, it decreases more slowly than the output voltage. Since the emitter of Q_2 is connected to the output voltage, it must also be decreasing faster than the base voltage. This causes Q_2 to conduct even harder further limiting the output current.

If the load current increases past a certain threshold, the circuit will "foldback" the output current. That is, even if the output is shorted directly to ground, the current will be limited to a value which is **less** than the maximum normal operating current. This reduction of output current under overload conditions is a very desirable characteristic. Since the pass transistor will have the full input voltage across it when the output is shorted to ground, it is prone to high power dissipation. In fact, the constant current

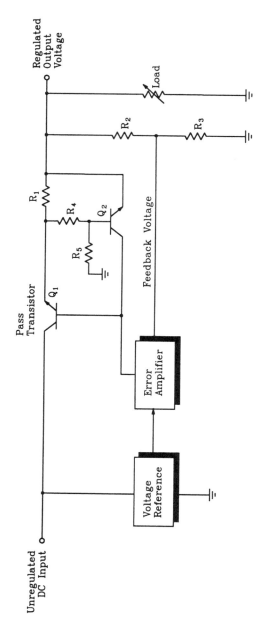

Figure 6.20 Foldback current limiting actually decreases the output current under overload conditions.

6.6 OVER-VOLTAGE PROTECTION

Some applications require a regulator circuit with over-voltage protection to protect the load against regulator malfunctions. That is, under normal operating conditions, the output of the regulator should stay at the regulated level. However, if the regulator were to fail (e.g., emitter-to-collector short in the series pass transistor), then the output may increase significantly over the regulated value and potentially cause damage to the load circuitry.

Figure 6.21 shows a common method of over-voltage protection which is built around a silicon-controlled rectifier (SCR). Under ordinary conditions, the voltage drop across R_4 is less than the base-emitter turn-on voltage of Q_2, and the circuit works identical to the unprotected circuit discussed earlier.

If a regulator malfunction causes the output voltage to rise above a threshold set by the ratio of R_4 and R_5, then Q_2 turns on and provides gate current for the SCR which causes it to fire. When an SCR has fired and is in the forward conducting state, the voltage drop across it is about 1 volt. Thus the base voltage of Q_1 is quickly dropped to about 1 volt. Since Q_1 is essentially an emitter follower, the output voltage will be dropped to a few tenths of a volt. This condition will continue as long as the SCR remains in conduction. To reset the SCR, the anode current must fall below a minimum value called holding current. In the circuit shown in Fig. 6.21, the main power source must be momentarily turned off to reset the SCR.

Capacitor C_1 is a transient suppressor and prevents accidental firing of the SCR during initial turn on of the regulator or as a result of a noise pulse.

Some power supply designs return the anode of the SCR directly to the unregulated DC input **with no limiting resistance.** If an over-voltage condition occurs and the SCR fires, the main supply is essentially shorted to ground via the SCR. This activates the current-limiting features of the main supply (often a fuse in the primary of the supply transformer). When the SCR is connected in this way, the circuit is called a "crowbar" circuit since it essentially throws a short circuit (like a steel crowbar) directly across the power supply.

6.7 POWER-FAIL SENSING

An op amp can be configured as a voltage comparator circuit and used to sense an *impending* power failure. This is commonly used in computer systems to protect the computer from erratic operation caused by power loss. If an impending power failure is detected, the computer quickly transfers all of the critical data to a permanent storage area that does not require power. Once power has been restored to the computer, it retrieves the stored data from the permanent memory and resumes normal operation.

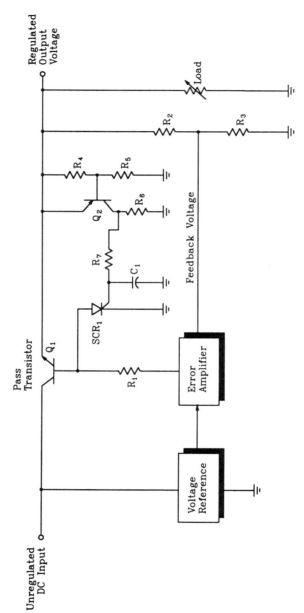

Figure 6.21 A series regulator circuit with over-voltage protection.

Sec. 6.8 Troubleshooting Tips for Power Supply Circuits

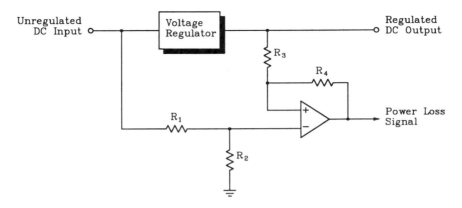

Figure 6.22 An op amp voltage comparator can be used to detect an impending power loss.

Figure 6.22 shows how an op amp can be used to detect an impending power failure and send a signal to a computer in time to save the critical data before the power actually goes away.

Under normal conditions, the inverting ($-$) input of the voltage comparator is more positive than the noninverting input. This condition is true even under conditions of minimum unregulated voltage. If a primary power loss occurs, the unregulated DC voltage will, of course, drop to zero. However, the filter capacitors (usually quite large) in the power supply will prevent the unregulated DC supply from decaying instantaneously. The regulator will continue to supply a constant voltage until the unregulated input voltage has decayed past a certain minimum point. Thus, up to a point, the voltage on the ($-$) input to the comparator will be decaying while the voltage on the ($+$) input remains constant. When the voltage on the ($-$) input passes the lower threshold of the voltage comparator, the output quickly changes states signaling a coming power loss. A computer monitoring this signal, can then take the necessary action to protect critical data. Resistors R_3 and R_4 provide hysteresis for the comparator.

The amount of time between primary power interruption and the point where the regulated output begins to drop is called *hold-up time* and is generally tens or even hundreds of milliseconds. Since a computer executes in the microsecond range, there is plenty of time to save the critical information **after** the unregulated input has started to decay but **before** the regulated output begins to drop.

6.8 TROUBLESHOOTING TIPS FOR POWER SUPPLY CIRCUITS

Power supply circuits are considered by some technicians and engineers to be simple, fundamental, nonglamorous, and even boring. However fundamental the *purpose* of power supplies may be, the troubleshooting of a defective supply is not always a simple task. What complicates the troubleshooting of a regulator circuit is the closed-loop nature of the system. A defect in any part of the loop can upset the voltages at all other points in the loop thus making it difficult to distinguish between cause and effect.

Nevertheless, armed with a thorough understanding of circuit operation and guided by systematic troubleshooting procedures, a defective regulator circuit can be quickly and effectively diagnosed. The following sequential steps will provide the basis for a logical, systematic troubleshooting procedure applicable to voltage regulator circuits.

1. **Observe the symptoms.** Due to the potentially high-power levels available in a supply, visible signs of damage are common. DO NOT, however, simply replace a burned component and reapply power. In many cases, the burned component is the result of a malfunction elsewhere in the supply. Nevertheless, detecting the burned component will help you narrow the range of possibilities.

   Symptom observation also includes taking careful note of the output symptoms. Is the output voltage too high, too low, zero, unregulated? Did the user of the equipment say how the problem was caused (e.g., an accidental short on the output)?

2. **Verify that the input to the regulator is correct.** If it is not correct, the regulator may not be the cause. On the other hand, if the problem is *no* input and the unregulated supply shows signs of damage, then suspect a short in the regulator circuit. In these cases, it is often helpful to disconnect the regulator circuit and get the unregulated supply back to normal as a first step. A simple way to disconnect series-regulator circuits is to remove the pass transistor. This is a particularly simple task for socket mounted power transistors.

3. **Check for possible short circuits.** Once the unregulated input voltage is shown to be correct, we can concentrate on the regulator portion of the supply. If the regulator was disconnected during step two and you have reason to believe a short circuit exists in the regulator, then DO NOT reconnect the regulator and apply full power. If you do and a short does exist, the newly repaired unregulated source will be damaged again. A better approach is to connect the regulator to the unregulated supply via a current meter. Use a variable autotransformer to supply the AC power to the unregulated power supply. Slowly increase the AC input voltage while monitoring the current meter. If a short exists in the regulator, then the current meter will exceed normal values with a very low-input voltage. If this is the case, you must rely on your theory of operation and an ohmmeter as your major tool.

4. **Open the regulator loop.** If the full supply voltage can be applied safely, but the regulator still doesn't work properly, then you can add a voltmeter to your tool kit. Since the regulator is inherently a closed-loop system, it is often difficult to distinguish between cause and effect. If the loop can be easily broken (e.g., removing a wire from the wiper arm of a potentiometer, removing a socket-mounted transistor, etc.), then this can help isolate the problem. After the loop has been opened, you can inject your own "good" voltage at the open-loop point from an external DC supply. The system can then be diagnosed using the split-half method, signal tracing, and so on, like any other open-loop system.

Chap. 6   Review Questions   293

5. **Force the circuit to known extremes.** If it is impractical to open the loop of the regulator, then you can sometimes force a condition at one point in the loop and watch for a response in another area. Your understanding of the operation of the components between the forced point and the monitiored point can lead you to the problem. A good example of forcing a condition is to either short the emitter-base circuit of a transistor to force it to cutoff, or to short the emitter-collector circuit to simulate a saturated condition. Be sure to examine the circuit carefully before shorting these elements, but in most circuits neither of these shorts will cause damage (see item 6), but they do force the circuit to go to one of two extremes. The extreme change will be passed through the rest of the circuit if everything is normal. A defect, however, will not respond to the change and thus reveal its identity to an alert technician or engineer.

6. **Use special care with switching regulators.** Here the regulator transistor has been selected on the assumption it is switching from full on to full off, and therefore, dissipates minimum power. If any portion of the regulator circuit causes the switching to stop and the pass transistor is in the ON state, it will almost certainly be damaged. This has two important ramifications. First, if your diagnosis reveals that no switching signal is being applied to the switching transistor, then suspect a bad transistor *after you correct the switching signal problem*. Second, you should never intentionally stop the oscillation in a switching supply by shorting components as described in step 5.

7. **Substitute the load.** Another method that can be helpful in isolating some power supply defects is to remove the load. This eliminates the possibility that a malfunction in the system is causing the supply to appear defective. If the supply is shown to be defective, then substituting an equivalent resistance in place of the system circuitry can simplify troubleshooting of the actual power supply. Additionally, it removes the possibility of causing damage to the system circuitry if the output of the supply were to become excessively high while troubleshooting the problem. (For example, a test probe may slip and cause a momentary short circuit.)

REVIEW QUESTIONS

1. List the three basic classes of voltage regulator circuits.
2. If the DC output voltage of a shunt-regulator circuit varies between 11 and 12.5 volts as the input line voltage varies from 110 to 130 volts, what is the percent of line regulation for the regulator?
3. If series-regulator circuit provides 25 volts DC under no load conditions, but drops to 24.3 volts when a full load is applied, what is the percent of load regulation for the regulator circuit.
4. If each of the following rectifier/regulator circuits requires 1.2 amps of current from the 120 VAC input line, which one would probably deliver the highest current to a 12 VDC load: series, shunt, switching? Explain your choice.

5. Refer to Fig. 6.8. If the input voltage is 16.8 VDC and the regulated output voltage is 8.9 volts DC, what is the power dissipation of Q_1 with a 300-milliampere load connected?
6. Refer to Fig. 6.8. If the reference voltage had a defect which caused it to go to +6 volts, what is the effect on output voltage (increase, decrease, remains the same)? Explain your answer.
7. Refer to Fig. 6.8. If resistor R_2 increased in value, what relative effect would this have on output voltage?
8. Refer to Fig. 6.9. What is the effect on output voltage if resistor R_2' increases in value?
9. Refer to Fig. 6.12. What is the effect on the current flow through Q_1 if resistor R_2 is decreased in value? Explain your answer.
10. If resistor R_3 in Fig. 6.12 were changed to 40 ohms (and no components were damaged), what would happen to the value of reference voltage? What would happen to the value of output voltage ($+V_{REG}$)? What would happen to the value of current through Q_1? Would the current through R_1 change?
11. Refer to Fig. 6.19. What is the effect on the current flow through the load resistor (assume a constant value of load resistance) if resistor R_2 is increased in value?
12. Refer to Fig. 6.19. What happens to the value of voltage dropped across the emitter-collector circuit of Q_1 if resistor R_3 is increased in value?
13. Refer to Fig. 6.20. What is the effect on circuit operation if transistor Q_2 develops an emitter-to-collector short?
14. Refer to Fig. 6.21. What is the effect on circuit operation if capacitor C_1 develops a short circuit?
15. Refer to Fig. 6.21. What is the effect on circuit operation if resistor R_4 becomes open?

CHAPTER 7

SIGNAL PROCESSING CIRCUITS

Most of the circuits presented in this chapter serve to condition analog signals for subsequent input to another circuit. Many of the circuits in the chapter could be categorized as waveshaping or conditioning circuits. We will, for example, examine circuits that can rectify low amplitude signals, limit the maximum excursion of signals, and change the DC level of waveforms. Many of the circuits are quite simple in terms of component count, but they play important roles in overall systems design.

7.1 CONCEPT OF THE IDEAL DIODE

Several of these circuits behave as though they had perfect or ideal diodes. Figure 7.1 contrasts the forward-biased characteristics of a perfect diode with that of a typical silicon diode.

Figure 7.1(a) shows a simple series circuit driven by an AC source. The output is taken across the diode. Basic circuit theory tells us that when the diode is reverse biased, it acts as a very high impedance (i.e., essentially an open circuit). In the case of the circuit in Fig. 7.1(a), we can expect to see nearly the full input voltage across the diode during times of reverse bias.

When the diode in Fig. 7.1(a) is forward biased, we would expect it to act as a short circuit (or at least a very low impedance). In this case, there would be very little voltage across it. Figure 7.1(b) shows the output waveforms that we could expect. We see that the ideal diode has **no** voltage across it when it is forward biased. This is in contrast with the silicon-diode waveform which has a 0.7 volt drop during forward-biased times. For purposes of this chapter, we shall limit our comparisons to the forward-biased performance of the diodes. Therefore, the waveforms in Fig. 7.1(b) show similar waveforms for both ideal and silicon diodes during reverse bias.

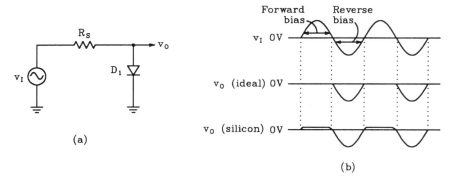

Figure 7.1 Comparison of silicon and ideal diodes.

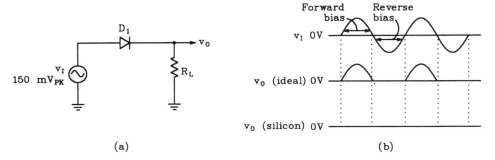

Figure 7.2 The forward voltage drop of a silicon diode can prevent rectification of small signals.

With regard to effect on circuit operation, is the 0.7 volt drop across the forward-biased silicon diode important? Does it adversely affect the circuit's performance? Well, many times we ignore the 0.7 volt drop when analyzing or even designing circuits and get along quite well. But, consider the circuit shown in Fig. 7.2 very closely. What is the peak voltage of the output voltage (v_O)? Contrast the output voltage for both ideal and silicon diodes.

The ideal diode, of course, produces a peak output which is equal to the input peak. The silicon diode, on the other hand, will drop 0.7 volts when it conducts leaving us with a maximum output voltage of $V_{PK} - 0.7$. Is that a problem? No, as long as the input signal is fairly large. However, what if the peak input signal is only 150 millivolts? Can you see, that the ideal diode will still produce the expected output waveform? The silicon diode, however, will have *no* output since the input never goes high enough to forward bias the junction. Thus, the silicon diode acts as a high impedance throughout the input cycle. In this case, that 0.7 volt difference between an ideal and a real diode makes the difference of whether or not the circuit will even work.

Figure 7.3 shows how an ideal half-wave rectifier can be made by placing silicon diodes in the feedback loop of an op amp. During the positive half cycle of the input

Sec. 7.2 Ideal Rectifier Circuits

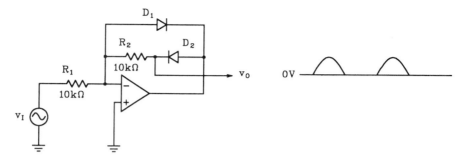

Figure 7.3 A half-wave rectifier that simulates an ideal diode.

waveform, diode D_1 is forward biased making the feedback resistance very low. Thus, the gain of the inverting amplifier is computed as

$$A_V = -\frac{R_F}{R_{IN}} \approx \frac{0}{10 \times 10^3} = 0$$

Additionally, since diode D_2 is reverse biased, no current will flow through R_2 so the output signal (v_O) will be zero.

On the negative half cycle, diode D_2 is forward biased completing the feedback loop through R_2. Since the current through R_2 is identical to the current through R_1 (ignoring bias currents), the voltage drop across R_2 will be identical to the input voltage. The left end of R_2 is connected to a virtual ground while the right end provides the inverted rectified output signal.

It is important to see that the 0.7 volt drop across the diodes has no effect on the output signal. Even if the input were only a few tenths of a volt, the circuit would still produce a full amplitude output signal. This same principle is applied to several of the circuits that follow. By including the diode in the feedback loop of the op amp, we make the *effects* of its nonideal forward voltage drop disappear.

7.2 IDEAL RECTIFIER CIRCUITS

Both half- and full-wave ideal rectifier circuits can be made with standard silicon diodes and an op amp. Figure 7.4 shows a dual half-wave rectifier. It is similar to the circuit presented in Fig. 7.3 with the addition of R_3 and R_4. The two outputs, one positive and one negative, can be used independently or combined in another op amp to produce a full-wave, precision-rectified signal.

7.2.1 Operation

Since the op amp in Fig. 7.4 is operated with closed-loop negative feedback, the inverting ($-$) input pin is a virtual ground point. Thus, for practical purposes, one end of resistors R_1, R_2, and R_3 are connected to ground. This means that the total input voltage will be

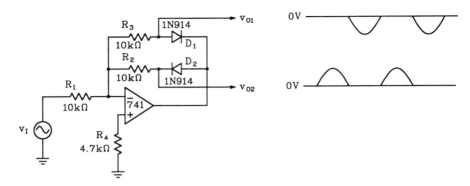

Figure 7.4 A dual, ideal diode, half-wave rectifier circuit delivering both positive and negative outputs.

developed across R_1 causing an input current to flow through R_1. Since negligible current will flow in or out of the ($-$) terminal, all of the input current continues through either R_2 or R_3.

On positive input alternations, electron current will leave the output terminal of the op amp, flow through D_1, R_3, R_1, and out to the positive source. Since resistors R_1 and R_3 are equal values, they will develop equal voltages. Thus, we can expect the voltage across R_3 to be the same as the input voltage. Since one end of R_3 is grounded, the other end (v_{O_1}) provides a signal equal in amplitude to the positive alternation of the input signal. However, because of the direction of current flow through R_3, the polarity of the v_{O_1} is inverted from v_I and produces negative half-wave waveforms.

During the negative alternation of the input cycle, electron current leaves the source, flows through R_1, R_2, D_2, and into the output of the op amp. Here again, since resistors R_1 and R_2 have the same current and are equal in value, they will have equal voltage drops. Thus, we will expect to see a signal at v_{O_2} that is the same amplitude as the input. Additionally, as a result of the direction of the current flow, the v_{O_2} output will provide a positive half-wave signal.

7.2.2 Numerical Analysis

The numerical analysis of the dual, half-wave rectifier circuit shown in Fig. 7.4 is fairly straightforward. We will determine the following characteristics:

1. maximum unclipped output signals
2. voltage gain
3. maximum input without distortion
4. highest frequency of operation

Maximum output signal. Outputs v_{O_1} and v_{O_2} are one diode drop away from the output of the op amp when their respective diode is conducting. Therefore, the maximum

Sec. 7.2 Ideal Rectifier Circuits

amplitude that we can expect at outputs v_{O_1} and v_{O_2} are 0.7 volts less than the maximum output of the op amp. That is

$$\boxed{v_{O_1}(\text{max}) = -V_{SAT} + 0.7} \quad (7.1)$$

Similarly, the maximum amplitude for the v_{O_2} output is computed as

$$\boxed{v_{O_2} = +V_{SAT} - 0.7} \quad (7.2)$$

The manufacturer's data sheet provides us with worst case and typical values for the maximum output swing. The minimum output swing for loads of 2 kilohms or more is listed as ±10 volts with a typical swing listed as ±13 volts. The typical value for loads of greater than 10 kilohms is ±14 volts. For purposes of our current analysis, let us use ±13 volts as the maximum output swing with ±15 volts V_{CC}. The maximum amplitudes for v_{O_1} and v_{O_2}, then, are computed as

$$v_{O_1} = -13 + 0.7 = -12.3 \text{ V},$$

and

$$v_{O_2} = +13 - 0.7 = +12.3 \text{ V}$$

Voltage gain. The voltage gain on either half cycle is found using the basic gain equation for inverting amplifiers, Eq. (2.6). In our present case, both half cycles will have the same gain since resistors R_2 and R_3 are the same value. We shall compute the voltage gain of the amplifer using R_2 as the feedback resistor.

$$A_V = -\frac{R_F}{R_I} = -\frac{R_2}{R_1} = -\frac{10 \times 10^3}{10 \times 10^3} = -1$$

For most applications, resistors R_2 and R_3 would be equal, but it is not necessary if unequal gains are desired.

Maximum input without distortion. We compute the highest input signal that we can have without distorting the output by applying our basic amplifier gain equation, Eq. (2.1). As mention in the preceding section, both half cycles have the same gain in this particular circuit. Let's compute the maximum input by using the maximum positive output signal. That is

$$A_V = \frac{v_{OUT}}{v_{IN}},$$

or

$$v_{IN} = \frac{v_{OUT}}{A_V}$$

$$v_I = \frac{12.3}{-1}$$
$$= -12.3 \text{ V}$$

Highest frequency of operation. The amplifier, in this case, is basically amplifying sinewaves. In fact, the actual output terminal of the op amp will have a sinusoidal waveform present. Of course the circuit can be used with nonsinusoidal inputs as well. The bandwidth considerations of the amplifier are identical to those discussed in Chap. 2. The upper frequency limit will be established by the finite bandwidth of the op amp or by the limits imposed by the op amp slew rate. Since the amplifier is configured for unity voltage gain, we can expect the bandwidth limitation to be similar to the unity-gain bandwidth of the op amp. The manufacturer's data sheet lists this limit as 1.0 megahertz.

The slew rate will likely impose a lower limit unless we apply only very low-amplitude signals. Assuming that we intend to apply maximum amplitude signals at some time, the highest frequency that can be amplified without slew-rate distortion is computed using Eq. (2.11).

$$f_{SRL} = \frac{\text{slew rate}}{\pi \, v_O(\max)}$$
$$= \frac{0.5 \text{ V}/\mu s}{3.14 \times 26}$$
$$= 6.12 \text{ kHz}$$

where $v_O(\max)$ is the maximum swing on the output of the op amp, and the 0.5 volts per microsecond slew rate is given by the manufacturer for the 741 op amp. So, if we expect to rectify maximum amplitude signals, we will be limited to 6.12 kilohertz or less. On the other hand, if the application uses only lower amplitude signals, then a wider bandwidth can be expected.

7.2.3 Practical Design Techniques

Let us now design a dual, half-wave rectifier circuit that satisfies the following design goals:

1. peak input voltage ± 250 millivolts to ± 5.0 volts
2. peak output voltages ± 375 millivolts to ± 7.5 volts
3. minimum input impedance   3000 ohms
4. highest input frequency   25 kilohertz

Sec. 7.2 Ideal Rectifier Circuits   301

Compute the required voltage gain. The design goals give no indication that we are to design for unequal gains on the two different half cycles. The required amplifier gain, then, is computed by applying the basic gain equation, Eq. (2.1).

$$A_V = \frac{v_O}{v_I} = \frac{375 \times 10^{-3}}{250 \times 10^{-3}} = 1.5$$

Of course, we could also have used the higher values of 5 volts and 7.5 volts for input and output voltage, respectively. This calculation gives us the absolute value of required voltage gain. By virtue of the circuit configuration, we know it will be an inverting gain (i.e., $A_V = -1.5$).

Select R_1. You will recall from our discussions of amplifiers in Chap. 2 that the input impedance of an inverting amplifier is determined by the value of input resistor. In our present case, resistor R_1 must be large enough to satisfy the minimum input impedance requirement. Additionally, there is little reason to go beyond a few hundred kilohms. For the present design example, let us select R_1 as 20 kilohms.

Compute R_2 and R_3. In our current design, resistors R_2 and R_3 will be equal since the gains for the two half cycles must be equal. We compute the value of R_2 (or R_3) by applying a transposed version of the inverting amplifier gain equation, Eq. (2.6).

$$A_V = -\frac{R_2}{R_1},$$

or

$$\begin{aligned} R_2 &= -A_V \times R_1 \\ &= -(-1.5) \times 20 \times 10^3 \\ &= 30 \text{ k}\Omega \end{aligned}$$

If this were not a standard value, then we would have to choose a close value or some combination of resistors to equal the required value.

Select D_1 and D_2. These diodes are fairly noncritical and can be one of many different diode types. There are two primary factors to consider when selecting these diodes

1. average forward current
2. peak inverse voltage

The highest reverse voltage that will be applied to either of the diodes is $\pm V_{SAT}$ (depending on the diode being considered). In most cases (including the present case) the two saturation voltages are equal. Therefore, we will need to select a diode with a peak inverse voltage rating greater than

$$V_{PIV}(\text{rating}) = V_{SAT} \qquad (7.3)$$

Of course, in our particular case, this means we will need diodes with reverse breakdown ratings of over 13 volts. This should be a simple task.

The average rectified current that flows through a particular diode is computed with Eq. (7.4).

$$I_{AVG} = 0.318 \frac{v_I(\text{peak})}{R_1} \qquad (7.4)$$

For our present application, the average forward current is computed as

$$I_{AVG} = 0.318 \frac{5}{30 \times 10^3} = 53 \ \mu A$$

Again, this rating is so low that most any diode should be capable of handling this current. For purposes of our present design, let us use the common 1N914A diodes due to cost and availability considerations.

Compute R_4. Resistor R_4 helps to minimize the output offset voltage that is caused by the op amp bias currents which flow through R_1, R_2, and R_3. Its value should be equal to the parallel combination of R_1 and either R_2 or R_3. In the present case, we compute R_4 as

$$R_4 = R_1 \| R_2 = \frac{1}{\frac{1}{20 \times 10^3} + \frac{1}{30 \times 10^3}} = 12 \ k\Omega$$

Select the op amp. Two of the more important characteristics that must be considered when selecting the op amp are

1. unity-gain bandwidth
2. slew rate

The minimum unity-gain bandwidth requirement can be estimated with Eq. (2.16) by substituting our actual closed-loop gain for A_{OL}. That is

$$A_{CL} = \frac{f_{UG}}{f_{IN}},$$

Sec. 7.2 Ideal Rectifier Circuits

or

$$f_{UG} = A_{CL} f_{IN}$$
$$= 1.5 \times 25 \text{ kHz}$$
$$= 37.5 \text{ kHz}$$

The minimum slew rate for the op amp is computed by applying Eq. (2.11). The required slew rate for our present design is computed as

$$f_{SRL} = \frac{\text{slew rate}}{\pi \, v_O(\text{max})},$$

or

$$\text{slew rate} = \pi f_{SRL} v_O(\text{max})$$
$$= 3.14 \times 25 \times 10^3 \times 15$$
$$= 1.18 \text{ V}/\mu\text{s}$$

The bandwidth requirement is easy to satisfy with nearly any op amp. The 1.18 volts per microsecond slew rate, however, exceeds the capabilities of the standard 741 op amp. The MC1741SC op amp, however, will work well for this application. Let us choose to use this device.

Figure 7.5 shows the final schematic of our dual, half-wave rectifier circuit. Its performance is indicated by the oscilloscope displays in Fig. 7.6. Figures 7.6(a) and 7.6(b) show the positive and negative outputs, respectively, with a 250-millivolt$_{PK}$ input signal at a frequency of 1.0 kilohertz. Notice the effect of the 0.7 volt forward voltage drop of diodes D_1 and D_2 is nonexistent. Figures 7.6(c) and 7.6(d) show the circuit's response to a 25-kilohertz signal with maximum input voltage. Finally, the actual performance of the circuit is contrasted with the original design goals in Table 7.1.

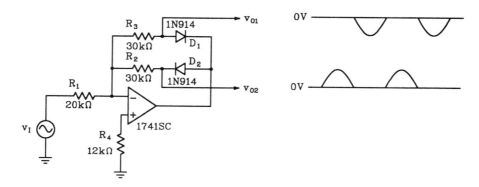

Figure 7.5 Final configuration of a dual, half-wave rectifier circuit designed for 25 kilohertz operation.

304   Signal Processing Circuits   Chap. 7

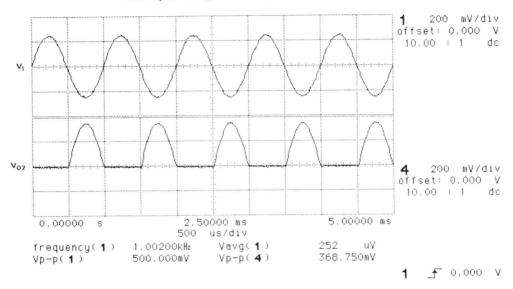

(a)

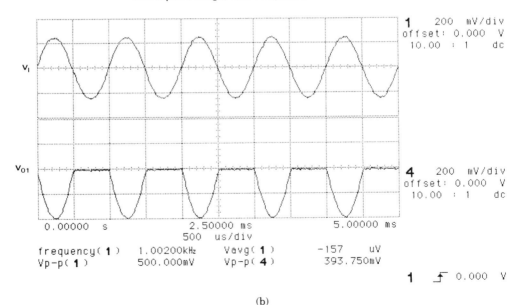

(b)

Figure 7.6   Oscilloscope displays showing the performance of the circuit shown in Fig. 7.5. (Test equipment courtesy of Hewlett-Packard Company)   (continued)

Sec. 7.2 Ideal Rectifier Circuits

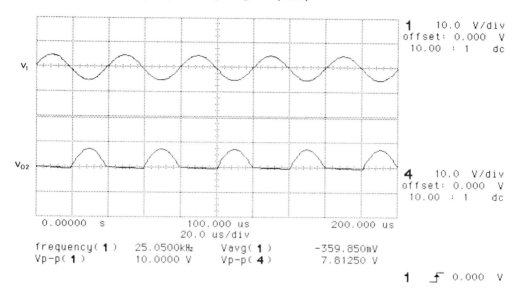

(c)

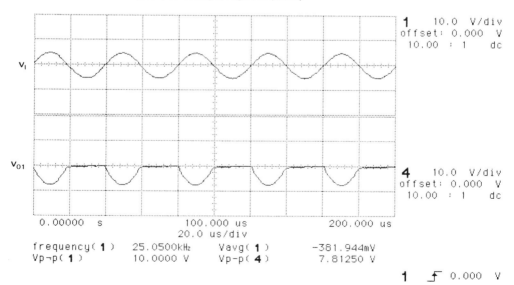

(d)

Figure 7.6c and d

TABLE 7.1

	DESIGN GOAL	MEASURED VALUE
Input voltage	250 mV–5 V_{PK}	250 mV–5 V_{PK}
Output voltage	375 mV–7.5 V_{PK}	368 mV–7.8 V_{PK}
Input impedance	≥ 3.0 kΩ	20 kΩ
Input frequency	0–25 kHz	0–25 kHz

7.3 IDEAL, BIASED CLIPPER

You may recall from basic electronics theory that a biased clipper or limiter circuit has no effect on the input signal as long as it is less than the clipping or reference voltage. Under these conditions, the input and output waveforms are identical. If, however, the input voltage exceeds the clipping level of the circuit, then the output waveform is clipped or limited at the clipping level. Figure 7.7 shows a basic, biased, shunt clipper, and its associated waveforms. We can make both series and shunt, and both biased and unbiased clippers with an op amp just as we can with simple diode circuits. The difference, however, is that the op amp version eliminates the effect of the diode's forward voltage drop (0.7 volts). This is a very important consideration when processing low-amplitude signals.

7.3.1 Operation

Figure 7.8 shows the schematic diagram of an op amp version of the biased, shunt-clipper circuit. The basic purpose is similar to the simple diode clipper shown in Fig. 7.7, but since the effects of the diode's forward voltage drop has been eliminated, it performs like an ideal clipper circuit. Potentiometer P_1 is used to establish the reference voltage or clipping level on the (+) input pin. Capacitor C_1 prevents fluctuations in the clipping level.

Let us first consider the operation of the circuit for input voltages that are less than the value of the reference voltage. Under these conditions, the inverting (−) input is always less positive than the noninverting (+) input of the op amp. Therefore, the output of the op amp will remain at a positive level. The positive voltage on the output of the op amp reverse biases diode D_1 and essentially opens the feedback loop. This allows the output of the op amp to go to +V_{SAT} and remain there. With diode D reverse biased (essentially open), the op amp circuit has no effect on the output signal. Figure 7.9 shows the equivalent circuit under these conditions. We would expect v_O to be identical to v_I as long as we avoid excessive loading. The equivalent circuit shown in Fig. 7.9 does reveal a disadvantage of the shunt-clipper circuit; it has a fairly high output impedance. This means that it will either load easily or we will have to follow it with a buffer amplifier.

Sec. 7.3 Ideal, Biased Clipper

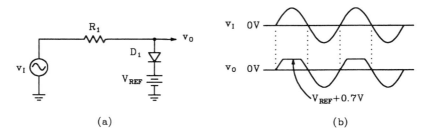

Figure 7.7 A simple, biased, shunt clipper circuit. The output signal cannot go above $V_{REF} + 0.7$ volts.

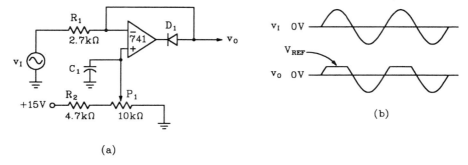

Figure 7.8 An ideal, biased, shunt clipper circuit. The forward voltage drop of D_1 has no effect on circuit operation.

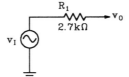

Figure 7.9 The equivalent circuit for the clipper circuit shown in Fig. 7.8 for input signals that are less than $+V_{REF}$.

Now let us consider circuit operation (Fig. 7.8) for input signals that exceed the reference voltage. Under these conditions, the output of the op amp will start to move in the negative (i.e., less positive) direction. This causes diode D_1 to become forward biased and closes the feedback loop. With a closed loop, we know that the voltage on the $(-)$ input will be equal to the voltage on the $(+)$ input (i.e., $+V_{REF}$). Since the output is taken from this same point, the output will, therefore, be equal to $+V_{REF}$ under these conditions. That is, as long as the input voltage is more positive than $+V_{REF}$, diode D_1 will remain forward biased, and the voltage on the $(-)$ input and at the output of the circuit will remain at the $+V_{ref}$ level.

If diode D_1 is reversed, then the circuit will clip the negative excursion of the input signal. Similarly, if the polarity of the reference voltage is reversed, the clipping will occur below zero.

7.3.2 Numerical Analysis

Let us now extend our understanding of circuit operation to include the numerical analysis of the circuit shown in Fig. 7.8. We shall compute the following characteristics:

1. minimum and maximum clipping levels
2. highest frequency of operation
3. maximum input voltage swing
4. input impedance
5. output impedance

Clipping levels. The clipping level in the circuit is determined by the voltage on the wiper arm of P_1. When the wiper arm is in the extreme right position, the reference voltage (and clipping level) will be zero volts since the wiper arm will be connected directly to ground.

The maximum clipping level will occur when the wiper arm is moved to the leftmost position. Under these conditions, the reference voltage is determined by applying the basic voltage divider formula. That is

$$+V_{REF} = V_{APP} \frac{P_1}{P_1 + R_2}$$

$$= +15 \frac{10 \times 10^3}{10 \times 10^3 + 4.7 \times 10^3}$$

$$= 10.2 \text{ V}$$

Highest frequency of operation. Since the feedback loop is essentially open-circuited for a majority of the input cycle, the highest frequency of operation is more dependent on slew rate than the bandwidth of the op amp. As the frequency increases, the output will begin to develop some degree of overshoot. That is, the output will rise beyond the clipping level momentarily then quickly drop to the desired level. This overshoot is caused when the output of the op amp switches slower than the input signal is rising. Thus, we continue to see the full input waveform at the output until the amplifier actually switches. The switching time, of course, is determined by the slew rate of the op amp. There is no precise maximum frequency of operation. The upper limit is determined by the degree of overshoot considered acceptable for a particular application. For purposes of our analysis and subsequent design, we shall consider a 1 percent overshoot to be acceptable. With this in mind, we can estimate the highest frequency of operation as follows:

$$\boxed{f_{MAX} = \frac{\text{slew rate}}{100(+V_{SAT} - V_{REF}(\min))}} \qquad (7.5)$$

Sec. 7.3   Ideal, Biased Clipper

For the circuit shown in Fig. 7.8, we estimate the highest frequency of operation (for 1% overshoot) as

$$F_{MAX} = \frac{0.5 \text{ V}/\mu\text{s}}{100(13 - 0)} = 385 \text{ Hz}$$

We can get higher operating frequencies by either accepting a higher overshoot or by selecting an op amp with a higher slew rate.

Maximum input signal.   During the time that D_1 (Fig. 7.8) is reverse biased, the full input signal is felt on the ($-$) input of the op amp. The manufacturer's data sheet indicates that the maximum voltage that should be applied to this input is equal to the supply voltage. So, in the case of the circuit in Fig. 7.8, the peak input voltage should be limited to 15 volts.

Input impedance.   The instantaneous input impedance will vary depending on whether D_1 is forward or reverse biased. During the time it is forward biased (worst-case input impedance), the input impedance is determined by the value of R_1 since the ($-$) input is a virtual AC ground point during this time because of the filtering action of C_1. When diode D_1 is reverse biased, the input impedance is the sum of R_1 and the input impedance of the following stage. Since the first value will always be lower, we shall compute minimum input impedance as

$$\boxed{Z_{IN}(\text{min}) = R_1} \qquad (7.6)$$

For the circuit shown in Fig. 7.8, the minimum input impedance is simply

$$Z_{IN}(\text{min}) = R_1 = 2.7 \text{ k}\Omega$$

Output impedance.   The output impedance also varies depending upon the conduction state of D_1. If diode D_1 is conducting, then the output impedance is nearly the same as the output impedance of the op amp itself which is a very low value. On the other hand, when D_1 is reverse biased, the output impedance is equal to the value of R_1. Since this latter value is always higher, we shall use it to estimate the output impedance.

$$\boxed{Z_{OUT}(\text{max}) = R_1} \qquad (7.7)$$

By using this value in the design of subsequent stages, we are assured that the signal will couple faithfully between stages on both alternations. In the case of the circuit shown in Fig. 7.8, the maximum output impedance is simply

$$Z_{OUT}(\text{max}) = R_1 = 2.7 \text{ k}\Omega$$

7.3.3 Practical Design Techniques

We are now ready to design an ideal, biased shunt-clipper circuit similar to the one shown in Fig. 7.8. We shall design to achieve the following performance goals:

1. clipping levels +3 to −3 volts
2. polarity of clipping negative peaks clipped
3. input frequency 100 hertz to 3 kilohertz
4. minimum input impedance 8 kilohm

Compute the reference voltage divider. Since the design requires bipolar (±3 volts) clipping levels, we choose to use a voltage divider like that shown in Fig. 7.10. The first step is to select a readily available potentiometer. Its value is not critical but choices between 1 and 50 kilohms would be typical. If the potentiometer is too small, the power consumption is unnecessarily high. If the value is too high, then the nonideal op amp characteristics become more noticeable. Let us choose to use a 10 kilohm potentiometer for P_1.

The values of R_2 and R_3 are computed with Ohm's Law. By inspection, we can see that these resistors have 12 (i.e., 15−3) volts across them. Additionally, they will have the same current flow as P_1. This current is also computed with Ohm's Law by dividing the voltage across P_1 (6 volts) by the value of P_1 (10 k). Combining all of this into equation form gives us

$$R_2 = \frac{P_1(V^+ - V_{CU})}{V_{CU} - V_{CL}} \qquad (7.8)$$

where V^+ is the +15 source, V_{CU} is the upper clipping level and V_{CL} is the lower clipping level. Similarly, the equation for R_3 is

$$R_3 = \frac{P_1(V^- - V_{CL})}{V_{CL} - V_{CU}} \qquad (7.9)$$

Applying these equations to our present design gives us the following results:

$$R_2 = \frac{10 \times 10^3 (15 - 3)}{3 - (-3)} = 20 \text{ k}\Omega,$$

and

$$R_3 = \frac{10 \times 10^3 [-15 - (-3)]}{-3 - 3} = 20 \text{ k}\Omega$$

Select R_1. The minimum value for R_1 is determined by the minimum required input impedance (8 kΩ in the present case). The upper limit depends on the op amp selected, desired output impedance, and required immunity to nonideal characteristics. It would be unusual to choose R_1 as anything greater than a few tens of kilohms since the resulting high-output impedance of the circuit would make it difficult to interface. Let us select R_1 as a 9.1 kilohm standard value.

Select the op amp. The op amp can be selected by assuming that the amplifier is passing strictly sinusoidal waveforms. Chapter 2 discusses this in more detail, but the

Sec. 7.3 Ideal, Biased Clipper

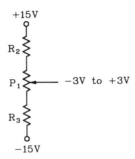

Figure 7.10 A voltage divider is used to provide the variable reference voltage.

primary consideration is op amp slew rate. The required slew rate can be estimated by applying Eq. (7.5). That is

$$\text{slew rate} = 100 f_{MAX} [+V_{SAT} - V_{REF}(\min)]$$
$$= 100 \times 3 \times 10^3 (+13 - -3)$$
$$= 4.8 \text{ V}/\mu s$$

This, of course, exceeds the 0.5 volts per microsecond rating of the standard 741. It does, however, fall within the 10 volts per microsecond slew-rate limit of the MC1741SC. Let us design around this device.

Select D_1. Diode D_1 must have a reverse breakdown voltage that is twice the value of the supply voltage. In the present case, it must withstand 30 volts. The current rating for D_1 is more difficult to determine since it is partially determined by the input impedance of the circuit being driven by the clipper. Since we are dealing with low currents, and since all of the diode current must be supplied by the op amp, it is reasonable to select a diode with a current rating that is greater than the short-circuit op amp current. The manufacturer's data sheet lists the maximum short-circuit current for the MC1741SC as ± 35 milliamps.

There are many diodes that will perform well in this application. Let us select the common 1N914A diode. This should be adequate under any probable circuit conditions.

Compute C_1. Capacitor C_1 helps to insure that the reference voltage on the (+) input of the op amp remains constant. Its value is not critical, and may even be omitted in many applications. A reasonable value can be computed by insuring that the reactance of C_1 is less than 10 percent of the resistance of the smaller of R_2 or R_3 at the lowest input frequency. In our present case, these resistors are both 20 kilohms. Therefore, we compute the value of C_1 by applying the basic equation for capacitive reactance.

$$X_c = \frac{1}{2\pi f C},$$

or

$$C = \frac{1}{2\pi f X_c}$$

$$= \frac{1}{6.28 \times 100 \times 2000}$$

$$= 0.796 \ \mu F$$

We shall select a standard value of 1.0 microfarad for our circuit.

The completed schematic of our shunt-clipper design is shown in Fig. 7.11. The actual performance of the circuit is indicated by the oscilloscope displays in Fig. 7.12. The actual circuit performance is compared with the original design goals in Table 7.2.

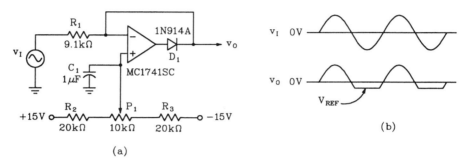

Figure 7.11 A biased, shunt clipper designed for variable clipping levels.

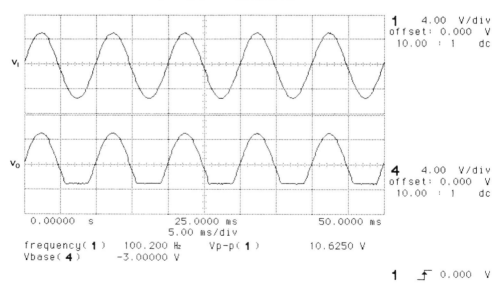

Figure 7.12 Oscilloscope displays showing the actual performance of the shunt clipper shown in Fig. 7.11. (Test equipment courtesy of Hewlett-Packard Company)   (continued)

Sec. 7.3 Ideal, Biased Clipper

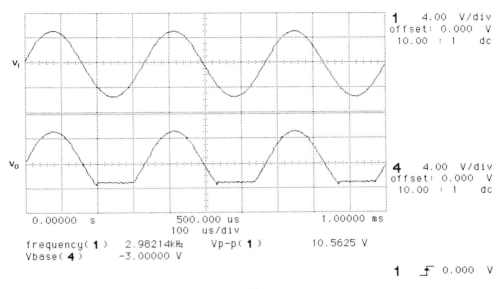

(b)

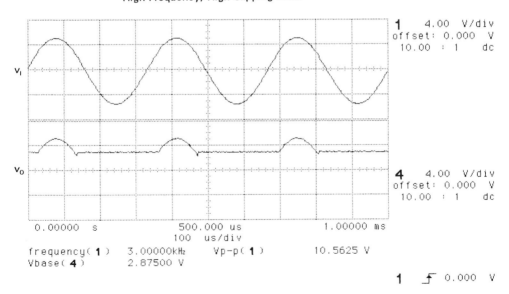

(c)

Figure 7.12b and c

(continued)

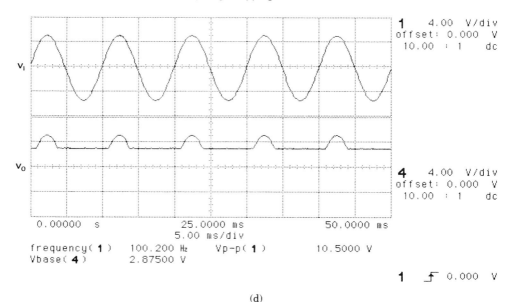

(d)

Figure 7.12d

TABLE 7.2

	DESIGN GOAL	MEASURED VALUE
Clipping levels	−3V, +3V	−3V, +2.875V
Input frequency	100 Hz–3 kHz	100 Hz–3 kHz
Input impedance	≥8 kΩ	≥9.1 kΩ

7.4 IDEAL CLAMPER

Figure 7.13 shows a basic diode-clamper circuit. Its purpose is to shift the average or DC level of the input signal without altering the waveshape. Alternatively, an application may require that the peaks of the signal be shifted to some new reference level. In either case, this is accomplished by clamping the peaks of the signal to a fixed reference level. Either positive or negative peaks may be clamped, to either a positive or negative reference level.

The operation of the simple clamper shown in Fig. 7.13 is best understood by starting when the input signal is at its negative peak (−10 V). At this instant, the 5-volt reference and the 10-volt source are series aiding. Since the only resistance in the circuit

Sec. 7.4   Ideal Clamper

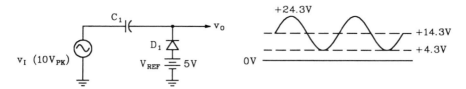

Figure 7.13 A simple, biased clamper circuit is used to shift the DC or reference level of the input signal.

is the forward resistance of the diode, capacitor C_1 quickly charges to 14.3 volts (negative on the left). That is, it charges to the combined voltage of the series-aiding source and reference voltage minus the forward voltage drop across D_1.

During all other portions of the input cycle, the diode will be reverse biased and effectively removed from the circuit. The output waveform then is being taken across an open circuit (reverse-biased diode) and will be identical to the "applied" signal. But what is the applied signal? It is now v_1 in series with the charge on C_1. Since C_1 has no discharge path, it acts like a battery. Therefore, the output waveform will be the same as the input waveform, but will have a positive 14.3 volts added to it. As the input signal passes through its zero voltage point, the output signal will be +14.3 volts. When the input is at +10 volts, the output will be at a +24.3-volt level. And when the input returns to the negative peak (−10 volts), the output will be at +4.3 volts. Any charge that has been lost on the capacitor due to leakage will be replaced when the input passes through its negative peak. As indicated by the resulting output waveform in Fig. 7.13, the output signal appears to have its negative peaks clamped to the reference voltage (less a diode drop). Also, notice that the DC level of the output is now at +14.3 volts DC. For this reason, the circuit is sometimes called a DC restorer circuit.

If the diode is reversed, the positive peaks will be clamped to the reference voltage. The reference voltage itself can be either positive or negative (even variable).

7.4.1 Operation

The operation of the basic clamper shown in Fig. 7.13 relies on the charging of a capacitor through a diode. For low-amplitude signals, the diode drop (0.7 volts) becomes significant. In fact, the circuit cannot be used at all if the peak input signal is below 0.7 volts since the diode cannot be forward biased. The circuit shown in Fig. 7.14 is called an ideal, biased clamper since it performs as though the diode were ideal (i.e., no forward voltage drop). This means the circuit can be used to clamp signals in the millivolt range.

Let's start our examination of the circuit in Fig. 7.14 as the input reaches the negative peak. This will cause the output of the op amp to go in a positive direction and will forward bias D_1. The diode places a near short-circuit around the op amp and essentially converts the circuit into a voltage follower with reference to the (+) input. This means that the output of the op amp will be at approximately the same voltage as the reference voltage (actually 0.7 volts higher). The (−) input of the op amp will also be at the reference voltage level. Capacitor C_2 charges to the difference in potential

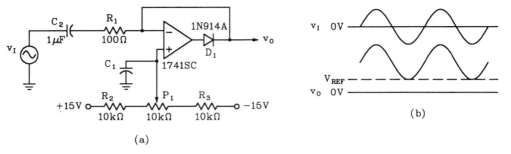

Figure 7.14 An ideal, biased clamper circuit.

between v_I and the reference voltage felt on the ($-$) input of the op amp. This action is similar to that described for the circuit in Fig. 7.13 but notice the absence of a diode drop. The rate of charge for C_2 is limited by the value of R_1.

For the remainder of the input cycle, the diode is reverse biased. This effectively disconnects the op amp from the circuit so the output will be the same as the input plus the voltage across C_2. Again, C_2 has no rapid discharge path so it will act as a DC source and provide the clamping action described previously.

In order for the clamper to be practical, it must drive into a very high-impedance circuit. For this reason, the clamper circuit shown in Fig. 7.14 is nearly always followed by a voltage follower circuit.

7.4.2 Numerical Analysis

For the purpose of numerically analyzing the behavior of the ideal, biased clamper circuit shown in Fig. 7.14, let us determine the following characteristics:

1. range of reference voltage adjustment,
2. maximum input voltage,
3. frequency range,
4. input impedance,
5. output impedance.

Range of reference voltage. Both upper and lower limits for the reference voltage (at the wiper arm of P_1) may be found by applying the basic voltage divider equation. The minimum voltage occurs when P_1 is moved toward the minus 15-volt source and is computed as

$$V_{REF}(\text{min}) = V^- + V_T \left(\frac{R_3}{R_1 + R_2 + P_1} \right)$$

Sec. 7.4   Ideal Clamper   317

$$= -15 + 30\left(\frac{10 \times 10^3}{(10 \times 10^3) + (10 \times 10^3) + (10 \times 10^3)}\right)$$

$$= -5 \text{ V}$$

The maximum reference voltage is found in a similar manner. That is

$$V_{REF}(\text{max}) = V^+ - V_T\left(\frac{R_2}{R_1 + R_2 + P_1}\right)$$

$$= +15 - 30\left(\frac{10 \times 10^3}{(10 \times 10^3) + (10 \times 10^3) + (10 \times 10^3)}\right)$$

$$= +5 \text{ V}$$

Maximum input voltage.   On the positive peak of the input signal, the ($-$) input of the op amp will be at a voltage that is higher than v_I(peak) by the amount of charge on C_2. The value of the voltage on the ($-$) input is equal to the peak-to-peak amplitude of the input signal plus the reference voltage.

The manufacturer's data sheet tells us that the maximum voltage on the ($-$) input pin of the op amp is equal to the supply voltage ($+15$ volts). Therefore, the maximum input signal is given as

$$\boxed{v_I(\text{peak-to-peak}) = V^+ - V_{REF}(\text{max})} \qquad (7.10)$$

If the diode were reversed, then V^- and V_{REF} (min) would be used to calculate the maximum input signal. In our particular circuit, the maximum input signal is computed as

$$v_I(\text{peak-to-peak}) = 15 - 5 = 10 \text{ volts peak-to-peak},$$

or

$$v_I(\text{RMS}) = \frac{10}{2 \times 1.414} = 3.54 \text{ volts (RMS)}$$

Frequency range.   During the majority of the input cycle capacitor C_2 is slowly leaking off charge. There are three paths for this discharge current

1. op amp bias current for the ($-$) input,
2. reverse-leakage current through D_1,
3. current through the input impedance of the following stage.

Since all of these currents are variable with voltage, temperature, frequency, and so on, the computation of a minimum operating frequency is not straightforward. Once the actual current values are known for a particular circuit under a certain set of conditions (presumably worst case), then the lower frequency limit will be determined by the amount of allowable discharge for C_2. That is, C_2 ideally maintains its full charge at all times unless the input signal changes amplitude. If C_2 is allowed to discharge excessively during a cycle, the circuit will begin to clip the negative peaks of the signal, and the output amplitude will begin to drop.

We can obtain a rough estimate of the lower cutoff frequency if we can estimate the effective discharge resistance (R_x) seen by C_2. This can be approximated by finding the parallel resistance of the ($-$) input, the diode's reverse resistance, and the input resistance of the following stage. All of these values are computed using DC parameters. For purposes of our present example, let us assume the output of the clamper is driving a standard 741 configured as a voltage follower. First, we compute the reverse resistance of the diode. The manufacturer's data sheet indicates that the diode will have about 25 nanoamperes of reverse current with 20 volts applied. From this we can estimate the reverse resistance as

$$R_{D_1} \approx \frac{20}{25 \times 10^{-9}} = 800 \text{ M}\Omega$$

The effective DC resistance of the ($-$) input can be estimated from the bias current data provided by the manufacturer. The data sheet indicates that the bias current will be about 300 nanoamperes (at 0 °C). We can assume a worst-case voltage equal to the maximum peak-to-peak input (10 volts in our case). Thus, Ohm's Law will allow us to estimate the effective resistance of the ($-$) input as

$$R_{(-)} \approx \frac{10}{300 \times 10^{-9}} = 33.3 \text{ M}\Omega$$

Finally, the approximate DC resistance of the ($+$) input of a standard 741 connected as a voltage follower (not shown in Fig. 7.14) can be computed. The data sheet indicates a maximum bias current of 800 nanoamperes. Again, we shall assume a voltage equal to the highest peak-to-peak input. Our estimate for the input resistance of the follower stage is then

$$R_{VF} \approx \frac{10}{1500 \times 10^{-9}} = 6.67 \text{ M}\Omega$$

The value of R_X is simply the parallel combination of these three estimated resistances. That is

$$R_X = R_{D_1} \| R_{(-)} \| R_{VF}$$
$$= \frac{1}{\frac{1}{800 \times 10^6} + \frac{1}{33.3 \times 10^6} + \frac{1}{6.67 \times 10^6}}$$
$$= 5.5 \text{ M}\Omega$$

Sec. 7.4   Ideal Clamper   319

Finally, the lower frequency limit of the clamper can be estimated with Eq. (7.11).

$$\boxed{f_L = \frac{16.7}{R_x C_2}} \quad (7.11)$$

This equation is derived by considering the tilt characteristics of a square-wave input. For the circuit being considered, the low-frequency limit is estimated as

$$f_L = \frac{16.7}{R_x C} = \frac{16.7}{5.5 \times 10^6 \times 1 \times 10^{-6}} = 3 \text{ Hz}$$

The high-frequency limit of the circuit shown in Fig. 7.14 is not well defined and is dependent on the application. To understand the effects of increasing input frequencies, consider that the input capacitor charges on the negative peak of the input waveform. This occurs when the output of the op amp goes positive and forward biases diode D_1. As frequency is increased, however, the output of the op amp begins to experience increasing delays. That is, the action on the output of the op amp occurs **after** the corresponding point on the input. This delay is partially caused by the internal phase shift of the op amp and partially by the effects of slew-rate limiting. In any case, the result is that C_2 is charged at some point after the negative peak. This means it won't be able to fully charge. This effect is evident on the output waveform as a reduction in the average DC level. The higher the input frequency, the worse this effect becomes. This effect is clearly illustrated in Fig. 7.15. The sinusoidal waveform is the input signal. The second waveform is the actual output pin of the op amp. The capacitor charges during the upper most portion of this latter waveform. Figure 7.15 (a) shows the circuit response to low-input frequencies (100 Hz). Notice that the capacitor charging time occurs at the negative peak of the input signal. Figure 7.15 (b) shows the same circuit with a higher input frequency (100 kHz). The charging point for the capacitor (positive peak on the output waveform) is now delayed and occurs **after** the negative peak of the input.

Input Impedance. The input impedance of the circuit varies with the input frequency and with the state of the circuit. As frequency increases, the reactance of C_2 decreases and lowers the input impedance. Additionally, during diode D_1's conduction times, the ($-$) input of the op amp is essentially an AC ground point. However, during the remainder of the cycle, this same point is at a high-impedance level.

For purposes of our present analysis, we shall consider the absolute minimum input impedance to be equal to the value of R_1.

Output impedance. The output impedance also varies with frequency and state of the circuit. During the charging time of C_2, the output impedance is quite low. During the remainder of the cycle, however, the impedance is determined by C_2 and R_1. In the present circuit, the maximum output impedance at the lower frequency limit (3 Hz) is approximately

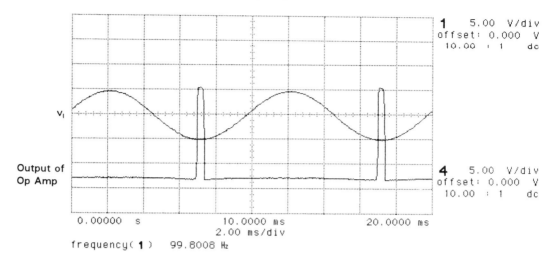

(a)

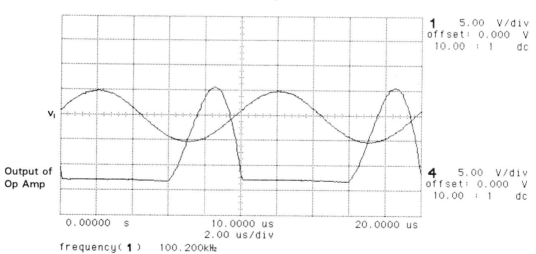

(b)

Figure 7.15 Oscilloscope displays showing the effects of internal delays and slew rate limiting on the operation of the biased clamper circuit. (Test equipment courtesy of Hewlett-Packard Company)

Sec. 7.4   Ideal Clamper   321

$$Z_O = \sqrt{X_{C2}^2 + R_1^2} = \sqrt{(53.08 \times 10^3)^2 + 100^2} \approx 53.08 \text{ k}\Omega$$

For a reliable operation, the circuit must drive into a very high impedance. This requirement is normally met by using a voltage follower buffer immediately after the clamper circuit. In any case, the input impedance of the following stage should be at *least* 10 times the output impedance of the clamper at the lowest input frequency.

7.4.3 Practical Design Techniques

Let us now design an ideal, biased clamper circuit that performs according to the following design goals:

1. minimum input impedance   1.1 kΩ kilohm
2. input voltage range   500 millivolt–2 volts (peak)
3. reference levels -1 through $+2$ volts
4. minimum input frequency   20 hertz

Computation and selection of all components other than C_2 is similar to the methods described previously for the biased, shunt clipper and will not be repeated here. The value for capacitor C_2 can be determined by applying Eq. (7.11).

$$F_L = \frac{16.7}{R_X C_2},$$

or

$$C_2 = \frac{16.7}{R_x F_L}$$

$$= \frac{16.7}{5.5 \times 10^6 \times 20}$$

$$= 0.152 \text{ }\mu\text{F}$$

where R_X was computed in a preceding paragraph. We shall use a standard 0.15 microfarad capacitor for C_2.

The final design for the ideal, biased clamper is shown in Fig. 7.16. Its performance is indicated by the oscilloscope displays in Fig. 7.17. Figure 7.17 (a) shows the circuit response for a minimum amplitude, minimum frequency input signal, and a minimum reference level. The output signal is just beginning to clip on the negative peaks. If this is critical in a particular application, then simply increase the size of C_2. Figure 7.17(b) shows the other extreme. That is, maximum input signal and maximum reference level at a higher frequency (5 kHz). Table 7.3 contrasts the actual measured performance of the circuit with the original design goals.

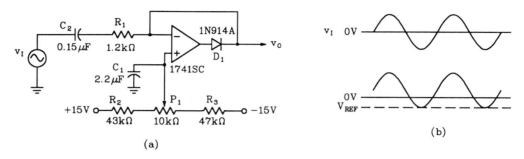

Figure 7.16 Final schematic of an ideal, biased clamper circuit design.

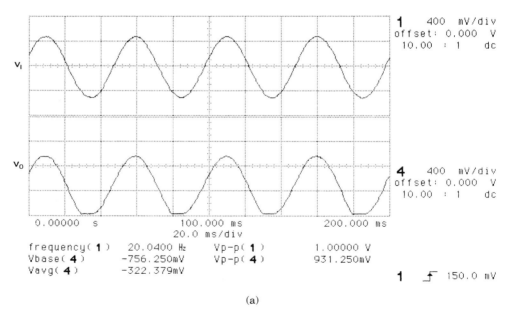

Figure 7.17 Oscilloscope displays showing the performance of the clamper circuit shown in Fig. 7.16. (Test equipment courtesy of Hewlett-Packard Company)   (continued)

Sec. 7.5 Peak Detectors   323

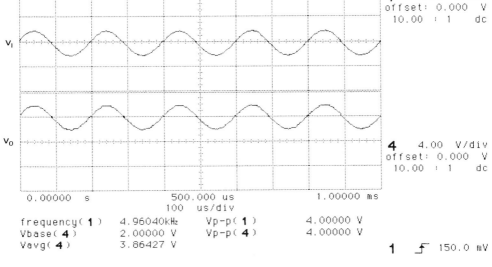

(b)

Figure 7.17b

TABLE 7.3

	DESIGN GOAL	MEASURED VALUES
Minimum input impedance	1.1 kΩ	1.2 kΩ
Input voltage range	500 mV–2 V (peak)	500 mV–2 V (peak)
Reference levels	-1V–$+2$ V	-0.76V–$+2$V
Minimum input frequency	20 Hz	20 Hz

7.5 PEAK DETECTORS

Many times it is necessary to develop a DC voltage that is equal to the peak amplitude of an AC signal. This technique is used for many applications including test equipment, ultrasonic alarm systems, and music synthesizers. As with the other circuits presented earlier in this chapter, the peak detector simulates an ideal diode by including the diode in the feedback loop of an op amp.

7.5.1 Operation

Figure 7.18 shows the schematic diagram of an ideal peak detector circuit. As the dotted lines in the figure indicate, the circuit is essentially an ideal clipper (an inverting clipper was discussed earlier in this chapter), followed by a parallel resistor and capacitor, and driving a voltage follower (discussed in Chap. 2). You will recall from the discussion on ideal clippers that the output of the clipper portion of the circuit will be a positive half-wave signal that is equal in amplitude to the peak of the input signal. Since C_1 is connected to this same point, it will be charged to this peak voltage.

The time constant for charging C_1 is very short and primarily consists of C_1 and the forward resistance of the diode. Thus, C_1 charges almost instantly to the peak output of the clipper circuit. When the output of the clipper starts to decrease (as it goes beyond the 90° point), diode D_1 becomes reverse biased. This essentially isolates capacitor C_1 and leaves the charge trapped. The only discharge path for C_1 is through R_5 and via leakage or op amp bias currents. In any case, the time constant is much longer than the charge time constant so C_1 holds its charge and presents a steady input voltage to A_2 that is equal to the peak amplitude of the input signal. A_2 of course, is simply a buffer amplifier and prevents unintentional discharging of C_1 caused by loading from the following circuit.

Resistor R_5 is the primary discharge path for C_1. If the input signal reduces its average (i.e., long-term) amplitude, then C_1 must be able to discharge to the new peak level. If the R_5C_1 time constant is too short, then the voltage on C_1 will not be constant and will have a high value of ripple. On the other hand, if the R_5C_1 time constant is too long, then the circuit cannot respond quickly to changes in the input amplitude. This characteristic is called fast attack (since C_1 responds quickly to amplitude increases) and slow decay (since C_1 is slow to respond to decreases in signal amplitude).

Resistor R_3 limits the current into the (+) input of A_2 when power is disconnected from the circuit. Without this resistor, the input circuitry for A_2 may be damaged as C_1 discharges through it. For capacitors smaller than 1 microfarad, resistor R_3 can normally

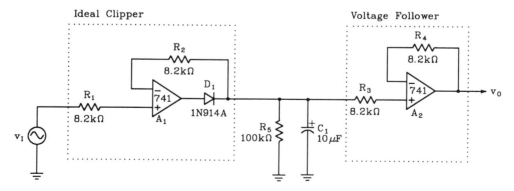

Figure 7.18 An ideal, peak detector circuit develops a DC output that is equal to the peak input voltage.

Sec. 7.5 Peak Detectors

be omitted. Resistor R_4 is to minimize the effects of bias currents in A_2. As in past circuits, we try to keep the resistance equal for both op amp inputs.

Resistor R_2 limits the current into the ($-$) input of A_1 when power is removed from the circuit. Again, this current comes from the discharge of C_1. Resistor R_1 is to minimize the effects of bias currents in A_1 and should be the same size as R_2.

7.5.2 Numerical Analysis

The basic numerical analysis of the clipper and buffer amplifier portions of the circuit (both voltage follower circuits) are presented in Chap. 2. These analyses will not be repeated here. Two additional characteristics to analyze are

1. lower frequency limit,
2. response time.

Lower frequency limit. The lower frequency limit is that frequency which causes the ripple voltage to exceed the maximum allowable level (determined by the design requirements). It can be estimated by applying the basic discharge equation for capacitors. That is

$$f = \frac{1}{R_5 C_1 \ln\left[\dfrac{E - E_0}{E - e_C}\right]} \tag{7.12}$$

where E_0 is the initial charge of the capacitor (V_{peak}), E is the voltage to which the capacitor will discharge (assumed to be zero), and e_C is the minimum allowable voltage on the capacitor. For purposes of this discussion, the lower frequency limit will be considered to be that frequency which causes the ripple voltage across C_1 to be 1 percent of the DC voltage. Having made this definition, we can apply a simplified equation to determine the lower frequency limit. That is

$$f_L = \frac{100}{R_5 C_1} \tag{7.13}$$

In the case of the circuit in Fig. 7.18, we estimate the lower frequency limit as

$$f_L = \frac{100}{100 \times 10^3 \times 10 \times 10^{-6}} = 100 \text{ Hz}$$

Response time. Response time is the term that describes how quickly C_1 can respond to decreases in the amplitude of the input signal. Here again, this can be computed from the basic discharge equation, Eq. (7.12). If, however, we assume the capacitor is charged to peak and discharges toward an eventual value of zero, then we can use the simplified form, Eq. (7.14).

$$t_R = R_5 C_1 \ln\left[\frac{v_{PK}(\text{old})}{v_{PK}(\text{new})}\right] \tag{7.14}$$

where $v_{PK}(\text{old})$ is the peak input signal amplitude before the decrease, and $v_{PK}(\text{new})$ is the peak input amplitude after the decrease in signal amplitude. For example, let us determine how quickly the circuit shown in Fig. 7.18 can respond if the input signal drops from 2.5 volts peak to 1.2 volts peak. We apply Eq. (7.14) as follows:

$$t_R = 100 \times 10^3 \times 10 \times 10^{-6} \ln\left[\frac{2.5}{1.2}\right] = 734 \text{ milliseconds}$$

7.5.3 Practical Design Techniques

Let us now design an ideal peak detector circuit similar to the one shown in Fig. 7.18. It should satisfy the following design goals:

1. input frequency range   300 to 3000 hertz
2. peak input voltage   1 to 5 volts
3. response time $\leq$ 200 milliseconds
4. ripple voltage $\leq$ 3 percent

Select the clipper op amp. The minimum unity-gain bandwidth is the same as the upper input frequency since A_1 is essentially operated at a closed-loop gain of one (when the rectifier conducts).

The minimum slew rate for the op amp is computed by applying Eq. (2.11). On the negative alternation of the input cycle, the output of A_1 will go to $-V_{SAT}$ since D_1 will be reverse biased and the op amp will be operating with open-loop gain. On the positive alternation, the output may have to go as high as 5.7 volts. That is, v_{PK} of the input plus the forward drop of D_1. Thus, the maximum output swing for purposes of determining the slew-rate requirement is $+5.7 - (-13) = 18.7$ volts. The required slew rate for our present design is computed as

$$f_{SRL} = \frac{\text{slew rate}}{\pi \, v_O(\text{max})},$$

or

$$\text{slew rate} = \pi f_{SRL} v_O(\text{max})$$
$$= 3.14 \times 3000 \times 18.7$$
$$= 0.176 \text{ V}/\mu\text{s}$$

Sec. 7.5   Peak Detectors   327

Both bandwidth and slew-rate requirements are easily satisfied with the standard 741 op amp. Let us choose to use this device.

Select the buffer amplifier.   The buffer amplifier has even less stringent requirements since it is amplifying a DC signal. We will not be concerned about bandwidth or slew-rate limitations. If the application is critical with regard to DC drift, then we can select an op amp to minimize this characteristic. For the present design, however, let us employ the basic 741 device.

Select D_1.   The peak inverse voltage of D_1 will be equal to the difference between $-V_{SAT}$ and the maximum peak input voltage. This difference in potential will exist when C_1 has charged to the maximum peak voltage and the output of A_1 swings to $-V_{SAT}$ on the negative alternation. In equation form, the peak inverse voltage of the diode is determined with Eq. (7.15).

$$V_{PIV}(\text{rating}) \geq V_{PK}(\text{max}) - (-V_{SAT}) \qquad (7.15)$$

In the case of the circuit being designed, the PIV rating of the diode is computed as

$$V_{PIV} = 5 - (-13) = 18 \text{ V}$$

The average current for D_1 is nearly negligible since it only conducts enough to recharge capacitor C_1, and because C_1 loses very little charge between consecutive cycles. The instantaneous current through D_1 however might be considerably higher when power is first applied to the system and C_1 is being charged initially. The safest practice is to insure that D_1 can survive the short-circuit current of A_1.

For our present design, let us use a 1N914A diode for D_1. This easily meets both the PIV and instantaneous current requirements.

Compute R_5 and C_1.   There are two conflicting circuit parameters that affect the choice of values for R_5 and C_1: allowable ripple voltage across C_1, response time. It is possible to establish values for these parameters in the initial design goals that cannot be physically implemented. In general, a faster response time leads to greater ripple.

For design purposes, we shall independently compute the required values for R_5 and C_1 to satisfy the ripple and the response time criteria. We will then make a judgment as to the optimum choice of values.

The *minimum* required RC time constant based on the ripple specification can be found by applying Eq. (7.12). (We could use Eq. (7.13) if the ripple goal were 1%).

$$f_L = \frac{1}{R_5 C_1 \ln\left[\dfrac{E - E_0}{E - e_C}\right]},$$

or

$$R_5C_1 = \frac{1}{f_L \ln\left[\dfrac{E - E_0}{E - e_C}\right]}$$

$$= \frac{1}{300 \ln\left[\dfrac{0 - 5}{0 - 4.85}\right]}$$

$$= 109.4 \text{ milliseconds}$$

Any RC time constant which is longer than this minimum value will satisfy the ripple goal.

The *maximum* R_5C_1 time constant based on the response time design goal can be found by applying Eq. (7.14).

$$t_R = R_5C_1 \ln\left[\frac{v_{PK}(\text{old})}{v_{PK}(\text{new})}\right],$$

or

$$R_5C_1 = \frac{t_R}{\ln\left[\dfrac{V_{PK}(\text{old})}{V_{PK}(\text{new})}\right]}$$

$$= \frac{200 \times 10^{-3}}{\ln\left[\dfrac{5}{1}\right]}$$

$$= 124.3 \text{ ms}$$

Any RC time constant which is less than this value will satisfy the response time requirement. At this point, we must choose values for R_5 and C_1 such that the RC time constant falls within the above window (i.e., 109.4 ms $\leq R_5C_1 \leq$ 124.3 ms). Additionally, we don't want to use resistor values smaller than a few kilohms nor larger than the low megohms. In the present case, let us select a standard value of 1.0 microfarad for C_1. The limits for R_5 can then be computed as

$$R_5(\text{min}) = \frac{109.4 \times 10^{-3}}{1 \times 10^{-6}} = 109.4 \text{ k}\Omega,$$

and

$$R_5(\text{max}) = \frac{124.3 \times 10^{-3}}{1 \times 10^{-6}} = 124.3 \text{ k}\Omega$$

Let us select a standard value of 120 kilohms for R_5.

Sec. 7.5 Peak Detectors

Compute R_1, R_2, R_3, and R_4. For our purposes, the exact values of these four resistors are noncritical, and any value in the range of 2 to 100 kilohms will suffice. It is, however, important that $R_1 = R_2$ and that $R_3 = R_4$. Let us arbitrarily choose all four resistors to be 10 kilohms.

This completes the design of our ideal peak detector circuit. The final schematic is shown in Fig. 7.19. The waveforms which indicate its performance are presented in Fig. 7.20. Figures 7.20(a) and 7.20(b) show the output response for minimum and

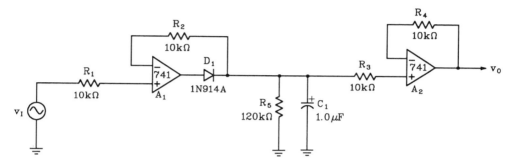

Figure 7.19 Final design of an ideal, peak detector circuit.

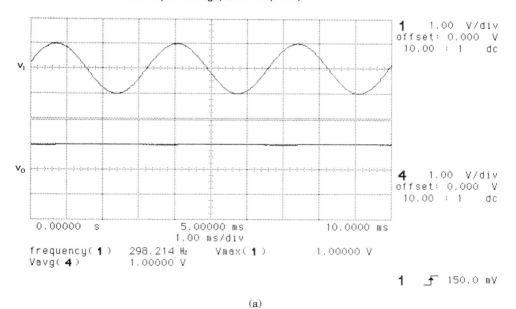

(a)

Figure 7.20 Oscilloscope displays showing the actual performance of the peak detector circuit shown in Fig. 7.19. (Test equipment courtesy of Hewlett-Packard Company) (continued)

330   Signal Processing Circuits   Chap. 7

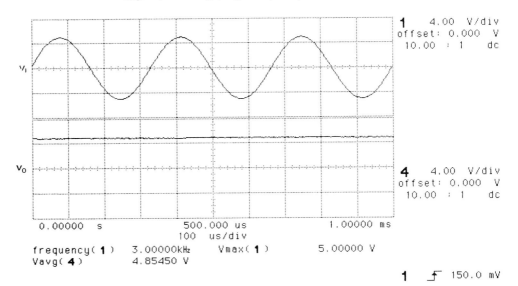

(b)

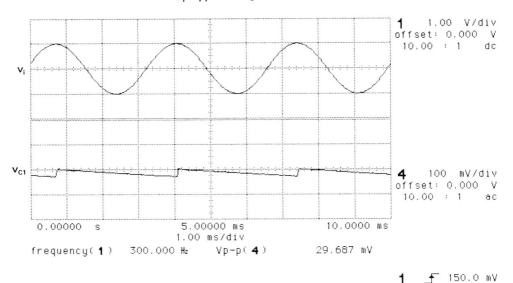

(c)

Figure 7.20b and c

(continued)

Sec. 7.6   Integrator

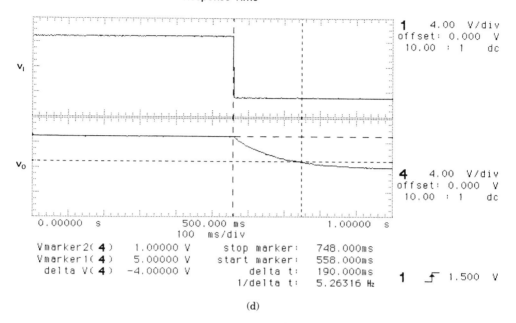

(d)

Figure 7.20d

TABLE 7.4

	DESIGN GOAL	MEASURED VALUE
Input voltage	1–5 volts peak	1–5 volts peak
Frequency range	300–3000 Hz	300–3000 Hz
Ripple voltage	≤3%	2.97%
Response time	≤200 ms	190 mS

maximum input conditions. Figure 7.20(c) illustrates the ripple voltage across C_1. The response time is being measured in Fig. 7.20(d). Here a square-wave input was applied and the time for the capacitor to discharge from 5 volts to 1 volt was measured. The design goals are compared with the measured results in Table 7.4.

7.6 INTEGRATOR

The integrator is one of the fundamental circuits studied in basic electronics. Its op amp counterpart is also an important circuit for many signal-processing applications. As you may recall, an integrator produces an output voltage which is proportional to *both* the duration and amplitude of an input signal. For example, if the input were a pulse waveform, then output would be a voltage that was proportional to the amplitude and pulse width of the input signal. In essence, the integrator computes the area (height × width) of the input signal. This corresponds to the mathematical operation in calculus called integration.

The integrator may also alter the shape of the input waveform. For example, a square wave will be converted to a triangular wave in the process of being integrated. When integrated, a triangle wave will produce a parabolic waveform that very closely approximates a sinewave. In the case of a sinewave input, the output will still be sinusoidal but may be shifted in phase and reduced in amplitude. For sinewave inputs, the integrator acts as a simple low-pass filter.

7.6.1 Operation

Figure 7.21 shows the schematic of an op amp integrator circuit. If you mentally open capacitor C_1 (which is true for DC signals anyway), you will see that the circuit is a simple inverting amplifier circuit. Resistors R_2 and R_1 determine the voltage gain of the circuit, and resistor R_3 is to minimize the effects of bias current. Recall that the ($-$) input of the op amp is a virtual ground point.

Now suppose that a step voltage is applied to the input terminal. This will cause a current to flow through resistor R_1. Since the current flowing in or out of the ($-$) input is negligible, we shall assume that all of the current flowing through R_1 continues through R_2 and C_1. Now, the voltage across R_2 and C_1 can only increase as fast as capacitor C_1 can take on a charge. When the input voltage first makes a change in amplitude, the current resulting from this voltage change is routed **entirely** through C_1. (Since the voltage across R_2 does not change instantly, neither can the current.) As the capacitor accumulates a charge, the current through R_2 begins to change. The circuit, however, is designed to insure that the current through R_2 is never allowed to be a substantial part of the capacitor current. For all practical purposes, as long as the input voltage is constant, the capacitor current is constant. As long as the input returns to its original state before the capacitor has had time to accumulate excessive voltage, this discussion is valid.

With a constant current through the capacitor, we will generate a linear ramp of voltage across it (and therefore at the output of the op amp). With a square-wave input,

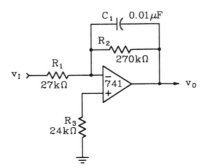

Figure 7.21 A basic op amp integrator circuit.

Sec. 7.6 Integrator

the capacitor will periodically charge and discharge with equal, but opposite polarity, currents. This, of course, produces a triangle wave at the output.

Resistor R_2 is included in the circuit to reduce the gain at low frequencies (DC in particular). Without R_2, the bias currents (small as they are) would eventually charge C_1 and cause an undesired DC offset in the output. This offset may even cause the amplifier to go into saturation. In terms of AC circuit theory, we insure that the reactance of C_1 is less than 10 percent of the value of R_2 at the lowest frequency of operation. This insures that the majority of the current will be used to charge and discharge C_1.

7.6.2 Numerical Analysis

Let us analyze the circuit shown in Fig. 7.21 and determine the following circuit characteristics:

1. lowest frequency of operation,
2. highest frequency of operation.

Many other characteristics (e.g., input and output impedance) are analyzed in the same way as a simple inverting amplifier (Chap. 2).

Lowest frequency of operation. The lower frequency limit of the integrator circuit shown in Fig. 7.21 is that frequency which causes the capacitive reactance of C_1 to be equal to one tenth of R_2. This is computed with the basic capacitive reactance equation.

$$X_C = \frac{1}{2\pi f C},$$

or

$$f = \frac{1}{2\pi C X_C}$$

$$= \frac{1}{6.28 \times 0.01 \times 10^{-6} \times 27 \times 10^3}$$

$$= 590 \text{ Hz}$$

This is not an ultimate limit, however, as the frequency is reduced below this value, the operation of the circuit becomes progressively less like an integrator and more like an inverting amplifier. Finally, at DC, it **is** an inverting amplifier.

Highest frequency of operation. The upper frequency limit is dependent on the characteristics of the op amp. In particular, the upper operating frequency will be the lower of the frequencies that are established by the bandwidth or slew rate of the op amp. Both of these considerations are discussed in greater detail in Chap. 2. For many applications, however, (including worst-case design consideration) the upper limit will be set by the slew rate of the op amp. This is computed with Eq. (2.11) as

$$f_{SRL} = \frac{\text{slew rate}}{\pi \, v_o(\text{max})}$$

$$= \frac{0.5 \text{ V}/\mu\text{s}}{3.14 \times 26}$$

$$= 6.12 \text{ kHz}$$

7.6.3 Practical Design Techniques

We shall now design an op amp integrator that will perform according to the following design goals:

1. input frequency   300 hertz to 20 kilohertz
2. input impedance ≥ 1000 ohms
3. input voltage   2 to 6.5 volts peak

Select the op amp. The first criteria for op amp selection is the upper frequency limit. That is, both slew rate and bandwidth considerations must allow the circuit to operate at the upper frequency specified in the design goals. Unless the circuit is specifically designed to handle small-amplitude signals, it will be the slew rate that limits the upper frequency of operation. In the present case, we shall determine the required op amp slew rate by applying Eq. (2.11).

$$f_{SRL} = \frac{\text{slew rate}}{\pi \, v_o(\text{max})},$$

or

$$\text{slew rate} = f_{SRL} \, \pi \, v_o(\text{max})$$

$$= 20 \times 10^3 \times 3.14 \times 26$$

$$= 1.63 \text{ V}/\mu\text{s}$$

This exceeds the 0.5 volts per microsecond slew rate of the standard 741 but is well within the capabilities of the MC1741SC.

Another op amp characteristic that is generally considered when designing op amp integrators is the bias current. In general, the lower the frequency of operation, the more problems caused by bias currents. If very low frequencies of operation are needed, it would be wise to select an op amp that has particularly low-bias currents. For purposes of the present design, however, let us opt to use the MC1741SC device.

Compute R_1. It is important that the input current to the circuit be much greater than the op amp bias current. Let us choose the input current to be at least 1000 times

Sec. 7.6 Integrator

the worst case bias current. The manufacturer's data sheet lists the maximum bias current as 800 nanoamperes. We shall establish our input current at 1000 × 800 nanoamperes or 800 μA.

We can now compute the value of R_1 using Ohm's Law.

$$R_1 = \frac{2}{800 \times 10^{-6}} = 2.5 \text{ k}\Omega$$

As long as this value exceeds the minimum input impedance requirement then it may be used as calculated. Otherwise, increase it to satisfy the impedance requirements. If a substantial increase is needed in order to establish the correct input impedance, then an op amp with a lower bias current should be selected and R_1 recalculated. For our design, let us use a standard 2.4-kilohm resistor for R_1.

Compute C_1. The basic electronics equations for charge (Q = CE and Q = It) can be set equal to each other and manipulated to give us our equation, Eq. (7.16), for C_1.

$$\boxed{C_1 = \frac{i_I}{2V_{SAT}f_L}} \tag{7.16}$$

where f_L is the lowest input frequency and i_I is the *maximum* input current. The maximum input current is computed with Ohm's Law as

$$i_I(\max) = \frac{v_I(\max)}{R_1} = \frac{6.5}{2.4 \times 10^3} = 2.7 \text{ mA}$$

For our current design, we compute C_1 as

$$C_1 = \frac{2.7 \times 10^{-3}}{2 \times 13 \times 300} = 0.346 \text{ μF}$$

We shall use a standard 0.33 microfarad capacitor for C_1. A low-leakage type of capacitor should be chosen.

Compute R_2. Resistor R_2 is chosen to have a resistance of at *least* 10 times the capacitive reactance of C_1 at the lowest input frequency. We simply apply the basic capacitive reactance equation to compute R_2.

$$R_2 = \frac{10}{2\pi f C_1} = \frac{10}{6.28 \times 300 \times 0.33 \times 10^{-6}} = 16 \text{ k}\Omega$$

Compute R_3. Resistor R_3 is computed as the parallel combination of R_1 and R_2. That is

$$R_3 = R_1 \| R_2 = \frac{1}{\frac{1}{2.4 \times 10^3} + \frac{1}{16 \times 10^3}} \approx 2 \text{ k}\Omega$$

The completed integrator design is shown in Fig. 7.22 with the circuit waveforms presented in Fig. 7.23. Figure 7.23(a) shows the response of the circuit at 300 hertz. Notice the linearity of the ramp waveform. Figures 7.23(b) and 7.23(c) show the circuit response to 20 kilohertz signals. The integrator action has essentially eliminated the observable waveform, but comparison of Fig. 7.23(a) to 7.23(c) will clearly show the circuit's response to changes in duty cycle. Figure 7.23(d) illustrates the circuit's response to a 6.5 volt peak input signal.

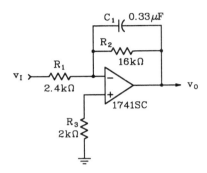

Figure 7.22 Final schematic of an integrator circuit design.

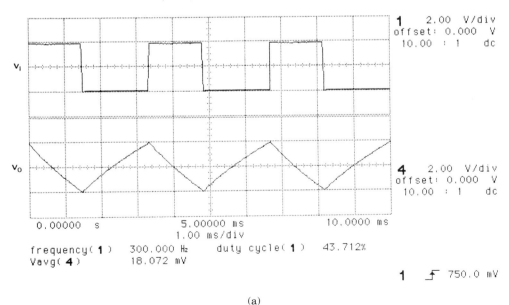

(a)

Figure 7.23 Oscilloscope displays showing the performance of the integrator circuit shown in Fig. 7.22. (Test equipment courtesy of Hewlett-Packard Company) (continued)

Sec. 7.6 Integrator

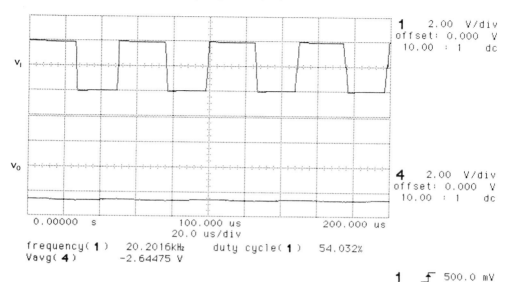

(b)

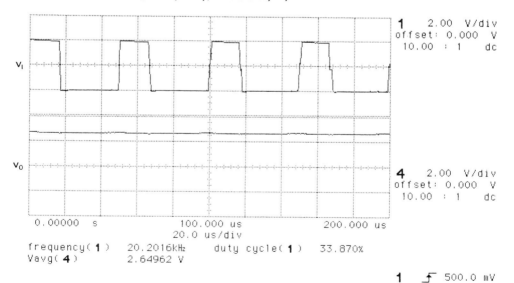

(c)

Figure 7.23b and c (continued)

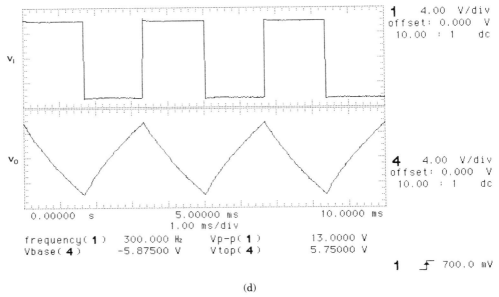

(d)

Figure 7.23d

7.7 DIFFERENTIATOR

The differentiator is another fundamental electronic circuit and is the inverse of the integrator circuit. In terms of mathematics, it produces an output signal that is the first derivative of the input signal. In more intuitive terms, the instantaneous output voltage is proportional to the instantaneous rate of change of input voltage. If, for example, we apply a linear ramp voltage to the input of a differentiator, we will expect the output to be a DC level since the rate of change of input voltage is a constant value. Similarly, if we apply a sine wave to the differentiator, the output will also be sinusoidal in shape, but it will be shifted in phase by approximately 90° since the maximum rate of change of a sine wave occurs as it passes through the 0° and 180° points.

7.7.1 Operation

Figure 7.24 shows the schematic diagram of an op amp differentiator circuit. From basic electronics, we know that the current through a capacitor is directly proportional to the rate of change of applied voltage. This is evident from the equation for capacitive current.

Sec. 7.7 Differentiator

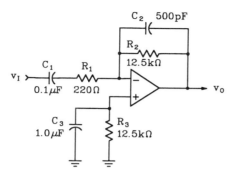

Figure 7.24 A differentiator produces an output voltage that is proportional to the rate of change of input voltage (i.e., $v_{O(t)} = k\, dv/dt$).

$$i_C = C\frac{\Delta v}{\Delta t} = C\frac{dv}{dt}$$

It is also evident from the capacitive reactance equation

$$X_C = \frac{1}{2\pi fC}$$

which shows that the opposition to current flow decreases as the frequency (rate of change of voltage) increases. In the case of the circuit shown in Fig. 7.24, we can expect capacitor C_1 to have greater currents for input voltages which *change* levels more quickly. Any current which flows through the capacitor must also flow through R_1 due to the series connection. Since no significant current flows in or out of the ($-$) input, we can conclude that the current through R_2 will also be the same as the input current and will be proportional to the rate of change of input voltage. The left end of R_2 is connected to a virtual ground point, therefore, the voltage across it **is** the output voltage of the op amp and is determined by the rate of change of input voltage.

The differentiator circuit is inherently unstable and prone to oscillation since the input impedance decreases with increasing frequency. Recall that the gain for an inverting op amp is determined by the ratio of the impedance in the feedback path to the input impedance. Since the input impedance decreases with frequency, it will cause the gain to increases at high frequencies. Even though the actual input signal may be a relatively low frequency, there are always high-frequency noise signals present. If the gain were allowed to increase excessively at high frequencies, these noise signals would interfere with the desired output and may well cause oscillation in the circuit. To prevent the gain from steadily increasing at high frequencies, we include capacitor C_2 in the feedback path. This capacitor tends to bypass resistor R_2 at noise frequencies thus reducing the circuit gain and improving the circuit stability. Additionally, resistor R_1 works to increase the stability by insuring that the input impedance has a practical minimum limit regardless of the frequency.

Resistor R_3 reduces the effects of the op amp bias current. As with previous circuits, we make R_3 equal to R_2 so that the DC resistance in both op amp terminals is the same.

Capacitor C_3 simply bypasses R_3 at high frequencies which further minimizes the circuit's response to noise frequencies.

7.7.2 Numerical Analysis

Since the output voltage of a differentiator circuit is determined by the rate of change of input voltage, we will want to know the maximum rate of change of input voltage that can be applied to the circuit without driving the circuit into saturation. We can estimate this rate by applying Eq. (7.17).

$$\boxed{\frac{\Delta v_I}{\Delta t} = \frac{V_{SAT}}{R_2 C_1}} \qquad (7.17)$$

For the circuit shown in Fig. 7.24, we estimate the maximum rate of change of input voltage as

$$\frac{\Delta v_I}{\Delta t} = \frac{13}{12.5 \times 10^3 \times 0.1 \times 10^{-6}} = 10,400 = 10.4 \text{ V/ms}$$

The input impedance is a rather complex issue since it varies with frequency and is affected by several components. The absolute minimum impedance, however, is established by R_1. In practice, though, the actual minimum impedance may never go as low as the value of R_1, but it is a good approximation for worst-case analysis.

The output impedance increases with frequency but can still be estimated as described in Chap. 2 for inverting amplifiers.

7.7.3 Practical Design Techniques

We shall now design a differentiator circuit that will satisfy the following design goals:

1. input waveform   triangle (dual ramp)
2. input voltage   ±2 volts
3. input frequency   2 kilohertz
4. output voltage   ±10 volts for the given input signal
5. op amp   741

Compute R_2. Resistor R_2 is selected to establish basic range of operation. A good rule of thumb for the initial selection of R_2 is given by Eq. (7.18).

$$\boxed{R_2 = \frac{25 v_O^+(\text{max})}{I_{SC}}} \qquad (7.18)$$

where $v_O^+(\text{max})$ is the highest expected output voltage, and I_{SC} is the short-circuit current rating of the op amp. For our present design, we compute R_2 as

Sec. 7.7 Differentiator   341

$$R_2 = \frac{25 \times 10}{20 \times 10^{-3}} = 12.5 \text{ k}\Omega$$

We shall select the nearest standard value of 12 kilohms.

Compute C_1.   The time constant for R_2C_1 is determined by the expected rate of change of input voltage as compared to the resulting output voltage. In equation form, we can compute the value of C_1 as

$$\boxed{C_1 = \frac{v_O^+(\max)\Delta t}{R_2 \Delta v_I}} \qquad (7.19)$$

Utilization of this equation requires us to know the rate of change of input voltage. The design specifications tell us that we will have a ramp voltage which goes from -2 volts to $+2$ volts and back at a frequency of 2 kilohertz. Thus, the Δv_I is 4 volts (i.e. -2V to $+2$V). The Δt is one half of the period of one input cycle. That is

$$\Delta t = \frac{T}{2} = \frac{1}{2F} = \frac{1}{2 \times 2000} = 250 \text{ }\mu\text{s}$$

We can now compute C_1 with Eq. (7.19) as

$$C_1 = \frac{10 \times 250 \times 10^{-6}}{12 \times 10^3 \times 4} = 0.052 \text{ }\mu\text{F}$$

Let's plan to use a standard 0.05 microfarad capacitor for C_1.

Compute R_1.   Resistor R_1 should equal the reactance of C_1 at a frequency higher than the normal operating frequency. In this way, R_1 has minimal effect for the normal input signal, but becomes effective for higher frequencies (noise). We can compute a reasonable value for R_1 with Eq. (7.20).

$$\boxed{R_1 = \sqrt{\frac{R_2}{\pi C_1 f_{UG}}}} \qquad (7.20)$$

where f_{UG} is the unity-gain frequency of the op amp. For the present application, we can compute the value of R_1 as follows:

$$R_1 = \sqrt{\frac{12 \times 10^3}{3.14 \times 0.05 \times 10^{-6} \times 1 \times 10^6}} = 276.5 \text{ }\Omega$$

We shall use the nearest standard value of 270 ohms.

Compute R_3.   Resistor R_3 is always equal to R_2 in order to maintain equal DC resistances in both of the op amp input terminals. Therefore, R_3 will also be a 12 kilohm resistor.

Compute C_3. The reactance of capacitor C_3 should be less than one tenth the resistance of R_3 at a frequency which causes the reactance of C_1 to be equal to the resistance of R_2. This insures that resistor R_3 will be effectively bypassed for all usable circuit frequencies. Since $R_2 = R_3$, we can express this in equation form as

$$\boxed{C_3 = 10C_1} \tag{7.21}$$

In our present circuit, we shall require a value of

$$C_3 = 10 \times 0.05 \times 10^{-6} = 0.5 \; \mu F$$

We shall use a standard value of 0.47 microfarad for C_3.

Compute C_2. In order to reduce the gain at high-noise frequencies and yet minimize the effect on normal circuit frequencies, we want to select capacitor C_2 such that its reactance is equal to R_2 at a frequency well above the highest normal operating frequency but yet well below the unity-gain frequency of the op amp. The following equation will provide a reasonable value for C_2.

$$\boxed{C_2 = \sqrt{\frac{C_1}{8 \pi R_2 f_{UG}}}} \tag{7.22}$$

Substituting values for our present circuit gives us

$$C_2 = \sqrt{\frac{0.05 \times 10^{-6}}{8 \times 3.14 \times 12 \times 10^3 \times 1 \times 10^6}} = 407 \; pF$$

We'll use the nearest standard value of 390 picofarad for C_2.

This completes the design of our differentiator circuit. The final schematic is shown in Fig. 7.25. The oscilloscope display in Fig. 7.26(a) shows the actual performance of the circuit under the conditions described in the original design goal. Figure 7.26(b) shows the circuit performance for a square-wave input signal. Since the rise and fall times are significantly faster than the ramp specified in the design goal, the input amplitude has to be much lower to avoid saturating the output. Notice that during periods of time

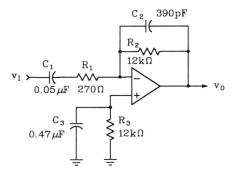

Figure 7.25 A differentiator circuit designed to produce a ±10 volt output for a 16-volt per millisecond input.

Sec. 7.7 Differentiator

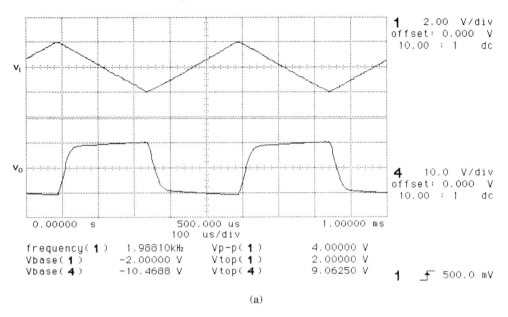

(a)

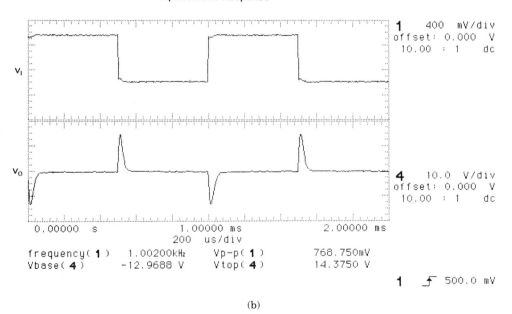

(b)

Figure 7.26 Oscilloscope displays showing the performance of the differentiator circuit shown in Fig. 7.25. (Test equipment courtesy of Hewlett-Packard Company)

TABLE 7.5

	DESIGN GOAL	MEASURED VALUE
Input frequency	2.0 kHz	2.0 kHz
Input voltage	±2 V	±2 V
Output voltage	±10 V	−10.5 V, +9.1 V

when the input signal is steady (i.e., *not changing states*) that the output is zero. Finally, Table 7.5 compares the original design goal with the measured performance.

7.8 TROUBLESHOOTING TIPS FOR SIGNAL PROCESSING CIRCUITS

Much of the troubleshooting procedures discussed with reference to basic amplifiers (Chap. 2) is applicable to the signal-processing circuits discussed in this chapter. A few additional techniques, however, may help isolate problems more quickly.

If the input connection to any of the circuits presented in Chap. 7 is connected directly to ground, then the output should go to its *normal* DC level. The normal value, of course, depends on the circuit being considered. In any case, if the output does go to the correct DC value with no input signal, then the problem is most likely caused by a defect in one of the AC branches. An AC branch will contain a series capacitor.

If the DC level on the output is abnormal when the input is shorted to ground, then apply the basic analytical techniques described in Chap. 2. When the DC output level is incorrect, more often than not the output will be saturated. Comparison of the polarity of the differential input voltage of the op amp with the output polarity will quickly reveal a defective op amp.

If you suspect an open, two-terminal component, you can momentarily parallel the suspected part with a known good one while monitoring the output. If the problem is corrected, you have located the defect.

REVIEW QUESTIONS

1. What is the name of the type of signal processing circuit that is used to shift the DC level of the input signal without altering its wave shape?
2. Refer to Fig. 7.2. In your own words, explain why the output waveform for the silicon diode is a constant zero volts.
3. Refer to Fig. 7.3. What is the effect on circuit operation if diode D_1 opens?
4. Refer to Fig. 7.3. What is the effect on circuit operation if resistor R_1 decreases in value?

Chap. 7   Review Questions   345

5. Refer to Fig. 7.5. While monitoring the output waveform on an AC-coupled oscilloscope, you momentarily short resistor R_4. Describe the effects, if any, that are noted on the oscilloscope display.
6. Refer to Fig. 7.5. Describe the effect on output wave shape if resistor R_3 is increased to 45 kilohms.
7. Refer to Fig. 7.8. What is the effect on output wave shape if resistor R_2 is increased in value?
8. Refer to Fig. 7.8. What is the effect on circuit operation if capacitor C_1 develops a short circuit?
9. Refer to Fig. 7.16. Describe the effect on output wave shape if resistor R_3 is increased in value.
10. Refer to Fig. 7.19. Describe the effect on circuit operation if resistor R_3 is changed to 4.7 kilohms.

CHAPTER 8

DIGITAL-TO-ANALOG AND ANALOG-TO-DIGITAL CONVERSION

The world of electronics can be neatly divided into two general classes based on the nature of the signal or circuit: digital and analog. Digital signals, devices, and circuits operate in one of two states at all times. These states may be high/low, on/off, up/down, 0 volts/5 volts, -5mA/$+5$ mA, or any other set of two-valued terms.

Analog signals, devices, and circuits, on the other hand, operate on a continuous range with an infinite number of values represented within a given range. An analog voltage for example, may be 1.5 volts or 1.6 volts. But it can also be an infinite number of values between these two numbers such as 1.55 volts, or 1.590754 volts.

A technician or engineer must generally be capable of working with both analog and digital devices and systems. This text will avoid revealing a prejudice toward one type of circuit or another. Similarly, the reader is encouraged to avoid developing such a prejudice. It is true that digital devices and techniques are steadily taking over operations and functions previously implemented by analog systems. But equally true, is the fact that the world in which we live is inherently analog. Quantities such as temperature, pressure, weight, speed, light intensity and color, volume, and all other similar quantities are analog in that they vary continuously and have an infinite number of possible values.

This chapter will focus on the circuits that interface analog systems with digital systems. Analog-to-digital (A/D) converters accept an analog signal at their input and produce a corresponding digital signal at the output. This output can then be processed and interpreted by a digital circuit (typically a microprocessor system). A digital-to-analog (D/A) converter, on the other hand, is used to convert the digital output from a microprocessor or other digital device into an equivalent analog signal. The analog signal is frequently used to control a real-world quantity (e.g., temperature, pressure, etc.).

The intent of this chapter, then, is to provide the reader with the concepts and terminology associated with A/D and D/A conversion. Additionally, several representative circuits will be presented which utilize operational amplifiers. An understanding

Sec. 8.1 D/A and A/D Conversion Fundamentals

of the operation of these fundamental circuits is important since they convey essential underlying principles. The actual implementation of the converter circuits, however, is another matter. Except for unique or very demanding applications (neither of which is targeted by this text), most A/D and D/A converter applications are resolved by utilizing an integrated circuit version of the A/D or D/A converter. The price and performance of these circuits makes them very difficult to beat by designing your own converter circuit.

8.1 D/A AND A/D CONVERSION FUNDAMENTALS

The concepts and terminology presented in this chapter are important to the reader whether designing a custom converter circuit or selecting an integrated version. In either case, the technician or engineer must be able to effectively evaluate the application and contrast it with the specifications of the converter circuit.

8.1.1 Analog-to-Digital Converters

Figure 8.1 illustrates the fundamental function of A/D conversion. The block labeled "A/D Converter" may be an integrated circuit or an array of op amps and other devices. In any case, it accepts the analog signal as its input and produces a corresponding digital output. The digital output is shown to consist of several lines. The number of lines varies with the *resolution* of the converter. Resolution describes the percentage of input voltage change required to cause a step change in the output. Table 8.1 shows the basic relationship between number of bits (lines) and equivalent resolution.

Suppose, for example, an 8-bit A/D converter was designed to accept 0- to 10-volt input signals. The 10-volt range would be divided into 256 discrete steps of 10/256 or about 39 millivolts per step. By contrast, a 4-bit A/D converter would have less resolution with each of the 16 output steps being equivalent to 6.25 percent of the full-scale input or 625 millivolts. Thus, the higher the resolution (i.e., number of bits in the converted output) the smaller the input change required to move to the next output step. Typical applications require resolutions of 8, 12, or 20 bits.

Since the analog input may take on any one of an infinite number of values, but the output must be resolved into a fixed number of discrete levels or steps, each output step inherently represents a range of input voltages. The process of forming discrete groups from the continuous input is *quantization*. The output, therefore, does not **exactly** represent a given input signal. Rather, it represents an approximation. The resulting error

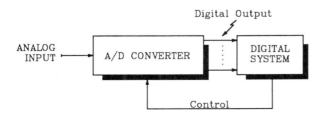

Figure 8.1 Analog-to-digital conversion makes an analog signal compatible with a digital system.

TABLE 8.1

RESOLUTION IN BITS	NUMBER OF STEPS	RESOLUTION AS PERCENT OF FULL SCALE (%)
1	2	50
2	4	25
3	8	12.5
4	16	6.25
5	32	3.125
6	64	1.5625
7	128	0.78125
8	256	0.390625

is called *quantization uncertainty* and is equal to $\pm 1/2$ the value of the least-significant bit of the converted output. In the example of the 8-bit converter described, this would equate to an uncertainty of $\pm 1/2 \times 39$ millivolts or about ± 19.5 millivolts. The magnitude of the quantization uncertainty is less with greater resolution. Therefore, if we need less quantization error, we must increase the number of bits in the output.

When the input signal is converted to a digital output, we would normally expect that steadily increasing values of input would produce equally spaced digital values in the output. Sometimes, however, the output may skip one or more steps or digital numbers. Similarly, the output may remain on a given step throughout a range that ideally includes two or more steps. This type of performance is generally caused by *linearity errors*. If the converter had no linearity problems, then the amount of input change to produce a change in the output would be consistent throughout the entire range of operation. When the amount of input change needed to reach the next step in the output varies, we call this variation nonlinearity.

Another characteristic of A/D converters describes the polarity of output changes when a steadily increasing input is applied. With an increasing input signal, we would expect (and want) a series of digital numbers in the output that are progressively larger. It is possible, however, for a particular output step to be smaller than the preceding step. That is, the magnitude of the digital output decreases on a particular step instead of increasing. This type of output response is called *nonmonotonic*. That is, a converter whose output is progressively higher for progressively higher inputs has the property of *monotonicity*.

Sometimes the intended range of input signals does not match the actual range. For example, the converter may be designed to accept a 0- to 10-volt input, but the actual device may be found to produce a maximum digital output for a 9.7-volt input. This discrepancy in full-scale operation is called *gain error* or *scaling error*.

Sec. 8.1 D/A and A/D Conversion Fundamentals

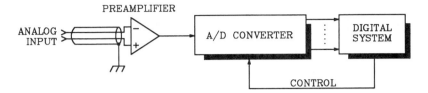

Figure 8.2 A preamplifier is used to boost low-level transducer signals to a level that is usable by the A/D converter. These amplifiers are frequently operated as differential amplifiers to reject common-mode noise.

The entire operational range of the A/D converter can be shifted up or down. For example, an A/D converter may be designed to produce a minimum digital output with 0 volts on the input. Actual measurement may reveal, however, that a DC offset voltage must be applied at the input in order to produce the minimum digital output. This error is called the *offset error* and is often expressed as a percentage of full-scale input voltage.

Accuracy is a term used to describe the overall performance of an A/D converter. It includes the combined effects of all errors and is a measure of the worst-case deviation from a given input signal and the equivalent value of its converted digital output.

The amount of time required to generate a particular digital number to represent a given analog input signal is called *conversion time*. Alternately, the number of these conversions that can be accomplished in one second is called the *conversion rate*.

8.1.2 Preamplifiers

Many A/D converter applications involve the conversion of transducer signals into corresponding digital numbers for subsequent processing by a microprocessor or programmable logic controller (PLC). Transducer signals are frequently very low level (current or voltage) and require amplification before they can be effectively applied to an A/D converter. Operational amplifiers are often used for this purpose. Additionally, special differential amplifiers called *instrumentation amplifiers* are frequently employed. Instrument amplifiers offer a very high rejection to common-mode signals (e.g., 60 Hz hum picked up on long cables), but offer high amplification to differential-mode signals (e.g., the actual transducer signal). These devices are discussed in greater detail in Chap. 11. Figure 8.2 shows the position of the preamplifier with respect to the A/D converter.

8.1.3 Sample-and-Hold Circuits

As mentioned previously, the conversion of an analog signal into an equivalent digital number requires a certain amount of time (conversion time). Since the analog signal may be changing values **during** the conversion process, substantial errors may be introduced. To eliminate this problem, we introduce a sample-and-hold (S/H) circuit between the preamplifier and the A/D converter. Figure 8.3 shows a block diagram for this case.

A S/H circuit is similar to the peak detector circuit presented in Chap. 7 with the exception that it is gated on and off. When the S/H circuit receives the *track* command,

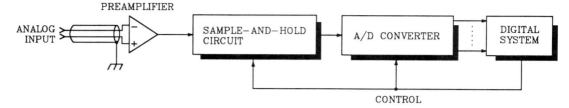

Figure 8.3 A sample-and-hold circuit is used to provide a steady input for the A/D conversion circuit.

it follows (i.e., samples) the input voltage. When a *hold* command is received, the S/H circuit opens its link to the input signal, and holds the most recently sampled value at its output. This output is held constant throughout the conversion time of the A/D converter. Once the conversion has been completed, the track command is issued, and the cycle repeats.

Once the track command has been received by the S/H circuit, it takes a certain amount of time for the output to match the present level of analog input. This delay is called the *acquisition time* of the S/H circuit. Similarly, there is a delay between the issuance of a hold command and the actual disconnecting of the S/H circuit from the input signal. This latter delay is called the *aperture time* of the S/H circuit.

The sampled input voltage is held constant by utilizing the charge on a capacitor. Although the capacitor has a very low discharge current, it does eventually leak off causing the S/H output to slowly decay or decrease. The rate at which this occurs is called the *droop rate* of the S/H circuit.

The more often a signal is sampled, the better the digital representation of the analog signal. If the input signal changes rapidly relative to the speed of the conversion process, then substantial portions of the input signal will be missed (i.e., go undetected). As an **absolute** minimum, the input signal must be sampled twice during each cycle. That is, the sampling rate must be at *least* twice the highest frequency component present in the input signal. While this may sound like a serious limitation, the use of a sample-and-hold circuit actually extends the highest usable frequency of an A/D converter by several thousand times.

Sample-and-hold circuits are available in integrated form. The SMP-10 is a sample-and-hold amplifier manufactured by Precision Monolithics. It offers a 3.5 microsecond acquisition time, a 50 nanosecond aperture time, and a 5.0 microvolt per millisecond droop rate.

8.1.4 Multiplexers

Many systems have several analog inputs that are monitored by a single computer or digital system. Each of these signals must be converted before the computer can process the signal. Since the A/D conversion circuitry can be quite expensive (relative to other subsystems), many designers opt to multiplex several analog inputs through a single A/D converter circuit. This technique is illustrated in Fig. 8.4.

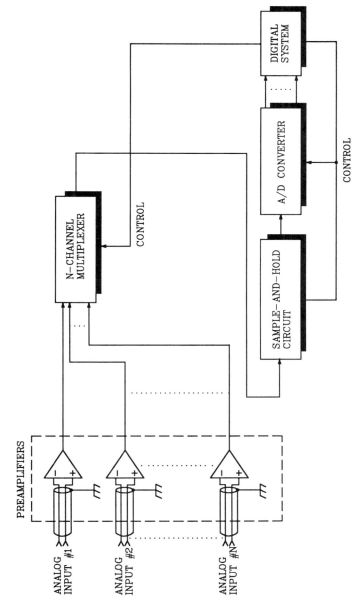

Figure 8.4 A multiplexer is used to route several analog inputs through a single A/D conversion circuit.

The multiplexer acts like a rotary switch that connects each of the analog inputs to the S/H circuit on a one-at-a-time basis. The position and timing of the "switch" is controlled by the computer or digital system. There should be total isolation between the channels of a multiplexer circuit. Sometimes, however, signal voltages from one channel will couple into another channel (generally via stray or internal capacitance). The resulting interference is called *crosstalk*.

This approach to system design can make it possible to use a higher performance (i.e., more expensive) sample-and-hold and A/D converter circuits by requiring only one such circuit for multiple inputs.

Multiplexers are available as integrated circuits. One such device is the MUX-16 manufactured by Precision Monolithics. This is a 16-channel device designed to select one of 16 analog input signals and to connect it through to a single analog output.

8.1.5 Digital-to-Analog Converters

Figure 8.5 shows the basic configuration for digital-to-analog (D/A) conversion. The digital system (frequently a microprocessor) computes the required value of analog signal and outputs an equivalent digital number. The D/A converter circuit then converts this digital number into an analog voltage or current for use by the external analog device.

Since the input to the D/A converter has a finite number of digital combinations, the resulting analog output also has a limited number of possible values (unlike pure analog signals which may have an infinite number of values). The greater the number of possible values, the closer the analog output will be to the ideal value. The number of possible levels is determined by the number of lines or bits in the digital number. More specifically, the number of states is computed as 2^N where N is the number of bits in the digital number. For example, an 8-bit D/A converter could be expected to produce 2^8 or 256 discrete output steps. If the full-scale range of the converter was 0 to 10 volts, then each step would be 10/256 or about 39 millivolts. If finer resolution were required, we would need more bits in the digital number. A converter with a 10-bit resolution, for example, would provide 2^{10} or 1024 steps with each step being equivalent to 10/1024 or about 9.8 millivolts.

Accuracy of a D/A converter describes the amount of error between the actual output of the converter and the theoretical output for a given input number. This rating inherently includes several other sources of error.

A certain amount of time is required for the output of a D/A converter to be correct once a particular digital number has been applied at the input. There are two major factors

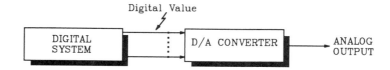

Figure 8.5 A digital-to-analog converter is used to make a digital signal (number) compatible with an analog system.

Sec. 8.1 D/A and A/D Conversion Fundamentals

which cause this delay. First, it takes time for the changes to pass through the converter circuitry. This delay is called *propagation time*. Second, the output of the D/A converter has a maximum rate of change called *slew rate*. This is identical to the slew-rate problems discussed with reference to op amps. The delay caused by slew-rate limiting and propagation time are collectively referred to as *settling time*. This is the total time required for the analog output to stabilize after a new digital number has been applied to the input.

The overall operating range of a D/A converter can be shifted up or down from the optimum point. This DC offset is called an *offset error*. In a somewhat similar manner, one end of the range can be correct but the other extreme is too high or too low. This latter error is called a *gain error* or *scaling error*.

As with A/D converters, we normally want to have a *monotonic* output. That is, the output should increase whenever the input number increases. However, it is possible for a D/A converter to have a reduction in analog output at a particular point in its range even though the digital input was increasing uniformly.

Figure 8.6 shows the performance of a low-quality D/A converter. Several of the potential problems described are present in the converted waveform. The input to the converter was a 4-bit down counter (e.g., 15, 14, 13, ..., 2, 1, 0, 15). The analog output should be 16 equally spaced, decreasing steps for each cycle producing a reverse sawtooth waveform. If you examine the waveform carefully, you can see the 16 distinct output levels. However, the steps are not equal in amplitude (linearity problems), the midpoint level actually increases instead of decreasing (nonmonotonic), and there are several *glitches* caused by switching transients. Although the performance indicated by

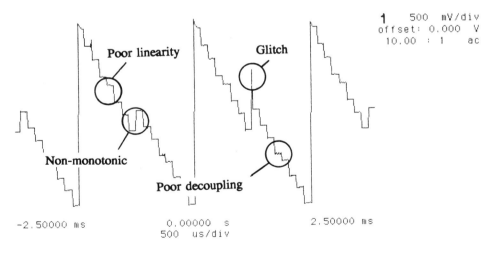

Figure 8.6 Oscilloscope display showing several imperfections in a low quality D/A converter. (Test equipment courtesy of Hewlett-Packard Company)

the waveform in Fig. 8.6 is certainly not representative of a practical D/A converter, it does provide an excellent example of several terms and definitions.

Let us now examine the actual circuitry for several of the more common methods of D/A and A/D conversion.

8.2 WEIGHTED D/A

Figure 8.7 shows the schematic diagram of a weighted digital-to-analog converter circuit built around a 741 op amp. You should recognize the configuration as being identical to the inverting summing amplifier discussed in Chap. 2. You may recall that the gain for each input is determined by the ratio of the feedback resistor and the respective input resistor. In the case of the circuit in Fig. 8.7, the individual gains are as follows:

$$A_{v1} = -\frac{R_F}{R_1} = -\frac{16 \times 10^3}{16 \times 10^3} = -1$$

$$A_{v2} = -\frac{R_F}{R_2} = -\frac{16 \times 10^3}{8 \times 10^3} = -2$$

$$A_{v3} = -\frac{R_F}{R_3} = -\frac{16 \times 10^3}{4 \times 10^3} = -4$$

$$A_{v4} = -\frac{R_F}{R_4} = -\frac{16 \times 10^3}{2 \times 10^3} = -8$$

For purposes of analysis, let us suppose that the two levels for the digital input signals are 0 and minus 1 volt. Table 8.2 shows the correlation between the digital number and the state of the four inputs.

You will likely recall from the discussion of inverting summing amplifiers, that the output voltage, at any given time, can be determined by simply adding the output voltages computed for each input individually. For example, if a digital number 5 were input to the circuit, the output voltage would be computed as follows:

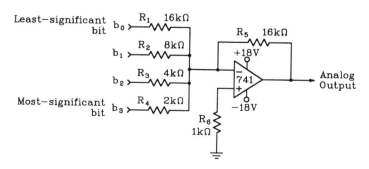

Figure 8.7 A weighted D/A converter with 4-bit resolution.

Sec. 8.2 Weighted D/A

TABLE 8.2

DIGITAL NUMBER	BINARY VALUE	b_3	b_2	b_1	b_0
0	0000	0V	0V	0V	0V
1	0001	0V	0V	0V	−1V
2	0010	0V	0V	−1V	0V
3	0011	0V	0V	−1V	−1V
4	0100	0V	−1V	0V	0V
5	0101	0V	−1V	0V	−1V
6	0110	0V	−1V	−1V	0V
7	0111	0V	−1V	−1V	−1V
8	1000	−1V	0V	0V	0V
9	1001	−1V	0V	0V	−1V
10	1010	−1V	0V	−1V	0V
11	1011	−1V	0V	−1V	−1V
12	1100	−1V	−1V	0V	0V
13	1101	−1V	−1V	0V	−1V
14	1110	−1V	−1V	−1V	0V
15	1111	−1V	−1V	−1V	−1V

$$v_{O_0} = A_{V_1}v_{I_0} = -1(-1) = +1 \text{ V}$$
$$v_{O_1} = A_{V_1}v_{I_1} = -2(0) = 0 \text{ V}$$
$$v_{O_2} = A_{V_2}v_{I_2} = -4(-1) = +4 \text{ V}$$
$$v_{O_3} = A_{V_3}v_{I_3} = -8(0) = 0 \text{ V}$$
$$\text{analog output} = v_{O_0} + v_{O_1} + v_{O_2} + v_{O_3} = +1 +0 +4 +0 = +5 \text{ V}$$

The scaling factor for the converter is such that each step in the output corresponds to one volt. This means that the analog voltage output will be the same **numerical** value as the digital input. This is not necessarily true for all converters. That is, the full-scale **digital** input for a 4-bit converter will always be 1111 (decimal 15). The full-scale output for the D/A converter shown in Fig. 8.7 is 15 volts, but it could just as easily be 5 volts, 10 volts, or any other number depending on the scale factor of the converter circuit.

For satisfactory performance, the input resistors must be very carefully selected (i.e., precision values) in order to maintain the correct ratios. If one or more resistors are the wrong value, the output will exhibit problems including poor linearity and/or lack of monotonicity. Even with careful selection of resistors, the simple weighted D/A converter is only useful for small numbers of bits since the ratio of the smallest resistor to the largest resistor quickly becomes impractical. That is, the ratio increases as 2^{N-1} where N is the number of bits in the input. For example, the resistor in the least-significant input of a 10-bit converter would be 2^{10-1} or 512 times larger than the resistor for the most-significant input.

A variation of the basic weighted D/A converter involves dividing the bits into two or more groups, and converting each group separately. The weighting resistors for each group are identical. The outputs from each of the individual converters can then be summed into a weighted, summing amplifier to produce the final output.

8.3 R2R LADDER D/A

One of the most popular methods for D/A conversion is shown in Fig. 8.8. It is called an R2R ladder D/A converter since the input network resembles the rungs on a ladder, and since the resistors in the input network are either equal (R) or have a 2:1 ratio (2R). One advantage of the R2R converter over the weighted converter previously discussed is immediately apparent; the resistors have a maximum of 2:1 ratio *regardless* of the number of bits being converted. This makes the matching of resistors much easier, and even makes the use of integrated resistors practical.

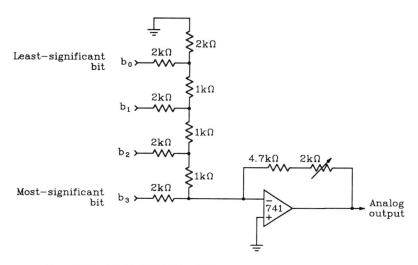

Figure 8.8 A 4-bit R2R ladder D/A converter utilizing a 741 op amp.

Sec. 8.3 R2R Ladder D/A

An easy way to analyze the operation of the circuit is to Thevenize the input circuit for one or more digital input numbers. This process was described in Chap. 1. Once the input circuit has been simplified with Thevenin's Theorem, you will be left with a simple inverting amplifier circuit whose input voltage is the Thevenin equivalent voltage and whose gain is determined by the ratio of feedback resistance to Thevenin equivalent input resistance. By performing several analyses with different input numbers, you will discover that the least-significant input produces the least effect on output voltage. The next input (b_1) has twice as much effect on output voltage. Similarly, bit b_2 has twice the effect of b_1 but only half the effect on output voltage as b_3. These variable effects are identical to the relative weights of the digits in a binary number.

The actual performance of an inexpensive R2R ladder D/A converter circuit similar to the one shown in Fig. 8.8 is revealed by the oscilloscope waveform in Fig. 8.9. Although the linearity is certainly less than optimum, it clearly illustrates the principles involved and would be adequate for many D/A applications. The linearity could be **greatly** improved by using precision resistors (rather than 5%), and by driving the digital inputs via analog switches (rather than directly from the output of a digital counter). Figure 8.9(a) shows the actual output of the converter. The 16 distinct output levels are easily seen in the waveform. Fig. 8.9(b) shows the same basic circuit after the output has gone through a simple low-pass filter. The abrupt changes in the output are now gone leaving us with a cleaner analog signal.

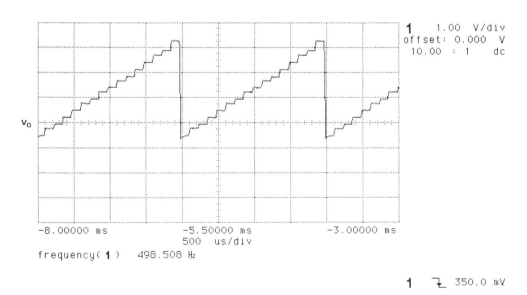

(a)

Figure 8.9 Oscilloscope displays showing the performance of the 4-bit D/A converter shown in Fig. 8.8. (Test equipment courtesy of Hewlett-Packard Company)   (continued)

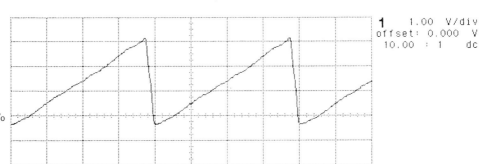

(b)

Figure 8.9b

Construction of your own D/A converters is feasible for noncritical applications. The low cost and high performance (e.g., laser-trimmed ladder resistors) available in integrated converters, however, make these devices the best choice for many applications. An example of such a device is the DAC03ADX, 10-bit D/A converter manufactured by Precision Monolithics.

8.4 PARALLEL A/D

Parallel A/D conversion (sometimes called *flash* conversion) is the fastest technique available and the simplest to understand. However, the practicality is limited to small numbers of bits since it requires $2^N - 1$ comparator circuits in order to produce an N-bit digital output. For example, to produce a 3-bit digital output (8 states) would require $2^3 - 1$ or 7 comparator circuits plus a significant amount of logic circuitry. Figure 8.10 shows the complete schematic diagram of a 3-bit parallel A/D converter circuit (including decoder logic).

The operation of the circuit is very straightforward. The voltage divider provides a stable reference for one input of each of the seven voltage comparators. Further, each voltage tap on the divider is 1.25 volts higher than the preceding one. This effectively divides the 10-volt range into 8 distinct ranges. These ranges and the corresponding

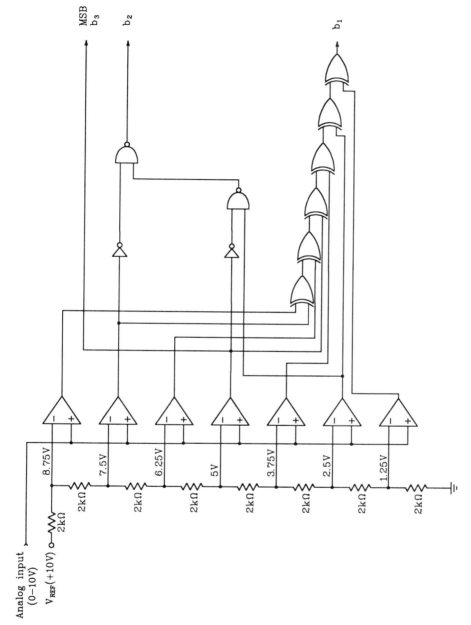

Figure 8.10 A 3-bit parallel A/D converter with a natural binary output.

TABLE 8.3

VOLTAGE RANGE	COMPARATOR STATES	DIGITAL RESULT
$0 \leq v_{IN} < 1.25$ V	0000000	000
1.25 V $\leq v_{IN} < 2.5$ V	0000001	001
2.5 V $\leq v_{IN} < 3.75$ V	0000011	010
3.75 V $\leq v_{IN} < 5$ V	0000111	011
5 V $\leq v_{IN} < 6.25$ V	0001111	100
6.25 V $\leq v_{IN} < 7.5$ V	0011111	101
7.5 V $\leq v_{IN} < 8.75$ V	0111111	110
8.75 V $\leq v_{IN} < 10$ V	1111111	111

comparator outputs are shown in Table 8.3. This Table also shows the converted digital output for each voltage range.

The converted digital output for the given converter is in standard binary format. Actual converter circuits, however, may use any one of a variety of codes including binary, gray code, excess-3, and others.

It is also possible to utilize the parallel converter in a hybrid configuration which gains some of the advantage of parallel conversion and yet avoids the geometrically increasing complexity normally associated with parallel conversion. This hybrid method essentially consists of applying the analog voltage to a small (4–7 bits) parallel converter. This converter generates the most-significant bits in the converted number. The digits are then reconverted to analog with a D/A converter and subtracted from the original input signal. The difference voltage is then converted with a second parallel converter to produce the least-significant bits of the digital result. This multistage method of parallel A/D conversion is faster than nonparallel methods but slower than a pure parallel approach. The complexity, however, is less than a pure parallel converter circuit. The PM-0820 CMOS high-speed 8-bit A/D converter circuit manufactured by Precision Monolithics is an example of a multistage parallel converter. It can deliver an 8-bit output in 1.5 microseconds.

8.5 TRACKING A/D

Figure 8.11 shows the schematic diagram of a tracking A/D converter. Here, the op amp plays a small but important role as a voltage comparator.

To understand the operation of the circuit, let us begin by assuming the counter is at zero and the analog input is at some positive voltage. The output of the counter (zero

Sec. 8.5 Tracking A/D

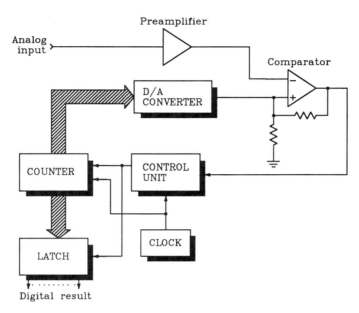

Figure 8.11 A tracking A/D converter will continuously follow (i.e., track) the analog input signal.

at the present time) is converted to an analog voltage by a D/A converter and applied to one input of a comparator. The other input is the amplified analog input signal. Under the given conditions, the output of the comparator will be low. The control unit interprets this comparator output to mean that the counter output is lower than the analog input so the counter is allowed to increment.

This situation continues on each subsequent clock pulse until the counter has incremented to a value that exceeds the analog input voltage. When this point is reached, the output of the D/A will be higher than the analog input voltage causing the comparator output to go to a high level. The control unit interprets this to mean the counter has exceeded the input, and directs the counter to begin counting down.

As the counter decrements, the output of the D/A becomes less. As soon as the D/A output falls below the analog input, the output of the comparator switches low again and causes the counter to start incrementing again. Thus, as the input changes, the counter automatically tracks it.

Every time the comparator *changes* state, the control unit transfers the counter value to a latch where it is accessible to other circuitry. This method is simple and inexpensive, but is not particularly fast (especially for large input changes). For example, let us assume that the clock is operating at 20 megahertz and the converter is designed to provide a 16-bit output. If the input signal makes a small (equivalent of one bit) change, then it will take the circuit 1/20 megahertz or 50 nanoseconds to provide a valid output. However, if the input made a full-range step change, then it would take the converter (1/20 MHz) $\times$ 2^{16} or 3.28 milliseconds to provide a valid result. This converter is best suited

362 Digital-to-Analog and Analog-to-Digital Conversion Chap. 8

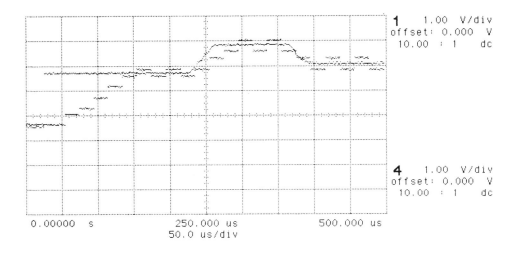

Figure 8.12 Oscilloscope display showing the operation of a tracking A/D converter circuit. (Test equipment courtesy of Hewlett-Packard Company)

for either slow signals or signals which make only small changes at any given time.

The oscilloscope display in Fig. 8.12 shows the performance of an actual tracking A/D converter circuit. The two superimposed waveforms are taken from the two inputs of the op amp voltage comparator (refer to Fig. 8.11). Near the left side of the screen, the analog input signal makes a large step change. It takes the counter and A/D circuit six clock pulses to climb to the new input level. At this point, the counter and A/D signal oscillate back and forth on either side of the analog signal. This oscillation in the least-significant bit will always occur since the counter must always count either up or down.

Near the center of the screen (in Fig. 8.12), the analog input makes another upward change. The counter and D/A output can be seen to track the change. Similarly, when the input makes a negative transition, the counter and D/A output continue to track the input signal. The response time of the tracking A/D converter is determined solely by the frequency of the input clock.

8.6 DUAL-SLOPE A/D CONVERSION

Figure 8.13 shows the schematic diagram of a basic dual-slope A/D converter. Let's first examine each of its subcircuits, and then analyze the overall operation of the circuit.

The heart of the circuit is an op amp, linear ramp generator circuit. Figure 8.14 shows the ramp generator isolated from the rest of the converter circuit. It is designed such that the charging current for capacitor C will always be constant. Basic circuit theory

Sec. 8.6 Dual-Slope A/D Conversion

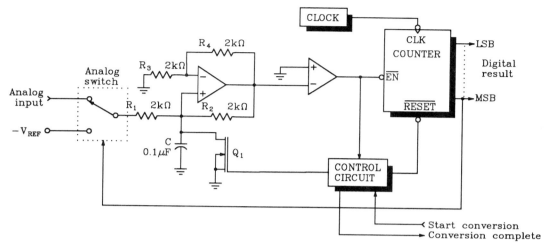

Figure 8.13 A basic dual-slope analog-to-digital converter circuit.

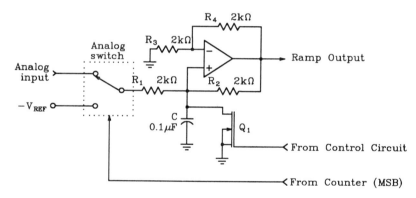

Figure 8.14 The linear ramp generator portion of the circuit shown in Fig. 8.13.

tells us that a constant charging current through a capacitor produces a linear ramp of voltage.

To understand the operation of the ramp generator circuit, let us assume that the capacitor is initially discharged (i.e., zero volts). This is the purpose of transistor Q_1. As long as Q_1 is saturated, capacitor C cannot accumulate a charge. Although the actual saturation voltage of Q_1 may be a few millivolts, let us assume it is truly zero volts for simplicity. Let us further assume (as an example), that the input voltage to the ramp generator is +5 volts. Now let's cutoff transistor Q_1 and allow capacitor C to begin charging. We shall compute the current through the capacitor at several times.

At the first instant after Q_1 is cutoff, the capacitor has 0 volts charge. Ohm's Law tells us that resistor R_1 will have a current of

$$I_{R_1} = \frac{E_{R_1}}{R_1} = \frac{5 - 0}{2 \times 10^3} = 2.5 \text{ mA}$$

The op amp is essentially a noninverting amplifier with respect to the capacitor voltage. The voltage gain is given by our basic equation for noninverting amplifiers.

$$A_V = \frac{R_F}{R_{IN}} + 1 = \frac{2 \times 10^3}{2 \times 10^3} + 1 = 2$$

The output voltage at this instant will be zero volts (i.e., 0×2). Resistor R_2 will have zero volts on both ends which means it has no current flow through it. We know that negligible current flows in or out of the (+) terminal of the op amp. Now, since 2.5 milliamperes of current is flowing through R_1, but no current flows to the op amp or through R_2, we can apply Kirchoff's Current Law to conclude that the entire 2.5 milliamperes must be flowing into capacitor C as a charging current. The direction of the electron current is from ground, up through capacitor C, and through R_1 to the positive 5 volt source. This establishes the initial slope of the charge on C. If we can maintain a constant current, we will maintain a linear slope across C.

Now, let us examine the circuit condition after capacitor C has accumulated 1 volt of charge (positive on top). The current through R_1 can now be computed as

$$I_{R_1} = \frac{E_{R_1}}{R_1} = \frac{5 - 1}{2 \times 10^3} = 2 \text{ mA}$$

With 1 volt on the capacitor and a voltage gain of 2, we can compute the output voltage of the op amp as

$$v_O = v_{IN} \times A_V = 1 \times 2 = 2 \text{ V}$$

The current through R_2 can be found with Ohm's Law since it has 1 volt on the left end and 2 volts on the right end.

$$I_{R_2} = \frac{E_{R_2}}{R_2} = \frac{2 - 1}{2 \times 10^3} = 0.5 \text{ mA}$$

Again, Kirchoff's Current Law will let us conclude that if 2 milliamperes are flowing to the left through R_1 and 0.5 milliampere is flowing to the right through R_2 then capacitor C must still be charging with a 2.5-milliampere current. Let's examine the circuit at one more point.

Suppose we let capacitor C accumulate a charge of 4 volts. The current through R_1 will then be

$$I_{R_1} = \frac{E_{R_1}}{R_1} = \frac{5 - 4}{2 \times 10^3} = 0.5 \text{ mA}$$

Sec. 8.6 Dual-Slope A/D Conversion

With +4 volts on the capacitor, the output voltage of the op amp must be

$$v_O = v_{IN}A_V = 4 \times 2 = 8 \text{ V}$$

The current through R_2 can be calculated as

$$I_{R_2} = \frac{E_{R_2}}{R_2} = \frac{8-4}{2 \times 10^3} = 2 \text{ mA}$$

Finally, we apply Kirchoff's Current Law to show that with 0.5 milliampere flowing right-to-left through R_1 and 2 milliamperes flowing left-to-right through R_2, there must surely be 2.5 milliamperes flowing upward through capacitor C. Since the current through capacitor C has remained constant at 2.5 milliamperes, we know that the voltage across it will be a linearly rising ramp. The slope of the ramp is given by the basic capacitor charge equations

$$\text{slope} = \frac{E_C}{T} = \frac{I_C}{C}$$

For the present case, the slope of the ramp across C is computed as

$$\text{slope} = \frac{I_C}{C} = \frac{2.5 \times 10^{-3}}{0.1 \times 10^{-6}} = 25 \text{ V/millisecond}$$

The output of the op amp will have a slope that is linear but twice as great since the amplifier has a voltage gain of 2. In either case, the slope of the ramp is determined by the charging current of C which is determined by the value of input voltage.

Now, let's analyze the overall operation of the dual-ramp A/D converter shown in Fig. 8.13. The input voltage to the ramp is switch selected as either the analog voltage to be converted (positive) or a fixed, negative reference voltage. Recall that the input voltage to the ramp circuit determines the slope of the ramp. The position of the analog switch is controlled by the state of the most-significant bit (MSB) of a counter. More specifically, if the MSB is low, then the switch will connect the analog input to the ramp generator. If the MSB of the counter is high, then the switch connects the negative reference voltage to the ramp generator input.

The counter is enabled (i.e., allowed to count) as long as the output of the ramp generator is positive. That is, as long as the ramp is above ground, the output of the comparator will be low and enable the counter. If the ramp ever goes below ground, then the output of the comparator will switch to a high state and disable the counter.

The control circuit provides the overall timing of circuit operation. Upon receipt of a *start conversion* signal from the main control system (generally a microprocessor), the control unit will reset the counter to zero and release (i.e., cutoff) Q_1. With the counter reset, the MSB will be zero and the analog switch will be connecting the analog input to the ramp generator circuit. As the counter counts up, the capacitor voltage (and the op amp output) will be linearly ramping up in a positive direction. This action is indicated in Figure 8.15 as t_1.

This action will continue until the counter reaches one half of its maximum count. At this point, the MSB of the counter will go high and cause the analog switch to move to the reference voltage position. With a negative input voltage applied to the ramp generator, the capacitor will begin to discharge. The discharge will be linear, and the rate will be determined by the value of the negative reference voltage. Eventually, the decreasing ramp will pass through zero volts. This causes the comparator to switch states and disables the counter. The control circuit senses this event and generates the *conversion complete* signal. This means that the digital result in the counter is now a valid representation of the analog input voltage.

We know that the initial slope (during time t_1 in Fig. 8.15) is determined by the value of the analog input voltage. The length of time for t_1, however, is fixed and is determined by the speed of the clock and the number of bits in the counter. Time t_2 in Fig. 8.15 is the amount of time required for the capacitor to linearly discharge to zero volts. The slope of t_2 is fixed and is determined by the negative reference. The time t_2, therefore, is variable and dependent on the value of voltage accumulated on capacitor C during time t_1. This voltage, of course, was determined by the value of analog input voltage. Since time t_2 is dependent on the value of analog input voltage, the number of counts registered in the counter will also be a function of the analog input voltage.

Figure 8.15 contrasts the results of two different analog input voltages. V_{C1} is the result of a higher input voltage. It takes a certain amount of time (t_2) to discharge the capacitor and stop the counter. A lower input voltage (V_{C2}) charges C to a lower voltage during the fixed time period t_1. Therefore, the discharge time (t_3) is shorter, and the counter will have a smaller count. The final converted result appears in the counter and ignores the MSB.

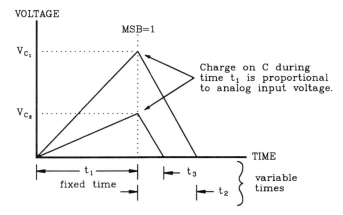

Figure 8.15 The positive slope of a dual-slope converter is determined by the value of analog input voltage. The slope of the negative ramp is determined by V_{REF}.

The dual-slope A/D conversion method is very popular in applications that do not require high-speed operation. It has distinct advantages that include high immunity to component tolerances, component drifts, and noise. This increased immunity stems from the fact that errors introduced during the positive slope will be largely offset by similar errors during the negative slope. The circuit offers total rejection of noise signals that are even multiples of the time period t_1 since the net effect of a **full** cycle of noise is zero.

Complete dual-slope converter systems are available in integrated form. A common application is for digital voltmeters. The analog portion of such a system is manufactured by National Semiconductor Corporation in the form of an LF12300 integrated circuit. Analog Devices has a patented improvement on the basic dual-slope converter called Quad-Slope conversion. This is used in the AD7550 13-bit A/D converter manufactured by Analog Devices.

8.7 SUCCESSIVE APPROXIMATION A/D

Figure 8.16 shows the functional block diagram of a successive approximation analog-to-digital converter system. This is probably the most widely used type of A/D converter since it is fairly simple and yet offers a relatively high speed of operation.

The analog input signal provides one input to a voltage comparator. The second input comes from the output of a D/A converter. The input to the D/A converter is provided by an addressable latch called the successive approximation register (SAR). Each bit of this register can be selectively set or cleared by the control unit. The control unit can be an internal portion of an integrated circuit, or it may be a complete microprocessor system. In any case, the overall operation of the successive approximation converter is described.

Let us suppose that the analog input voltage is 5.7 volts. Let us further suppose that the SAR and D/A converter are 4-bit devices that provide 0.625 volts per step at the output of the D/A converter. Finally, let us assume that the SAR is initially set to 1000 thus producing a 5-volt output for the D/A converter.

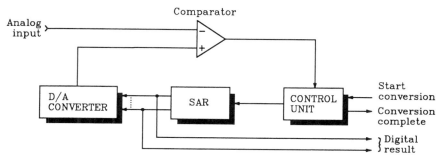

Figure 8.16 A functional block diagram of a successive-approximation analog-to-digital converter circuit.

Under these conditions, the output of the comparator will be low since the analog input (inverting input) is higher than the D/A output voltage (noninverting input). The control unit interprets this to mean that the SAR value is too low. The control unit then leaves the MSB (b_3) alone and also sets the next bit (b_2). The SAR now holds the binary value of 1100 which converts to 7.5 volts at the output of the D/A converter. Since this exceeds the value of analog input voltage, the comparator output will go high. The control unit interprets this to mean that the SAR value is too high.

The control unit then resets bit b_2 (since that is what caused the excessive value) and sets the next most significant bit (b_1). The SAR value of 1010 is now applied to the D/A converter to produce a comparator input of 6.25 volts. This is still higher than the analog input voltage so the comparator output remains high. The control unit again interprets this to mean that the SAR value is too large.

The control unit then resets bit b_1 and sets the next lower bit (b_0) thus yielding an SAR value of 1001. This converts to 5.625 volts at the output of the comparator. Since this is still lower than the analog input, b_0 will remain set. The *conversion complete* signal is now generated indicating a completed conversion. The result (1001) is available in the SAR.

In general, the control unit starts by setting the most-significant bit and monitoring the output of the comparator. This bit will then be left set or it will be reset as a function of the state of the comparator. In either case, the same process is applied to the next lower bit and so forth until the least-significant bit is left either set or reset. With each progressive step, the approximation gets closer. Regardless of the magnitude of the analog input voltage (within the limits of the converter), it will always take as many clock periods to convert the voltage as there are bits in the converted number. In this example, the successively better approximations were as listed in Table 8.4.

There is a striking similarity between the logic used during the successive approximation process and the logic applied during a split-half troubleshooting exercise. In both cases, each successive step reduces the number of options by one half.

There are many integrated forms of A/D converters which utilize the successive approximation technique. One such device is the ADC-910 which is a 10-bit converter manufactured by Precision Monolithics. It provides a 10-bit result in 6 microseconds and is compatible with microprocessors. It should also be pointed out, that all functions

TABLE 8.4

ANALOG INPUT (VOLTS)	SAR STATUS	D/A OUTPUT (VOLTS)	COMPARATOR OUTPUT
5.7	1000	5	Low
5.7	1100	7.5	High
5.7	1010	6.25	High
5.7	1001	5.625	Low

represented in Fig. 8.16, except for the comparator and D/A converter, can be implemented with software internal to a microprocessor. The speed, however, would generally be much slower than a dedicated converter.

REVIEW QUESTIONS

1. A certain A/D converter has a 12-bit resolution specification for the analog input range of 0 to 10 volts. What is the smallest voltage change that can be represented in the output?
2. If the binary output of an A/D converter does not numerically increase with every increase in input voltage, we say the converter is _____.
3. The total time required for an A/D converter to obtain a valid digital output for a given analog input is called _____.
4. Contrast and explain the terms *acquisition time* and *aperture time* with reference to sample-and-hold circuits.
5. Would a weighted D/A converter be the best choice for a 64-bit converter circuit? Explain your answer.
6. What type of A/D conversion circuit provides the fastest conversion times? What are its disadvantages?
7. If a certain successive approximation A/D converter requires 10 microseconds to resolve a 0- to 2-volt step change on the input, how long will it take to resolve a 0- to 5-volt step change?
8. Repeat quest. 7 for a tracking A/D converter.
9. The use of a sample-and-hold circuit greatly reduces the highest usable frequency for an A/D converter. (True or False)
10. Discuss the relationship between droop rate in a sample-and-hold circuit and the necessary conversion time of a subsequent A/D converter circuit.

CHAPTER 9

ARITHMETIC FUNCTION CIRCUITS

This chapter presents several circuits which are designed to perform mathematical operations including adding, subtracting, averaging, absolute value, and sign changing. Several other common, but more complex, circuits that perform mathematical functions are presented in Chapter 11.

9.1 ADDER

An adder circuit has two or more signal inputs, either AC or DC, and a single output. The magnitude and polarity of the output at any given time is the algebraic sum of the various inputs. In Chap. 2, we discussed an inverting adder circuit. It was called an inverting summing amplifier. By making the feedback resistor and all input resistors the same size, the circuit will provide a mathematically correct sum (i.e., no voltage gain). The following discussion will introduce the noninverting adder circuit.

9.1.1 Operation

Figure 9.1 shows the schematic diagram for a noninverting adder circuit. The input signals may be AC, DC, or some combination of signals. The op amp, in conjunction with R_1 and R_2, is a simple noninverting amplifier whose gain is determined by the ratio of R_2 to R_1. Whatever voltage appears on the (+) input will be amplified.

The voltage which appears on the (+) input is the output of a resistive network composed of R_3 through R_N and the associated input voltages. Since all input resistors are equal in value and connect together at the (+) input, we can infer that the relative effects of the inputs are identical. The absolute effect, of course, is determined by the gain of the amplifier. If R_2 is set to the correct value, then the gain of the op amp will be such that the output voltage corresponds to the sum of the input voltages.

Sec. 9.1   Adder

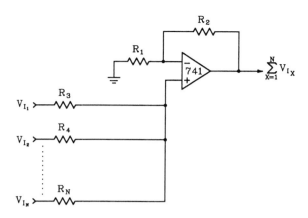

Figure 9.1   A noninverting adder circuit sums the instantaneous voltage at several inputs.

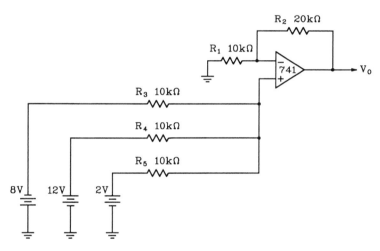

Figure 9.2   A 3-input noninverting adder with DC inputs.

9.1.2 Numerical Analysis

We shall now analyze the circuit shown in Fig. 9.2 to numerically confirm its operation. Let's first determine the behavior of the op amp and its gain resistors (R_1 and R_2). These components form a standard noninverting amplifier whose gain can be determined with the familiar gain equation.

$$A_V = \frac{R_F}{R_I} + 1 = \frac{R_2}{R_1} + 1 = \frac{20 \times 10^3}{10 \times 10^3} + 1 = 3$$

The voltage gain of a noninverting adder will always be the same as the number of inputs.

All that remains is to determine the value of voltage on the (+) input pin. A simple method for calculating this voltage is to apply Thevenin's Theorem to the network. This

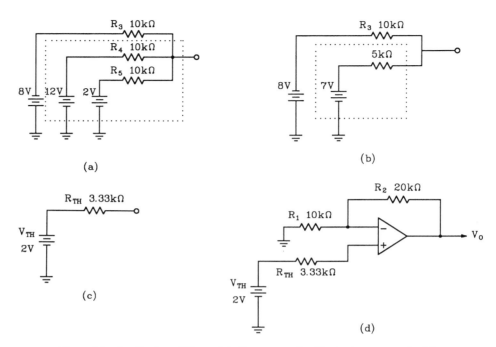

Figure 9.3 Application of Thevenin's Theorem to simplify the input network of the noninverting adder circuit.

will give us an equivalent circuit consisting of a single resistor and a single voltage source. The application of Thevenin's Theorem is shown sequentially in Fig. 9.3(a) through 9.3(c). Figure 9.3(a) shows the original input network. We shall apply Thevenin's Theorem to the portion of the circuit within the dotted box.

First we see that we have two opposing voltage sources in this part of the circuit which yield a net voltage of

$$V_X = 12 - 2 = 10 \text{ V}$$

Now, with two equal resistors in the circuit, there will be 5 volts dropped across each one. The equivalent Thevenin voltage, then, can be found with Kirchoff's Voltage Law.

$$V_{TH} = +12 - 5 = +7 \text{ V}$$

Alternately, we could apply Kirchoff's Voltage Law as

$$V_{TH} = +2 + 5 = +7 \text{ V}$$

By replacing the two upper sources with their internal resistances (assumed to be zero), we can determine the Thevenin resistance which is simply the value of R_5 and R_4 in parallel.

$$R_{TH} = R_4 \| R_5 = \frac{R_4 R_5}{R_4 + R_5} = \frac{10 \times 10^3 \times 10 \times 10^3}{(10 \times 10^3) + (10 \times 10^3)} = 5 \text{ k}\Omega$$

Sec. 9.1 Adder

The results of our first simplification are shown in Fig. 9.3(b). The Thevenin equivalents computed above are shown inside of the dotted box. We can reapply Thevenin's Theorem to the remaining circuit to obtain our final simplified circuit.

First we see that we have series-aiding voltage sources for an effective voltage source equal to

$$V_X = 7 + 8 = 15 \text{ V}$$

The portion of this effective voltage that drops across R_3 can be found with the voltage divider formula.

$$V_{R_3} = \left(\frac{R_3}{R_3 + 5 \text{ k}\Omega}\right) 15 = \left[\frac{10 \times 10^3}{(10 \times 10^3) + (5 \times 10^3)}\right] 15 = 10 \text{ V}$$

The resulting Thevenin voltage can now be found with Kirchoff's Voltage Law.

$$V_{TH} = -8 + 10 = +2 \text{ V}$$

We don't really need the value of Thevenin resistance for the remainder of the problem, but in the name of completeness we shall compute it as

$$R_{TH} = 5 \text{ k} \| 10 \text{ k} = \frac{5 \times 10^3 \times 10 \times 10^3}{(5 \times 10^3) + (10 \times 10^3)} = 3.333 \text{ k}\Omega$$

The fully simplified circuit is shown in Fig. 9.3(c). Figure 9.3(d) shows our equivalent circuit reconnected to the amplifier portion of the circuit. The output voltage can be easily computed by applying our basic gain equation.

$$V_O = V_I A_V = 2 \times 3 = 6 \text{ V}$$

This value confirms the correct operation of our adder which should provide the algebraic sum of its inputs (i.e., $+2 + 12 - 8 = +6$).

9.1.3 Practical Design Techniques

Let us now design a noninverting adder that will satisfy the following design goals:

1. accept four inputs $-10 < \text{sum} < 10$
2. minimum input impedance > 6 kilohms
3. frequency range   0 to 1 kilohertz

Select the value for R_1, and R_3 to R_6. The design of the noninverting adder circuit is very straightforward since all resistor values are the same with the single exception of the feedback resistor. Selection of the common resistor value is made by considering the following guidelines:

1. High-value resistances magnify the nonideal op amp characteristics and make the circuit more susceptible to external noise.
2. Low-resistor values present more of a load on the driving circuits.

Resistor selection will determine the input impedance presented to the various signal sources as expressed by Eq. (9.1).

$$R_{IN} = R\left(\frac{1}{N-1} + 1\right) \quad (9.1)$$

where R is the common resistor value and N is the number of inputs to the adder circuit. Since the design goal specifies a minimum input of 6 kilohms, let us substitute this value into Eq. (9.1) and solve for the value of R.

$$6000 = R\left(\frac{1}{3} + 1\right)$$

$$R = \frac{6000}{1.333} = 4.5 \text{ k}\Omega$$

This represents the smallest value that we can use for the input resistors and still meet our input impedance requirement. Let us choose to use 5.6 kilohms resistors for R_1, and for R_3 to R_6.

Calculate the feedback resistor. The value of the feedback resistor must be selected such that the voltage gain is equal to the number of inputs. In our case we will need a gain of 4. Since we already know the value of R_1, we can transpose our basic noninverting amplifier gain equation to determine the value of feedback resistor.

$$A_V = \frac{R_2}{R_1} + 1,$$

or

$$R_2 = R_1(A_V - 1)$$

Since voltage gain will always be equal to the number of inputs, this can also be written as

$$R_2 = R_1(A_V - 1) = R_1(N - 1) \quad (9.2)$$

In our particular case, we can compute the required value of R_2 as

$$R_2 = 5.6 \times 10^3(4 - 1) = 16.8 \text{ k}\Omega$$

If we expect the adder to generate the correct sum, it is essential to keep the resistor ratios correct. Therefore, since 16.8 kilohms is not a standard value, we will need to use a variable resistor for R_2 or some combination of fixed resistors (e.g., 15 kΩ in series with 1.8 kΩ).

Select an op amp. There are several nonideal op amp parameters which may affect the proper operation of the noninverting adder. An op amp should be selected to minimize those characteristics which are most important for a particular application. The

Sec. 9.2   Subtractor

various nonideal parameters that should be considered are discussed in Chap. 10. In general, if DC signals are to be added, then an op amp with a low offset voltage and low drift would likely be in order. For AC applications, bandwidth and slew rate are two important limitations to be considered.

Based on the modest gain/bandwidth requirements required for this particular application, let us use a standard 741 op amp. There are other more precision devices that can be substituted to optimize a particular characteristic (e.g., low noise). Many of these alternate devices are pin compatible with the basic 741.

The final schematic diagram of our noninverting adder design is shown in Fig. 9.4. The measured performance is contrasted with the original design goals in Table 9.1. It should be noted that the noninverting adder is particularly susceptible to component tolerances and nonideal op amp parameters (e.g., bias current and offset voltage). For reliable operation, components must be carefully selected and good construction techniques must be used.

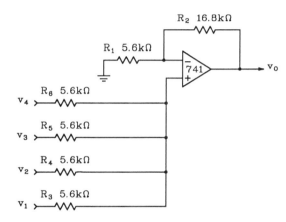

Figure 9.4 Final design of a 4-input noninverting adder circuit.

TABLE 9.1

INPUT VOLTAGES				OUTPUT VOLTAGE	
V_1	V_2	V_3	V_4	IDEAL	ACTUAL
−15.09 V	4.37 V	0 V	0 V	10.72 V	10.8 V
−8.1 V	−2.0 V	6.6 V	2.7 V	−0.8 V	−0.74 V
1.52 V	0 V	−3.1 V	−1.49 V	−3.07 V	−2.98 V

9.2 SUBTRACTOR

Another circuit which performs a fundamental arithmetic operation is the subtractor circuit. This circuit generally has two inputs (either AC or DC) and produces an output voltage that is equal to the instantaneous difference between the two input signals. This is, of course, the very definition of a difference amplifier which is another name for the subtractor circuit.

9.2.1 Operation

Figure 9.5 shows the schematic diagram of a basic subtractor circuit. A simple way to view the operation of the circuit is to mentally apply the Superposition Theorem (without numbers). If we assume that V_A is zero volts (i.e., grounded), then we can readily see that the circuit functions as a basic inverting amplifier for input V_B. The voltage gain for this input will be determined by the ratio of resistors R_1 and R_2. If we assume that the voltage gain is -1, then the output voltage will be $-V_B$ volts as a result of the V_B input signal.

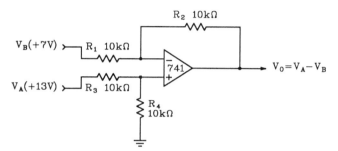

Figure 9.5 A subtractor circuit computes the voltage difference between two signals.

In a similar manner, we can assume input V_B is grounded. In this case, we find that the circuit functions as a basic noninverting amplifier with respect to V_A. The overall voltage gain for the V_A input will be determined by the ratio of R_1 and R_2 (sets the op amp gain) and the ratio of R_3 and R_4 which form a voltage divider on the input. If we assume that the voltage divider reduces V_A by half, and we further assume that the op amp provides a voltage gain of 2 for voltages on the $(+)$ input, then we can infer that the output voltage will be $+V_A$ volts as a result of the V_A signal.

According to the Superposition Theorem, the output should be the net result of the two individual input signals. That is, the output voltage will be $+V_A - V_B$. Thus we can see that the circuit does indeed perform the function of a subtractor circuit.

9.2.2 Numerical Analysis

The numerical analysis of the subtractor circuit is straightforward and consists of applying the Superposition Theorem. Let us first assume that the V_A input is grounded and compute the effects of the V_B input. The output will be equal to V_B times the voltage gain of the inverting amplifier circuit. That is

$$V_{O_B} = V_B \times A_V$$

$$= V_B \left(-\frac{R_2}{R_1} \right)$$

Sec. 9.2 Subtractor

$$= +7\left(-\frac{10 \times 10^3}{10 \times 10^3}\right)$$

$$= -7 \text{ V}$$

Now let us ground the V_B input and compute the output voltage caused by the V_A input. First, V_A is reduced by the voltage divider action of R_3 and R_4. The voltage appearing on the (+) input is computed with the basic voltage divider equation.

$$V'_A = V_A\left(\frac{R_4}{R_3 + R_4}\right)$$

$$= +13\left(\frac{10 \times 10^3}{(10 \times 10^3) + (10 \times 10^3)}\right)$$

$$= 6.5 \text{ V}$$

The voltage on the (+) input will now be amplified by the voltage gain of the noninverting op amp configuration. That is

$$V_{OA} = V'_A\left(\frac{R_2}{R_1} + 1\right)$$

$$= +6.5\left(\frac{10 \times 10^3}{10 \times 10^3} + 1\right)$$

$$= +13 \text{ V}$$

The actual output voltage will be the algebraic sum of the two partial outputs computed. That is

$$V_O = V_{OA} + V_{OB} = +13 + (-7) = +6 \text{ V}$$

This, of course, is the result that we would expect from a circuit which is supposed to compute the difference between two input voltages.

The input impedance for the inverting input (V_B signal) can be computed with Eq. (2.7). For the circuit shown in Fig. 9.5, we can compute the inverting input resistance as

$$Z_I = R_1 = 10 \text{ k}\Omega$$

The input impedance presented to the noninverting input is essentially the value of R_3 and R_4 in series. That is

$$Z_I \approx R_3 + R_4 = (10 \times 10^3) + (10 \times 10^3) = 20 \text{ k}\Omega$$

If resistors R_3 and R_4 are quite large, then a more accurate value can be obtained by considering that the input resistance of the op amp itself is in parallel with R_4. The input resistance of the 741 is specified as at *least* 300 kilohms at 25 °C. Taking this into account, we can recompute the input resistance of the noninverting input as

$$Z_I = R_3 + (R_4 \| R_{OP})$$
$$= (10 \times 10^3) + \left(\frac{300 \times 10^3 \times 10 \times 10^3}{(300 \times 10^3 + (10 \times 10^3)} \right)$$
$$= 19.68 \text{ k}\Omega$$

The small signal bandwidth of the circuit can be computed by applying Eq. (2.22) to the (+) input. For our present circuit, we can estimate the bandwidth as

$$\text{bandwidth} = \frac{\text{unity-gain frequency}}{\text{closed-loop gain}} = \frac{1 \times 10^6}{2} = 500 \text{ kHz}$$

The highest practical operating frequency may be substantially lower than this due to slew-rate limiting of larger input signals. If we assume that the output will be required to make the full output swing from $+V_{SAT}$ (+13V) to $-V_{SAT}$(−13V), then we can apply Eq. (2.11) to determine the highest input frequency that can be applied without slew-rate limiting.

$$f_{SRL} = \frac{\text{slew rate}}{\pi V_o(\text{max})} = \frac{0.5 \text{ V}/\mu\text{s}}{3.14 \times 26} = 6.12 \text{ kHz}$$

9.2.3 Practical Design Techniques

Now let's design a subtractor circuit that will satisfy the following design goals:

1. input voltages   0 to +5 volts
2. input frequency   0 to 10 kilohertz
3. input impedance   ≥ 2.5 kilohms

Compute R_1 to R_4. The value of R_1 is determined by the minimum input impedance of the circuit. Its value is found by applying Eq. (2.7).

$$R_1 = Z_I = 2.5 \text{ k}\Omega$$

Resistors R_2 through R_4 are set equal to R_1 in order to provide the correct subtractor performance. We shall choose to use standard values of 2.7 kilohms for these resistors.

Select the op amp. The primary considerations for op amp selection are slew rate and bandwidth. Unless the input signals are very small, it will be the slew rate that establishes the upper frequency limit. The required slew rate to meet the design specifications can be found by applying Eq. (2.11). The maximum output voltage change will be 10 volts and the frequency may be as high as 10 kilohertz.

$$\text{slew rate} = \pi f V_o(\text{max}) = 3.14 \times 10 \times 10^3 \times 10 = 0.314 \text{ V}/\mu\text{s}$$

Sec. 9.2 Subtractor

This is within the range of the standard 741 op amp. The unity-gain frequency required to meet the bandwidth requirements can be estimated with Eq. (2.22). In the present case, the minimum unity-gain frequency is computed as

$$f_{UG} = bw \times A_V = 10 \times 10^3 \times 2 = 20 \text{ kHz}$$

This is well below the 1.0 megahertz unity-gain frequency of the standard 741. Let us use this device in our design.

There are several other nonideal op amp parameters that could play a significant role in op amp selection. These factors are discussed in Chap. 10.

The schematic of our subtractor design is shown in Fig. 9.6. Its performance with two, in-phase AC inputs is shown in Fig. 9.7. Channels 1 and 2 of the oscilloscope are connected to the V_A and V_B inputs, respectively. Channel 4 shows the output of the op

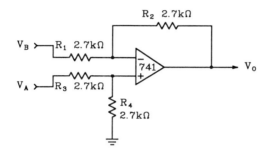

Figure 9.6 A subtractor circuit designed for 10 kilohertz operation with 0 to 5-volt input signals.

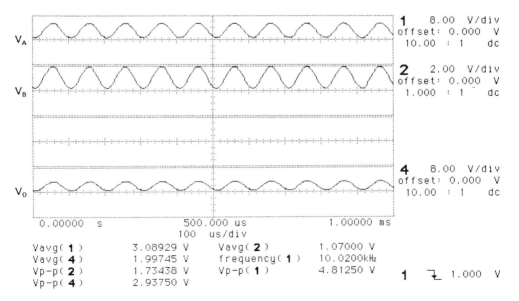

Figure 9.7 Oscilloscope display showing the actual performance of the subtractor circuit shown in Fig. 9.6. (Test equipment courtesy of Hewlett-Packard Company)

TABLE 9.2

INPUT VOLTAGES		OUTPUT VOLTAGE	
V_A	V_B	DESIGN GOAL	MEASURED VALUE
+4.56 V	+2.1 V	+2.46 V	+2.42 V
0 V	0 V	0 V	−0.05 V
+1.6 V	+3.99 V	−2.39 V	−2.43 V

amp which is the desired function ($V_O = V_A - V_B$). The response of the circuit to DC signals is listed in Table 9.2.

9.3 AVERAGING AMPLIFIER

The schematic diagram of an inverting averaging amplifier is shown in Fig. 9.8. Although this represents a separate mathematical operation, the configuration of the circuit is similar to the inverting adder or inverting summing amplifier discussed in Chap. 2.

Since its operation and design is nearly identical to the inverting summing amplifier, only a brief analysis is given here.

To understand its operation, let us apply Ohm's and Kirchoff's Laws along with some basic equation manipulation. Since the inverting input ($-$) is a virtual ground point, each of the input currents can be found with Ohm's Law.

$$I_1 = \frac{V_1}{R_1},$$

and

$$I_2 = \frac{V_2}{R_2},$$

and

$$I_N = \frac{V_N}{R_N}$$

Since negligible current flows in or out of the ($-$) input pin, Kirchoff's Current Law will show us that the sum of the input currents must be flowing through R_3. That is

$$I_3 = I_1 + I_2 + I_N,$$

or

$$I_3 = \frac{V_1}{R_1} + \frac{V_2}{R_2} + \frac{V_N}{R_N}$$

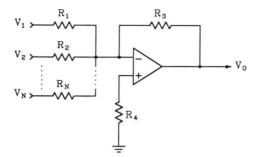

Figure 9.8 An N-input averaging amplifier circuit.

The voltage across R_3 will be equal to the output voltage since one end of R_3 is connected to a virtual ground point and the other end is connected directly to the output terminal. We can, therefore, conclude that

$$I_3 = -\frac{V_O}{R_3}$$

Now, for proper operation of the averaging circuit, all of the input resistors must be the same value; let us call this value R. Further, the feedback resistor (R_3) is chosen to be equal to R/N where N is the number of inputs to be averaged. Making these substitutions in our previous equations gives us the following:

$$\frac{V_1}{R} + \frac{V_2}{R} + \ldots + \frac{V_N}{R} = -\frac{NV_O}{R}$$

$$V_1 + V_2 + \ldots + V_N = -NV_O$$

$$-V_O = \frac{V_1 + V_2 + \ldots + V_N}{N}$$

This final expression for the output voltage should be recognized as the equation for computing arithmetic averages. That is, add the numbers (input voltages) together and divide by the number of inputs. The minus sign simply shows that the signal is inverted in the process of passing through the op amp circuit.

9.4 ABSOLUTE VALUE CIRCUIT

An op amp circuit can be configured to provide the absolute value (|**VALUE**|) of a given number. You will recall from basic mathematics that an absolute value function produces the magnitude of a number without regard to sign. In the case of an op amp circuit designed to generate the absolute value of its input, we can expect the output voltage to be equal to the input voltage without regard to polarity. So, for example, a +6.2-volt input and a −6.2-volt input would both produce the same (typically +6.2-volt) output.

9.4.1 Operation

There are several ways to obtain the absolute value of a signal. The schematic diagram in Fig. 9.9 shows one possible way. The first stage of this circuit is a dual, half-wave rectifier the same as that analyzed in Chap. 7. Recall that on positive input signals, the output goes in a negative direction and forward biases D_1. This completes the feedback loop through R_2. Additionally, the forward voltage drop of D_1 is essentially eliminated by the gain of the op amp. That is, the voltage at the junction of R_2 and D_1 will be the same magnitude (but opposite polarity) as the input voltage.

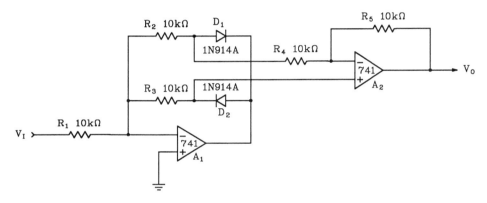

Figure 9.9 An absolute value circuit which computes the value of $|V_I|$.

When a negative input voltage is applied to the dual, half-wave rectifier circuit, the output of the op amp goes in a positive direction. This forward biases D_2 and completes the feedback loop through R_3. Diode D_1 is reverse biased. In the case of the basic dual, half-wave rectifier circuit, the voltage at the junction of R_3 and D_2 is equal in magnitude (but opposite polarity) to the input voltage. In the case of the circuit in Fig. 9.9, however, this voltage will be somewhat lower due to the loading effect of the current flowing through R_2, R_4, and R_5.

The outputs from the dual, half-wave rectifier circuit are applied to the inputs of a difference amplifier circuit. Since the two half-wave signals are initially 180° out of phase, and since only one of them gets inverted by amplifer A_2, we can conclude that the two signals appear at the output of A_2 with the same polarity. That is, both polarities of input signal produce the same polarity of output signal. This is, by definition, an absolute value function.

9.4.2 Numerical Analysis

Now let us extend our understanding of the absolute value circuit shown in Fig. 9.9 to include a numerical analysis of its operation. First notice that all resistors are the same

Sec. 9.4   Absolute Value Circuit

value. This will greatly simplify our algebraic manipulation. We shall analyze the circuit for both polarities of input voltage.

Figure 9.10 shows an equivalent circuit that is valid whenever V_I is positive. The ground on the lower end of R_3 is provided by the virtual ground at the ($-$) input of A_1. It is easy to see that we now have two inverting amplifiers in cascade. The output voltage, therefore, will be equal to the input voltage times the voltage gains of the two amplifiers. That is

$$V_O = V_I A_{V_1} A_{V_2} = V_I \left(-\frac{R_2}{R_1}\right)\left(-\frac{R_5}{R_4}\right) = V_I \left(\frac{R_2 R_5}{R_1 R_4}\right)$$

Since all resistors are the same value (R), we can further simplify the expression for V_O.

$$V_O = V_I \left(\frac{RR}{RR}\right) = V_I$$

In the case of positive input signals, the output voltage is equal to the input voltage.

Now let us consider the effects of negative input voltages. Figure 9.11 shows an equivalent circuit that is valid whenever V_I is negative. Kirchoff's Current Law, coupled with our understanding of basic op amp operation, will allow us to establish the following relationship:

$$I_1 = I_2 + I_3$$

Ohm's Law can be used to substitute resistance and voltage values.

$$-\frac{V_I}{R_1} = \frac{V_X}{R_3} + \frac{V_X}{R_2 + R_4}$$

Since all resistor values are equal (R), we can further simplify this latter expression as follows:

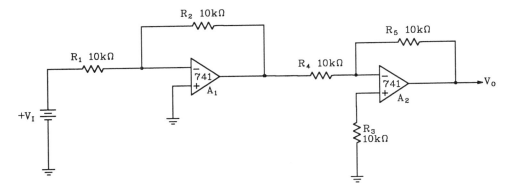

Figure 9.10 An equivalent circuit for the circuit in Fig. 9.9 during times when V_I is positive.

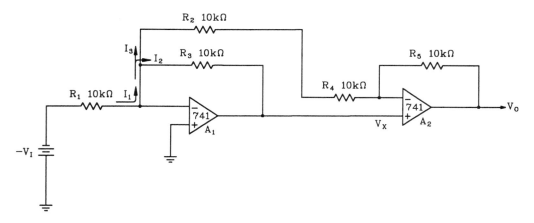

Figure 9.11 An equivalent circuit for the circuit in Fig. 9.9 during times when V_I is negative.

$$-\frac{V_I}{R} = \frac{2V_x + V_x}{2R} = \frac{3V_x}{2R},$$

or

$$-V_I = \frac{3V_x}{2}$$

We shall come back to this equation momentarily. For now, though, let us determine the output voltage in terms of V_x. Amplifier A_2 is a basic noninverting amplifier with respect to V_x since the left end of R_2 connects to ground (virtual ground provided by the inverting input of A_1). Therefore, the output voltage can be expressed using our basic gain equation for noninverting amplifiers.

$$V_O = V_x\left(\frac{R_5}{R_2 + R_4} + 1\right)$$

Since the resistors are all equal, we substitute R for each resistor and solve the equation for V_x.

$$V_O = V_x\left(\frac{R}{R + R} + 1\right)$$
$$= V_x\left(\frac{R + 2R}{2R}\right)$$
$$= \frac{3V_x}{2},$$

Sec. 9.4 Absolute Value Circuit

or

$$V_x = \frac{2V_o}{3}$$

If we now substitute this last expression for V_x into our earlier expression, we can determine the output voltage in terms of input voltage. That is

$$-V_I = \frac{3V_x}{2}$$

$$= \frac{3\left(\frac{2V_o}{3}\right)}{2}$$

$$V_o = -V_I$$

For negative input voltages, the output is equal in amplitude but the opposite polarity. This, coupled with our previous analysis for positive input voltages, means that the output voltage is positive for either polarity of input voltage and equal in amplitude to the input voltage. This is, of course, the proper behavior for an absolute value circuit where $V_o = |V_I|$.

The input impedance, output impedance, frequency response, and so on, are computed in the same manner as similar circuits previously discussed in detail. These calculations are not repeated here.

9.4.3 Practical Design Techniques

Since all resistor values are the same in the absolute value circuit, the calculations for design are fairly straightforward. Let us design an absolute value circuit that will perform according to the following design goals:

1. input voltage $-10V \leq V_I \leq 10V$
2. input impedance > 18 kilohms
3. frequency range   0 to 100 kilohertz

Select the value for R. The minimum value for all of the resistors is determined by the required input impedance. The maximum value is limited by the nonideal characteristics of the circuit (refer to Chap. 10), but is generally below 100 kilohms. The minimum value for R can be found by applying Eq. (2.7).

$$R = \text{input impedance} = 18 \text{ k}\Omega$$

Let us choose to use 22-kilohm resistors for our design.

Select the op amps. As usual, slew rate and small signal bandwidth are used as the basis for op amp selection. The required unity-gain frequency for A_2 can be computed with Eq. (2.22).

$$f_{UG} = bw \times A_V = bw\left(\frac{R_5}{R_2 + R_4} + 1\right)$$

$$= 100 \times 10^3\left(\frac{10 \times 10^3}{(10 \times 10^3) + (10 \times 10^3)} + 1\right)$$

$$= 150 \text{ kHz}$$

The bandwidth for A_1 will be somewhat lower since it has a voltage gain of -1. But in either case, the standard 741 will be more than adequate with regard to bandwidth.

The required slew rate for either amplifier can be computed with Eq. (2.11). The voltage and frequency limits are stated in the design goals.

$$\text{slew rate} = \pi f V_o(\text{max}) = 3.14 \times 100 \times 10^3 \times 20 = 6.28 \text{ V}/\mu\text{s}$$

This slew rate exceeds the capability of the standard 741 but does fall within the 10 volts per microsecond slew rate capability of the MC1741SC. Let us utilize this device in our design. It should be noted that in critical applications there are several additional nonideal op amp characteristics that should be evaluated before selecting a particular op amp. These characteristics are discussed in Chap. 10.

The final schematic diagram for our absolute value circuit is shown in Fig. 9.12. Its performance is evident from the oscilloscope displays in Fig. 9.13. In Fig. 9.13(a) the input is a sine wave. Since both polarities of input translate to a positive output, we have essentially built an ideal full-wave rectifier circuit. Fig. 9.13(b) shows the circuit's response to a very slow (essentially varying DC) triangle wave.

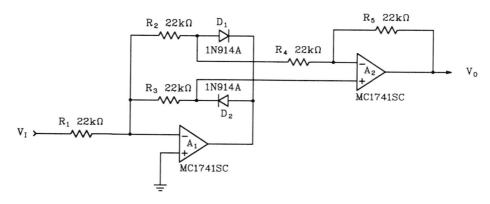

Figure 9.12 Final design of a 100-kilohertz absolute value circuit.

Sec. 9.4 Absolute Value Circuit

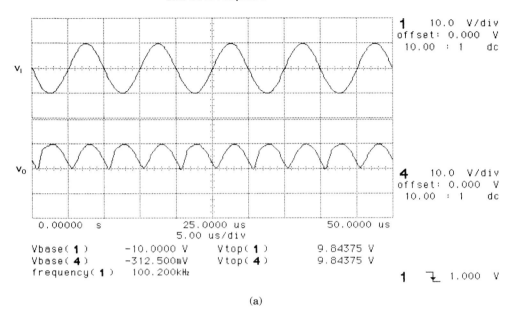

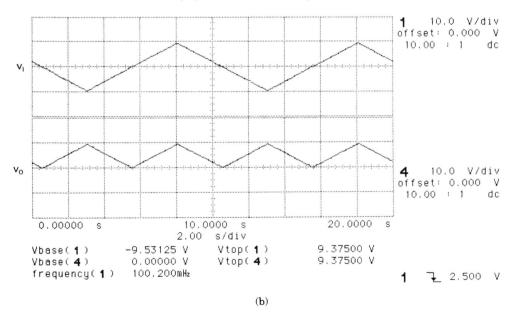

Figure 9.13 Oscilloscope displays showing the performance of the absolute value circuit shown in Fig. 9.12. (Test equipment courtesy of Hewlett-Packard Company)

9.5 SIGN-CHANGING CIRCUIT

A sign changing circuit is a simple, but important member of the arithmetic circuits family. There are many arithmetic operations which require sign or polarity changes. A sign changing circuit then is a circuit which can either invert a signal or, alternately, pass it through without inversion. One specific application which could utilize a sign changing circuit is a combination add/subtract circuit. You will recall the basic rule for algebraic subtraction, "change the sign of the subtrahend and then proceed as in addition." We could, therefore, route one of the adder inputs through a sign changing circuit that could invert or not invert the signal to subtract or add, respectively.

9.5.1 Operation

The schematic diagram of a sign changing circuit is shown in Fig. 9.14. The single-pole double-throw (SPDT) switch would generally be an analog switch that is controlled by another circuit (e.g., a microprocessor system). When the switch is in the upper position, the amplifier is configured as a basic inverting amplifier. The gain ($A_V = -1$) is determined by the ratio of R_1 and R_2. Resistor R_3 works to minimize the effects of bias currents.

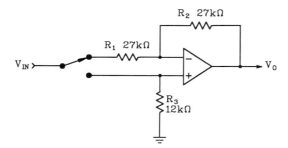

Figure 9.14 A sign-changing circuit allows selectable inversion.

When the switch is moved to the lower position, the circuit is configured as a simple voltage follower ($A_V = 1$). Resistor R_3 determines the input impedance of the circuit. The value of R_3 can be selected such that the input impedance offered by the sign changer is the same in both modes. This allows the circuit to present a constant load on the driving stage. Alternatively, R_3 can be chosen to minimize the effects of bias currents. Resistor R_1 is open-circuited in the noninverting mode and has no effect on circuit operation.

9.5.2 Numerical Analysis

The numerical analysis for this circuit must be done in two stages (one for each position of the switch). Each of these calculations is literally identical to the calculations presented in Chap. 2 for inverting and noninverting amplifiers. They are not repeated here.

9.5.3 Practical Design Techniques

Let us now design a simple sign changer circuit that satisfies the following design requirements:

1. input impedance ≥ 47 kilohms
2. frequency range   0 to 8.5 kilohertz

Compute R_1 and R_2. The value of R_1 is established by the input impedance requirement. Its value is computed with Eq. (2.7).

$$R_1 = Z_1 \geq 47 \text{ k}\Omega$$

If we plan to minimize the effects of bias currents in the circuit through proper selection of R_3, then R_1 should be chosen to be at least twice the minimum impedance value. Let us choose a standard value of 100 kilohms to insure that the impedance criteria will be met under all conditions.

Resistor R_2 will be the same value as R_1 since we want the circuit to have a voltage gain of -1.

Compute R_3. The value of R_3 should be approximately equal to the value of R_1 if we want to maintain a constant input impedance for both modes. On the other hand, if minimizing the effects of bias current is more important in a particular application, then R_3 will be selected to be equal to the parallel combination of R_1 and R_2. For purposes of our present example, let us compute R_3 to minimize the effects of bias current. Its value is calculated with Eq. (2.26).

$$R_3 = \frac{R_1 R_2}{R_1 + R_2} = \frac{100 \times 10^3 \times 100 \times 10^3}{(100 \times 10^3) + (100 \times 10^3)} = 50 \text{ k}\Omega$$

Use the nearest standard (5%) value of 51 kilohms for R_3. Note that by selecting R_1 to be at least twice the minimum input impedance, we guarantee that R_3 will also satisfy the requirements of minimum input impedance.

Select the op amp. We shall select the op amp on the basis of bandwidth and slew rate. The minimum unity-gain frequency is found as

$$f_{UG} = bwA_V = 8.5 \times 10^3 \times 1 = 8.5 \text{ kHz}$$

This value should be doubled (i.e., 17kHz) for critical applications. The minimum required slew rate (based on a full output swing) is calculated with Eq. (2.11).

$$\text{slew rate} = \pi f V_o(\text{max}) = 3.14 \times 8.5 \times 10^3 \times 26 = 0.694 \text{ V}/\mu\text{s}$$

The bandwidth specification can be satisfied by most any op amp. The required slew rate, however, exceeds the 0.5 volts per microsecond rating for the standard 741. Let us plan to use an MC1741SC device which satisfies both bandwidth and slew-rate requirements.

The schematic of our sign changing circuit is shown in Fig. 9.15. Its performance is evident from the oscilloscope displays shown in Fig. 9.16. The waveforms in Fig. 9.16(a) show the circuit's performance in the invert mode. Figure 9.16(b) shows the circuit operated in the noninvert mode.

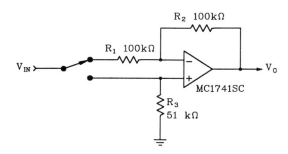

Figure 9.15 Final design of an 8.5-kilohertz sign-changing circuit.

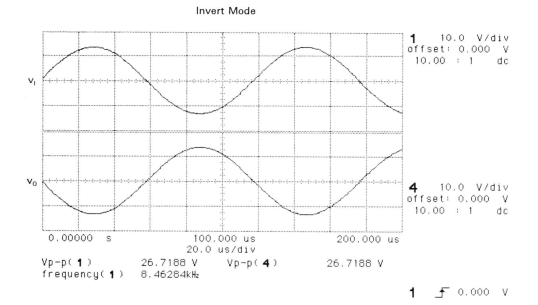

(a)

Figure 9.16 Oscilloscope displays showing the actual performance of the sign-changing circuit shown in Fig. 9.15. (Test equipment courtesy of Hewlett-Packard Company)

(continued)

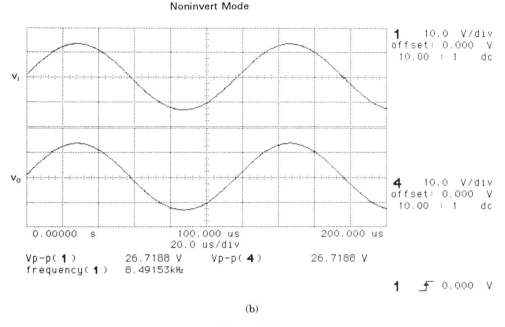

(b)

Figure 9.16b

9.6 TROUBLESHOOTING TIPS FOR ARITHMETIC CIRCUITS

Most of the arithmetic circuits presented in this chapter are electrically similar to other amplifier circuits previously studied. It is the component values and intended application that cause them to be classified as arithmetic circuits. With this in mind, all of the troubleshooting tips and procedures previously discussed for similar circuits are applicable to the companion arithmetic circuit. These will not be repeated here.

Proper operation of the arithmetic circuits presented in this chapter requires accurate component values. It is common to use precision resistors to obtain the desired performance. If a component changes value, its effects may be more noticeable in the arithmetic circuits than in similar, nonarithmetic circuits previously discussed. The very nature of the arithmetic circuit generally implies a higher level of required accuracy than a simple amplifier. Thus, the probability of problems being caused by component variations is greater with arithmetic circuits.

To diagnose a problem caused by a component variation, it is helpful to apply your theoretical understanding of the circuit to minimize the number of suspect components. By applying various combinations of signals and monitoring the output signal, you can usually identify a particular input that has an incorrect response. If, on the other hand, all of the inputs seem to be in error (e.g., shifted slightly), then the problem is most likely a component that is common to all inputs (e.g., the feedback resistor or op amp).

If your job content requires you to do frequent maintenance on arithmetic circuits, then it is probably worth your while to construct a test jig to aid in diagnosing faulty circuits. The test jig could consist of a number of switch-selectable voltages applied to several output jacks. The voltages should be accurate enough to effectively test the particular class of circuits being evaluated. This jig, coupled with a table showing the performance of a known good circuit can be used to very quickly isolate troubles in arithmetic circuits. Of course, this entire test fixture could easily be interfaced to a computer for automatic testing and comparison.

If the circuit seems to work properly in the laboratory, but consistently goes out of tolerance when placed in service, you might suspect a thermal problem. Nearly all of the components in all of the circuits are affected by temperature changes. Short of providing a constant temperature environment, your only options for improving performance under changing temperature conditions are listed

1. locate a defective component
2. substitute compatible components with tighter tolerances
3. redesign the circuit using a different technique

You can artificially simulate temperature changes to a single component by spraying a freezing mist on a particular component. Sprays of this type are available at any electronics supply store. Although every component that you spray may cause a shift in operation, an abrupt or dramatic or erratic response from a particular component may indicate a failing part.

There are numerous choices for all of the components in the circuits presented in this chapter. Improved immunity to temperature variations can often be obtained by simply substituting components with more stringent tolerances. Resistors with a 5 percent rating can be replaced, for example, with 1 percent resistors. Similarly, a general-purpose op amp could be replaced with a pin-compatible op amp having lower bias currents, noise, or temperature coefficients.

Finally, if items 1 and 2 do not resolve a particular thermal problem, then redesign may be in order. Frequently, there is a trade-off between circuit simplicity and circuit stability. To achieve rock-solid stability often requires a step increase in circuit complexity.

REVIEW QUESTIONS

1. Refer to Fig. 9.2. What is the effect on circuit operation if resistors R_3 to R_5 are all increased to 20 kilohms?
2. Refer to Fig. 9.2. Show how the circuit can be analyzed by applying the Superposition Theorem to the input circuit.
3. Refer to Fig. 9.5. What is the effect on circuit operation if resistor R_3 develops an open?

Chap. 9   Review Questions

4. Refer to Fig. 9.5. What is the effect on circuit operation if resistor R_2 develops an open?
5. Refer to Fig. 9.9. What is the effect on circuit operation if resistor R_3 is doubled in value?
6. Refer to Fig. 9.9. Describe the effect on circuit operation if resistor R_4 became open.
7. Would the circuit shown in Fig. 9.9 still operate correctly if resistors R_4 and R_5 were **both** doubled in value? Explain your answer.

CHAPTER 10

NONIDEAL OP AMP CHARACTERISTICS

For purposes of analysis and design in the preceding chapters, we have considered many of the op amp parameters to be ideal. For example, we generally assumed the input bias current to be zero, we frequently ignored output resistance, and disregarded any effects caused by drift or offset voltage. This approach not only greatly simplifies the analysis and design techniques, but it is a practical method for many situations. More demanding applications, however, require that we acknowledge the existence of certain nonideal op amp characteristics. This chapter will describe these additional considerations.

10.1 NONIDEAL DC CHARACTERISTICS

We shall classify the nonideal characteristics of op amps into two general categories: DC, AC. Let us first consider the effects of nonideal DC characteristics.

10.1.1 Input Bias Current

As briefly noted in Chap. 1, the first stage of an op amp is a differential amplifier. Figure 10.1 shows a representative circuit that could serve as the input stage of an op amp. Clearly, the currents that flow into or out of (depending on whether NPN or PNP transistors are used) the inverting ($-$) and noninverting ($+$) op amp terminals is actually base current for the internal transistors. For proper operation, we must always insure that both inputs have a DC path to ground. They cannot be left floating, and they cannot have series capacitors. These currents are very small (ideally zero), but may cause undesired effects in some applications.

Figure 10.2 can be used to show the effect of nonideal bias currents. This shows a basic op amp configured as *either* an inverting or a noninverting amplifier with the

Sec. 10.1 Nonideal DC Characteristics

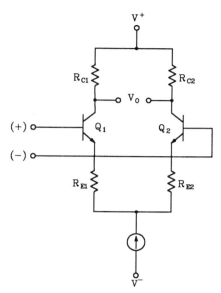

Figure 10.1 A representative input stage for a bipolar op amp.

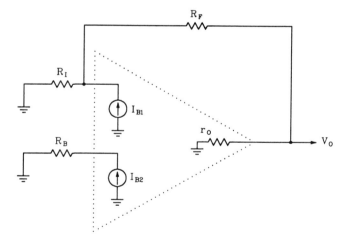

Figure 10.2 A model that can be used to determine the effects of bias current.

input signal removed (i.e., input shorted to ground). The direction of current flow for the bias currents and the resulting output voltage polarities are essentially arbitrary since different op amps have different directions of current flow. However, for a given op amp, both currents will flow in the same direction (i.e., either in or out). For our immediate purposes, let us assume that the arrows on the current sources indicate the direction of electron flow. We shall now apply Ohm's Law in conjunction with the Superposition Theorem to determine the output voltage produced by the bias currents. First, the noninverting bias current (with the inverting bias current set to zero) will cause a voltage drop across R_B with a value of

$$V_{R_B} = I_{B_2} \times R_B$$

This voltage will be amplified by the noninverting gain of the amplifier and appear in the output as

$$V_{O_2} = V_{R_B}\left(\frac{R_F}{R_I} + 1\right) = I_{B_2}R_B\left(\frac{R_F}{R_I} + 1\right)$$

It should be noted that this voltage will be negative in our present example since we have assumed the electron flow was out of the input terminals.

Now let's consider the effect of the bias current for the inverting input. According to the Superposition Theorem, we must set the bias current on the noninverting input to zero. Having done this, we see that since no current is flowing through R_B there will be no voltage across it. Therefore, the voltage on the (+) input will be truly zero or ground. Additionally, we know that the closed loop action of the amplifier will force the inverting pin to be at a similar potential. This means that the inverting pin is also at ground potential. Recall that we referred to this point as a virtual ground. In any case, with zero volts across R_I there can be no current flow through R_I. The entire bias current for the inverting input, then, must flow through R_F (by Kirchoff's Current Law). Since the left end of R_F is grounded and the right end is connected to the output, the voltage across R_F is equal to the output voltage. Therefore, the output voltage caused by the bias current on the inverting pin can be computed as

$$V_{O_1} = I_{B_1} \times R_F$$

Since current has been assumed to flow out of the (−) pin, we know that the resulting output voltage will be positive. Note that this is opposite the polarity of the (+) input.

Now, continuing with the application of the Superposition Theorem, we simply combine (algebraically) the individual voltages computed above to determine the net effect of the two bias currents. Since the polarities of output voltage caused by the two bias currents were opposite, the net output voltage must be

$$V_O = V_{O_1} - V_{O_2},$$

or

$$V_O = I_{B_1}R_F - I_{B_2}R_B\left(\frac{R_F}{R_I} + 1\right)$$

The manufacturer does not generally provide the value of the individual bias currents. Rather, the bias current (I_B) listed in the specification sheet is actually an average of both bias currents. In general, the two currents are fairly close in value. Therefore, if we assume that the two currents are equal, the preceding equation becomes

$$\boxed{V_O = I_B\left[R_F - R_B\left(\frac{R_F}{R_I} + 1\right)\right]} \tag{10.1}$$

Sec. 10.1   Nonideal DC Characteristics   397

So, to estimate the output voltage caused by the bias currents in a particular amplifier circuit, we apply Eq. (10.1). As an example, let us estimate the output voltage caused by the bias currents for the inverting amplifier circuit shown in Fig. 10.3.

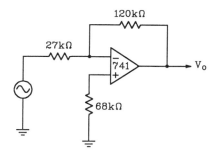

Figure 10.3   A basic inverting amplifier circuit.

By referring to the data sheets in the appendix, we can determine the maximum value of bias current for a 741 under worst-case conditions to be 1500 nanoamperes. Substituting this into Eq. (10.1) gives us

$$V_O = I_B \left[R_F - R_B \left(\frac{R_F}{R_I} + 1 \right) \right]$$

$$= 1500 \times 10^{-9} \left[(120 \times 10^3) - (68 \times 10^3) \left(\frac{120 \times 10^3}{27 \times 10^3} + 1 \right) \right]$$

$$= -375.33 \text{ mV}$$

Now, a good question to ask is, "How do we design our circuits to minimize the effects of op amp bias currents?" We would prefer to have zero volts in the output as a result of the bias currents, therefore, let us set Eq. (10.1) equal to zero, and then manipulate the equation.

$$0 = I_B \left[R_F - R_B \left(\frac{R_F}{R_I} + 1 \right) \right]$$

$$= R_F - R_B \left(\frac{R_F}{R_I} + 1 \right)$$

$$R_F = R_B \left(\frac{R_F}{R_I} + 1 \right)$$

$$R_B = \frac{R_F}{\frac{R_F}{R_I} + 1}$$

$$= \frac{R_F}{\frac{R_F + R_I}{R_I}}$$

$$R_B = R_F\left(\frac{R_I}{R_I + R_F}\right)$$

$$= \frac{R_F R_I}{R_F + R_I}$$

This final result should be recognized as an equation for two parallel resistances. You will recall that throughout the earlier chapters we always tried to set R_B equal to the parallel resistance of R_F and R_I. This is an important result.

$$\boxed{R_B = \frac{R_F R_I}{R_F + R_I} = R_F \| R_I} \qquad (10.2)$$

As a final example, let us replace R_B in Fig. 10.3 with the correct value and compare the results. The correct value for R_B is determined with Eq. (10.2).

$$R_B = \frac{R_F R_I}{R_F + R_I} = \frac{120 \times 10^3 \times 27 \times 10^3}{(120 \times 10^3) + (27 \times 10^3)} \approx 22 \times 10^3 = 22 \text{ k}\Omega$$

We now apply Eq. (10.1) to determine the resulting output voltage with the correct value of R_B.

$$V_O = I_B\left[R_F - R_B\left(\frac{R_F}{R_I} + 1\right)\right]$$

$$= 1500 \times 10^{-9}\left[(120 \times 10^3) - (22 \times 10^3)\left(\frac{120 \times 10^3}{27 \times 10^3} + 1\right)\right]$$

$$= 333 \text{ }\mu\text{V}$$

This, as you can see, is an improvement of over 1000 times. You should realize, however, that these calculations were based on the assumption that the two bias currents are identical. While they are close, they are not actually equal. The difference between these two bias currents is the subject of the next section.

10.1.2 Input Offset Current

The value of bias current listed in the manufacturer's data sheet is the average of the two individual currents. The value of current listed in the manufacturer's data sheet as input offset current is the difference between the two bias currents. The difference is always less than the individual currents. In the case of the standard 741, the worst-case input offset current is listed as 500 nanoamperes (compared to 1500 nA for bias current). The typical values for these currents at room temperature are 20 nanoamperes and 80 nanoamperes for input offset current and input bias current, respectively.

While deriving Eq. (10.1), we generated the following intermediate step:

$$V_O = I_{B_1}R_F - I_{B_2}R_B\left(\frac{R_F}{R_I} + 1\right)$$

Sec. 10.1   Nonideal DC Characteristics   399

Now, from this point, let us assume that R_B is selected to be equal to the parallel combination of R_F and R_I as expressed in Eq. (10.2). If we substitute this equality into this equation and then manipulate the equation, we can produce a useful expression.

$$V_O = I_{B_1}R_F - I_{B_2}\frac{R_F R_I}{R_F + R_I}\left(\frac{R_F}{R_I} + 1\right)$$

$$= I_{B_1}R_F - \left[\frac{I_{B_2}R_F^2 + I_{B_2}R_F R_I}{R_F + R_I}\right]$$

$$= I_{B_1}R_F - I_{B_2}R_F$$

$$= R_F(I_{B_1} - I_{B_2})$$

Since the quantity $I_{B_1} - I_{B_2}$ is the very definition of input offset current, we can make this substitution and obtain Eq. (10.3).

$$\boxed{V_O = R_F I_{IO}} \quad (10.3)$$

As long as we select the correct value for R_B, we can use the simplified equation, Eq. (10.3) to determine the output voltage caused by op amp input currents. In the case of the standard 741 shown in Fig. 10.3 (but with the correct value for R_B), the worst-case output voltage caused by offset current is

$$V_O = 120 \times 10^3 \times 500 \times 10^{-9} = 60 \text{ mV}$$

A more likely result can be found by using the typical value of offset current at room temperature. That is,

$$V_O = 120 \times 10^3 \times 20 \times 10^{-9} = 2.4 \text{ mV}$$

10.1.3 Input Offset Voltage

Input offset voltage is another parameter that is listed in the manufacturer's data sheet. Like the bias currents, input offset voltage produces an error voltage in the output. That is, with zero volts applied to the inputs of an op amp, we would expect to find zero volts at the output. In fact, we will find a small DC offset present at the output. This voltage is called the output offset voltage and is a result of the combined effects of bias current (previously discussed above) and input offset voltage.

The error contributed by input offset voltage is a result of DC imbalances within the op amp. The transistor currents (see Fig. 10.1) in the input stage may not be exactly equal due to component tolerances within the integrated circuit. In any case, an output voltage is produced just as if there were an actual voltage applied to the input of the op amp. To facilitate the analysis of the problem, we model the circuit with a small DC source at the noninverting input terminal (see Fig. 10.4). This apparent source is called the input offset voltage. It will be amplified and appear in the output as an error voltage. The output voltage caused by the input offset voltage can be computed with our basic gain equation.

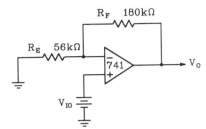

Figure 10.4 The input offset voltage contributes to the DC offset voltage in the output of an op amp.

$$V_O = V_{IO}\left(\frac{R_F}{R_I} + 1\right)$$

The manufacturer's data sheet for a standard 741 lists the worst-case value for input offset voltage as 6 millivolts. In the case of the circuit shown in Fig. 10.4, we could compute the output error voltage caused by the input offset voltage as follows.

$$V_O = 6 \times 10^{-3}\left(\frac{180 \times 10^3}{56 \times 10^3} + 1\right) = 25.29 \text{ mV}$$

The polarity of the output offset may be either positive or negative. Therefore, it may add or subtract from the DC offset caused by the op amp bias currents. The worst-case output offset voltage can be estimated by assuming that the output voltages caused by the bias currents and the input offset voltage are additive. In that case, the resulting value of output offset voltage can be found as

$$\boxed{V_{OO} = R_F I_{IO} + V_{IO}\left(\frac{R_F}{R_I} + 1\right)} \qquad (10.4)$$

Most op amps, including the 741, have provisions for nulling or canceling the output offset voltage. Appendix 4 shows the recommended nulling circuit for an MC1741SC. It consists of a 10-kilohm potentiometer connected between the offset null pins (1 and 5) of the op amp. The wiper arm of the potentiometer connects to the negative supply voltage. The amplifier is connected for normal operation (excluding any DC input signals), and the potentiometer is adjusted to produce zero volts at the output of the op amp. You should realize, however, that this only cancels the output offset voltage at one particular operating point. With temperature changes or simply over a period of time, the circuit may drift and need to be readjusted. Nevertheless, it is an improvement over a circuit with no compensation.

10.1.4 Drift

The term *drift* is a general term that describes the change in DC operating characteristics with time and/or temperature. The temperature drift for input offset current is expressed in terms of nA/°C, and the drift for input offset voltage is expressed in terms of μV/

Sec. 10.1   Nonideal DC Characteristics   401

°C. The temperature coefficient of each of these quantities varies over the temperature range. The variation may even include a change in polarity of the temperature coefficient. The maximum drift may be provided in tabular form by some manufacturers, but more meaningful data is available when the manufacturer provides a graph showing the response of input offset current and input offset voltage to changes in temperature. The data sheet in appendix 1 is essentially a compromise between these two methods. Here the manufacturer has provided the values of input offset current and input offset voltage at room temperature and at the extremes of the temperature range.

The usual way to reduce the effects of drift is to select an op amp that has a low temperature coefficient for these parameters. Additionally, in certain critical applications some success can be achieved by including a thermistor network as part of the output offset voltage compensation network.

10.1.5 Input Resistance

An ideal op amp has an infinite input resistance. Practical op amps, however, have a lower, but still very high, input resistance. The errors caused by nonideal input resistance in the op amp do not generally cause significant problems. What problems may be present can generally be minimized by insuring that the following conditions are satisfied:

1. The differential input resistance should be at *least* 10 times the value of feedback resistor for inverting applications.
2. The differential input resistance should be at *least* 10 times the values of the feedback and source resistances for noninverting applications.

In most cases, these requirements are easily met. In more demanding applications, the designer may select a FET input op amp. The MC34001 op amp made by Motorola is an example of a FET input op amp. It provides an input resistance of 10^{12} ohms and is pin compatible with the standard 741 device.

10.1.6 Output Resistance

Figure 10.5 shows a circuit model that can be used to understand the effects of nonzero output resistance in an op amp. Considering that the ($-$) input pin is a virtual ground point, we find that R_F and R_L are in parallel with each other. This combination forms a voltage divider with the series output resistance (R_O). The actual output voltage (V'_O) will be somewhat less than the ideal output voltage (V_O).

The numerical effect of R_O can be determined by applying the basic voltage divider equation.

$$V'_O = V_O \left(\frac{R_F \| R_L}{R_F \| R_L + R_O} \right)$$

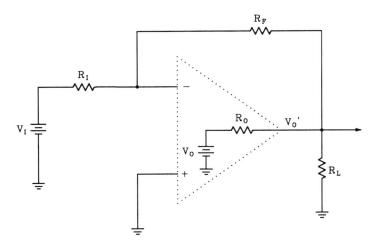

Figure 10.5 The effect of op amp output resistance (R_o) is to reduce the output voltage.

$$V'_O = V_O \left(\frac{\frac{R_F R_L}{R_F + R_L}}{\frac{R_F R_L}{R_F + R_L} + R_O} \right)$$

$$= \frac{V_O}{1 + R_O \left(\frac{R_F + R_L}{R_F R_L} \right)}$$

Now, as long as the condition

$$R_O \left(\frac{R_F + R_L}{R_F R_L} \right) \ll 1$$

is maintained, the actual output voltage will show very little loading. In practice, this condition is generally easy to satisfy. In most cases, the maximum output current of the op amp will be exceeded before the output resistance becomes a problem.

10.2 NONIDEAL AC CHARACTERISTICS

10.2.1 Frequency Response

The effects of limited bandwidth have been discussed in several of the earlier chapters with reference to specific circuits. In general, the open-loop DC gain of an op amp is extremely high (typically well over 200,000). However, as the frequency of operation increases, the gain begins to fall off. At some point, the open-loop gain reaches one. We call this the unity-gain frequency. The unity-gain frequency is also referred to as the gain-bandwidth product.

When the op amp is provided with negative feedback, the closed-loop gain is less than the open-loop gain. As long as the closed-loop gain is substantially less than the open-loop gain (at a given frequency), then the circuit is relatively unaffected by the reduced open-loop gain. However, at frequencies which cause the open-loop gain to approach the expected closed-loop gain, the actual closed-loop gain also begins to fall off.

We can estimate the highest operating frequency for a particular closed-loop gain as follows.

$$\text{bw} = \frac{\text{unity-gain frequency}}{\text{closed-loop gain}}$$

This is adequate for many, if not most, applications, but it ignores the fact that the closed loop gain falls off more rapidly as it approaches the open-loop curve. In fact, when the circuit is operated at the frequency computed, the response will be about 3 dB below the ideal voltage gain. If this 3 dB drop is unacceptable for a particular application, then the gain must be reduced (or use an op amp with a higher gain-bandwidth product).

10.2.2 Slew Rate

In order to provide high-frequency stability, op amps have one or more capacitors connected to an internal stage. The capacitor may be internal to the op amp (internally compensated), or it may be added externally by the designer (externally compensated). Section 10.2.4 discusses frequency compensation in greater detail. In either case, this compensating capacitance limits the maximum rate of change that can occur at the output of the op amp. That is, the output voltage can only change as fast as this capacitance can be charged and discharged. The charging rate is determined by two factors

1. charging current
2. size of the capacitor

The charging current is determined by the design of the op amp and is not controllable by the user. In the case of internally compensated op amps, the value of capacitance is also fixed. The user does have control over the capacitance values for externally compensated op amps. The smaller the compensating capacitor the wider the bandwidth and the faster the slew rate. Unfortunately, the price paid for this increased performance is a greater amplification of noise voltages and a greater tendency for oscillations.

Since slew rate is, by definition, a measure of volts per second, the severity of problems caused by limited slew rates is affected by both signal amplitude and signal frequency. We can determine the largest output voltage swing for a given slew rate and operating frequency by applying Eq. (2.11).

$$v_O(\text{max}) = \frac{\text{slew rate}}{\pi f}$$

Of course, we can also transpose this equation to determine the highest operating frequency for a given output amplitude.

10.2.3 Noise

The term *noise*, as used here, refers to undesired voltage (or current) fluctuations created within the internal stages of the op amp. Although there are many internal sources of noise and several types of noise, it is convenient to view them collectively as a single source connected to the noninverting input terminal. This approach to noise analysis is shown in Fig. 10.6.

The value of this equivalent noise source is provided by some manufacturers as *equivalent input noise*. The gain given to this noise voltage is computed with our basic noninverting amplifier gain equation. That is

$$v_O = V_N \left(\frac{R_F}{R_I} + 1 \right)$$

It is important to note, however, that R_I for the purposes of calculating noise gain is the total resistance from the inverting pin to ground. This has particular significance in the case of a multiple-input summing amplifier. In this case, the R_I value is actually the parallel combination of all input resistors. Thus, the noise gain of the circuit is higher than any of the individual gains.

Another important aspect of noise gain is apparent when capacitance is used in the input circuit (e.g., the differentiator circuit). By having a capacitance in series with the input terminal $(-)$, we cause the effective R_I to decrease at higher frequencies thus increasing the noise gain of the circuit.

The data sheet for the standard 741 provided in appendix 1 shows one manufacturer's method of providing noise specifications for an op amp. The graph labeled "Output Noise versus Source Resistance" is particularly useful. Here the manufacturer is indicating the total RMS output noise for various gains and various source resistances. The source resistor always generates thermal noise that increases with resistance and temperature. For low values of source resistance (below 1.0 kΩ), the op amp noise is the primary

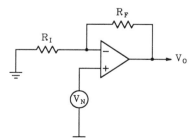

Figure 10.6 All of the internal noise sources in an op amp can be viewed as a single source (V_N) applied to the non-inverting input.

Sec. 10.2 Nonideal AC Characteristics

contributor to overall output noise. Thus, the curves remain fairly flat for various values of source resistance. As the source resistance is increased beyond 10 kilohms, its noise begins to swamp out the internal op amp noise, and we begin to see a steady rise in overall noise with increases in source resistance.

The graph labeled "Spectral Noise Density" also provides us with greater insight into the noise characteristics of the 741. This graph gives us an indication of the relative magnitudes of noise signals at various frequencies. In particular, notice that at frequencies above 1.0 kilohertz, the distribution of noise voltage is fairly constant. This flat region is largely caused by the noise generated in the source resistance. Much of the internal op amp noise, decreases with increasing frequency. By the time we reach 1.0 kilohertz, this internal noise contributes little to the overall noise signal. Below 1.0 kilohertz, however, the overall noise amplitude increases sharply as frequency is decreased. This increase in overall output noise is largely caused by increased internal op amp noise. This increasing noise level for decreasing frequencies can present problems in DC amplifiers.

There are several things we can do to minimize the effects caused by noise voltages. First, we can take steps to minimize the noise gain of the circuit. This means avoiding the use of large values of feedback resistance (R_F) and small values of input resistance (R_I). Unfortunately, the gain of the circuit for normal signals determines the ratio of these two components. However, if we bypass R_F with a small capacitor, we can cause the noise gain to decrease at frequencies beyond the normal operating range of the circuit. That is, the normal input signals will see the bypass capacitor as an open and will be unaffected. The noise signals, on the other hand, will see the bypass capacitor as a low impedance and reduce the overall noise gain of the circuit.

A second way to minimize the effects of noise in an op amp circuit is to insure that the resistance between the inverting input and ground, and between the noninverting input and ground, are equal. You will recall that this same procedure helped us minimize the effects of input offset current.

Third, since the noise generated by resistors increases with resistance (actually $\sqrt{R}$), we should avoid large values of resistance when noise is potentially a problem.

Finally, we can reduce the op amp's contribution to overall output noise by selecting an op amp that is optimized for low-noise operation. The OP-27 op amp manufactured by Motorola, for example, is called an "ultra-low" noise device. It generates only 3.0 nanovolts of RMS noise at 1.0 kilohertz.

10.2.4 Frequency Compensation

We know from the study of oscillators that the two primary criteria for oscillation are in-phase (or 360°) feedback and a gain of at least unity at the feedback frequency. The op amp has a varying gain depending upon the operating frequency, but the gain requirement for oscillation is certainly a possibility.

Additionally, the op amp inherently has a 180° phase shift between its inverting input and the output by virtue of its operation. Now if we provide an additional 180° phase shift **and** provide a loop gain of at least unity, we will have built an oscillator.

In an earlier chapter, we provided this extra phase shift externally at a specific frequency to construct an op amp oscillator. All of the internal stages of an op amp also have certain frequency and phase characteristics. As frequency increases, the cumulative phase shift of these internal stages also increases. If, at some frequency, the total phase shift reaches 360° (or 0°) and, **at the same time,** we have a loop gain of at least unity, the circuit will oscillate even though that may not be the intended behavior.

Principles of frequency compensation. So far in the text we have been primarily focused on the frequency characteristics of the op amp with only occasional reference to the phase characteristics. Figure 10.7 shows a simple RC circuit and graphs showing its frequency and phase performance. Table 10.1 shows this same data in tabular form.

There are several things to be observed from these data. First note that the break (or cutoff) frequency occurs at 10 kilohertz. At this frequency, $X_C = R$, the voltage gain is 0.707, and the phase shift is about 45°. Now notice that if we are at least a decade lower than the break frequency, the following occurs:

1. Voltage gain remains fairly constant (near unity).
2. The dB/decade drop in output voltage is very slight.
3. The phase shift is near 0° ($< 6°$).

As the cutoff frequency is passed, however, things change. By the time we are one decade above the cutoff frequency and continuing beyond, the following occurs:

1. Voltage gain decreases by a factor of 10 for each decade increase in frequency.
2. The dB/decade drop continues at about -20 dB per decade.
3. The phase shift is near 90° ($> 84°$).

TABLE 10.1

FREQUENCY (Hz)	VOLTAGE GAIN	dB per DECADE	PHASE SHIFT (degrees)
1	1	—	-0.0057
10	0.99999	-0.00000478	-0.0541
100	0.99995	-0.000429	-0.5701
1000	0.99505	-0.0427	-5.7
10,000	0.70763	-2.964	-44.96
100,000	0.09965	-17.046	-84.29
1,000,000	0.01001	-19.979	-89.43
10,000,000	0.001001	-20.022	-89.95

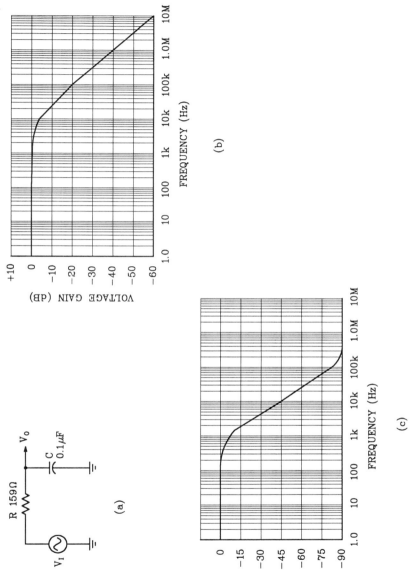

Figure 10.7 The frequency/phase response of a simple RC circuit.

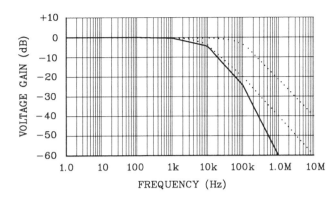

Figure 10.8 A two-stage RC filter has a cumulative effect.

If we were to add a second RC section to the circuit shown in Fig. 10.7, these effects would be magnified. That is, for frequencies well below **both** break frequencies, the circuit would behave as described for low frequencies. For frequencies greater than the break frequency of one RC section, but less than the second, the circuit response would be as described for high frequencies. Finally, when the operating frequency was higher than **both** break frequencies, the following effects would occur:

1. Voltage gain decreases by a factor of 100 for each decade increase in frequency.
2. The dB/decade drop continues at -40 dB per decade.
3. The phase shift is near 180°.

This two-stage response is shown graphically in Fig. 10.8 as a solid line. The dotted curves represent the responses of the individual RC sections. A similar cumulative effect would occur for each subsequent RC section that was added.

As briefly mentioned earlier in this section, each of the internal stages of an op amp have frequency and phase characteristics similar to the RC sections presented. Figure 10.9 shows the open-loop frequency response of a hypothetical amplifier **with no frequency compensation.**

This response presents us with "good news" and "bad news." The good news stems from the fact that we have a substantial increase in bandwidth relative to a compensated op amp (e.g., the standard 741 shown as a dotted curve). The bad news, however, is caused by the multiple break frequencies which will certainly cause greater than 180° of phase shift. This internal shift coupled with the inherent 180° shift from the inverting input terminal will make this particular op amp very prone to oscillation. Now, let's determine how prone.

If we superimpose the closed-loop gain response on the open-loop response originally shown in Fig. 10.9, we get Fig. 10.10. The closed-loop responses for several gains are shown. Now, the important characteristic for stability (i.e., no oscillations) is summarized in the following statement:

Sec. 10.2 Nonideal AC Characteristics 409

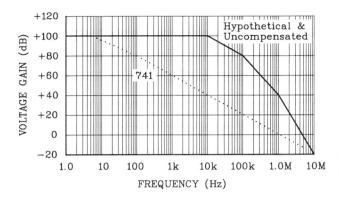

Figure 10.9 An uncompensated frequency response provides both "good" and "bad" news.

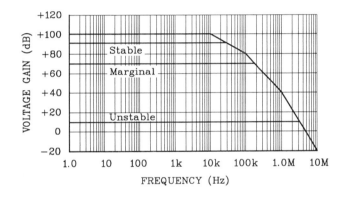

Figure 10.10 To insure against oscillations, the intersection of the closed-loop gain and the open-loop gain curves must occur with a net slope of less than 40 dB/decade.

> To insure against oscillations, the intersection of the closed-loop gain and the open-loop gain curves must occur with a *net* slope of less than 40 dB per decade.

If this rule is violated, unstable operation is assured. On the other hand, if this rule is faithfully followed, it is still possible to construct an unstable amplifier circuit. Nevertheless, it provides us with an excellent starting point. We shall consider exceptions at a later point in the discussion.

One method for verifying amplifier stability using this method relies on the use of simple sketches plotted on semilog graph paper. The logarithmic scale is used to plot frequency, and the linear scale is used to plot voltage gain expressed in dB. Done in this way, it is a simple matter to determine the net slope at the point of intersection between the closed- and open-loop gain curves. This method, although valid in concept, is often difficult to implement since the manufacturer may not provide adequate data regarding the uncompensated open-loop response. More often the manufacturer provides a set of curves that indicate the overall open loop response **with** frequency compensation (either internal or external). The designer must interpret these graphs to ascertain safe operating regions.

In any case, Fig. 10.10 clearly indicates that the tendency for oscillation becomes greater as the closed-loop gain of the circuit is reduced. The worst-case gain, with regard to stability, occurs for 100 percent feedback or unity voltage gain. If the circuit is stable for unity gains, then we can be assured of stability at all other higher gains.

Internal frequency compensation. Many general-purpose op amps (e.g., 741, MC1741SC) are internally compensated to provide stability for all gains down to and including unity. This is generally accomplished by adding a capacitor to one of the internal stages. This causes the overall response to have an additional roll-off characteristic (like adding another series RC section). The break frequency of this added circuit is chosen to be lower than all other break frequencies present in the output response defining it as the dominant network. Figure 10.11 illustrates how the response curve of an uncompensated op amp is shifted downward by the introduction of a compensation capacitor. Notice that the effects of each of the break frequencies is still present in the response, but that the amplifier gain falls below unity *before* the slope exceeds 40 dB/decade. While this does insure maximum stability, it is clearly detrimental to the bandwidth of the amplifier.

The internal capacitance, called a lag capacitor, can be connected to any one of several points within the op amp. Since larger values of capacitance are required for lower impedance points, it is common to connect the lag capacitor at a high-impedance point in the device. Additionally, by inserting the capacitor in one of the earlier stages, rather than the output stage, it has less of a slowing effect on the slew rate. Probably the most common value of internal compensating capacitance is 30 picofarad. It can be readily identified on the simplified schematic of the standard 741 included in appendix 1.

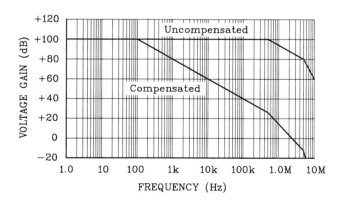

Figure 10.11 Adding a compensation capacitor increases stability but reduces the bandwidth of an amplifier.

Sec. 10.2 Nonideal AC Characteristics

External frequency compensation. Although the inclusion of an internal compensating capacitor greatly simplifies the use of an op amp and makes it less sensitive to sloppy designs, it does cause an unnecessarily severe reduction in the bandwidth of the circuit. Alternatively, the manufacturer may elect to bring out one or more pins for the connection of an external compensating capacitor. The value of the capacitor can be tailored by the designer for a specific application.

The extreme case, of course, is to put heavy compensation on the device to make it stable all the way down to unity gain. This makes the externally compensated op amp equivalent to the internally compensated one. However, many applications do not require unity gain. In these cases, we can use a smaller compensating capacitor which directly increases the bandwidth. So long as the closed-loop gain curve intersects the open-loop gain curve with a net slope of less than 40 dB, we will generally have a stable circuit.

The LM301A op amp is an externally compensated, general-purpose op amp. It's data sheet includes a graph that illustrates the effect on open-loop frequency response for compensating capacitors of 3 and 30 picofarads. A second graph shows the dramatic increase in large-signal frequency response obtained by using a 3-picofarad capacitor instead of a 30-picofarad. With a 30-picofarad capacitor, the full-power bandwidth is limited to about 7.5 kilohertz (nearly the same as a standard 741). By using a 3-picofarad compensating capacitor, however, the full-power bandwidth goes up to about 100 kilohertz. This can be attributed to an increased slew rate. It should be noted that 3 picofarads is a **very** small capacitance. This value can easily be obtained or even exceeded by stray wiring capacitance.

The negative side, however, is evident in the open-loop frequency response. The curve for 3 picofarads does not extend to unity gain (0 dB). That is, we must design the circuit to have substantial (e.g., 40 dB) gain to insure stability.

Feed-forward frequency compensation. The lowest, and therefore most dominant, break frequency in an uncompensated op amp is generally caused by the frequency response of the PNP transistors in the input stage. Although this stage does not provide a voltage gain, the attenuation to the signal increases with frequency. By connecting a capacitor between the inverting input and the input of the next internal stage, we can effectively bypass the first stage for high frequencies. This boosts the overall frequency response of the op amp. This technique is called *feedforward compensation*.

The manufacturer must provide access to the input terminal of the second stage in order to employ feedforward compensation. In the case of the LM301, feedforward compensation is achieved by connecting a capacitor between the inverting input and pin 1 (balance). Examination of the internal schematic will reveal that this effectively bypasses the NPN differential amplifier on the input and the associated level-shifting PNP transistors.

When feedforward compensation is used, a small capacitor must be used to bypass the feedback resistor to insure overall stability. The manufacturer provides details for each op amp that demonstrates the selection of component values.

The use of feedforward compensation produces all of the following effects:

1. increased small-signal bandwidth
2. increased full-power bandwidth
3. improved slew rate

10.2.5 Input Resistance

Calculation of input resistance, or more correctly input impedance, was presented in Chap. 2. In the case of a noninverting configuration, we found that the open-loop input resistance of the op amp is magnified when the feedback loop is closed. Eq. (2.29) was used to determine the effective input impedance once the loop was closed

$$Z_I = R_{OP} A_V \left(\frac{R_I}{R_F + R_I} \right)$$

This calculation, of course, produces a very high value which, for most applications, can be viewed as the ideal value of infinity.

In the case of an inverting configuration, the input impedance is generally considered to be equal to the value of the input resistor. That is, the impedance directly at the inverting input is generally considered to be zero (virtual ground). This approximation is satisfactory for most applications.

If a greater level of accuracy is needed, we can estimate the actual input impedance directly at the inverting input with Eq. (10.5).

$$\boxed{R_I(-) \approx \frac{R_F}{A_{OL}}} \tag{10.5}$$

The total input impedance for the amplifier circuit is simply the sum of the input resistor and the value computed with Eq. (10.5).

10.2.6 Output Resistance

The calculation and effects of output resistance, or impedance, was presented in Chap. 2. It was found that the closed-loop output resistance was substantially reduced from the open-loop value. Equation (2.15) was used to estimate the closed-loop output impedance for either inverting or noninverting configurations

$$R_O = \frac{\text{output impedance (open loop)} \times (R_I + R_F)}{A_{OL} R_I}$$

This is an adequate approximation for most applications. In the case of very low-open-loop gains (e.g., at higher frequencies), Eq. (10.6) provides a more accurate estimate of the closed-loop output impedance.

$$\boxed{R_O = \frac{\text{output impedance (open loop)} \times (R_I + R_F)}{R_I + R_F + A_{OL} R_I}} \tag{10.6}$$

In most cases, the finite output resistance has little effect on circuit operation. The maximum output current capability will generally limit the size of load resistor to a value that is still substantially larger than the output resistance of the op amp. Therefore, the voltage divider action of output resistance is minimal.

10.3 SUMMARY AND RECOMMENDATIONS

For some applications, many of the nonideal characteristics of op amps can be ignored without compromising the design. But, how do you know which parameters are important under what conditions? The answer to that question is quite complex, but the following will provide you with some practical guidelines.

10.3.1 AC-Coupled Amplifiers

If a particular amplifier is AC-coupled (e.g., capacitive-, optically, or transformer-coupled), then the nonideal DC characteristics can often be ignored. Any offsets caused by bias currents, drift, and so on, will be noncumulative. That is, the effects will be limited to the particular stage being considered and will not upset the operation of subsequent stages. Therefore, as long as the DC offset is not so great as to present problems (e.g., saturation on peak signals) in the present stage, it can probably remain uncompensated.

Frequency response and slew rate, on the other hand, are important in nearly every AC-coupled application. These parameters should be fully evaluated before selecting a particular amplifier for a given application.

Noise characteristics can often be ignored, but it depends on the application and on the amplitude of the desired signal relative to the noise signal. If the noise signal has an amplitude that is comparable to the desired signal, then the designer should take steps (discussed in an earlier section) to minimize the circuit noise response. On the other hand, if the primary signal is many times greater than the noise signal, then the design may not require any special considerations with regard to noise reduction.

10.3.2 DC-Coupled Amplifiers

DC-coupled amplifiers seem to present some of the more formidable design challenges. Depending on the specific application, a DC amplifier may be affected by literally all of the nonideal op amp characteristics. This would certainly be the case for a DC-coupled, low-level, wideband amplifier.

If, however, the input frequency is always very low (e.g., the output of a temperature transducer), then considerations regarding slew rate and bandwidth can often be disregarded. In these cases, the emphasis needs to be placed on the DC parameters such as DC offsets and drifts.

10.3.3 Relative Magnitude Rule

A good rule of thumb that is applicable to all classes of amplifiers and all of the nonideal characteristics of op amps involves the relative magnitude of the nonideal quantity compared to the desired signal. If the nonideal value is less than 10 percent of the desired signal quantity, then ignoring it will cause less than a 10 percent error. Similarly, keeping the nonideal value below 1 percent of the desired signal will generally keep errors within 1 percent even if the nonideal quantity is disregarded.

Consider, for example, the case of input bias current. If the input bias current is approximately 700 microamperes and the input signal current is expected to be 1.2 milliamperes, then to ignore bias current would be to make a significant error since the bias current is comparable in magnitude to the desired input current.

Now suppose the cumulative effects of bias current and input offset voltage for a particular amplifier are expected to produce a 75-millivolt offset at the output of the op amp. If the normal output signal is a 1-volt sine wave riding on a 5-volt DC level, then the undesired 75 millivolts offset could probably go unaddressed without producing any serious effects on circuit operation.

10.3.4 Safety Margins on Frequency Compensation

It is common to speak of *phase margin* and *gain margin* with reference to op amp frequency compensation. The terms describe the amount of safety margin between the designed operating point of the op amp and the point where oscillations will likely occur. The absolute limits (i.e., zero safety margin) occurs when the closed-loop gain reaches unity and has a phase shift of $-180°$. This is the dividing line between oscillation and stable operation.

Gain margin is defined as the difference between unity and the actual closed-loop voltage gain at the point where a $-180°$ phase shift occurs. To insure stable operation and to allow for variances in component values, the loop gain should fall to about one third or -10 dB by the time the phase shift has reached $-180°$. Similarly, the phase margin is the number of degrees between the actual phase shift and $-180°$ at the time the loop gain reaches unity. A safety margin of about 45° is recommended. If these safety margins are maintained and the capacitive loading on the output of the amplifier is light, then the circuit should be stable and perform as expected.

REVIEW QUESTIONS

1. The value of the difference between the two input bias currents of an op amp is provided by the manufacturer in the data sheet and is called _____.
2. Drift is a nonideal characteristic that primarily affects AC-coupled, DC-coupled op amp circuits.

Chap. 10   Review Questions   415

3. What type of op amp would you select if a very high value of input impedance was required?
4. Noise generation in op amp circuits can be reduced by selecting very large values of resistance. (True or False)
5. If the intersection of the closed-loop gain curve and the open-loop gain curve for an op amp amplifier occurs with a net slope of 60 dB per decade, will the amplifier be stable? Explain your answer.

CHAPTER 11

SPECIALIZED DEVICES

This chapter will provide an overview of specialized op amps and their applications. The focus will be on identifying the primary characteristics of the various devices and on illustrating potential applications for each device.

11.1 PROGRAMMABLE OP AMPS

One class of specialized op amp is the micropower programmable op amp. These devices utilize an external resistor to establish the quiescent operating current for the internal stages. That is, the internal stages are biased at a particular operating current by the selection of an external resistor. Several characteristics of the amplifier are altered by changes in the programming current (I_P)

1. DC supply current
2. open-loop voltage gain
3. input bias current
4. slew rate
5. unity-gain frequency
6. input noise voltage

Typical values of DC supply current range from less than 1 microampere to as high as 1 milliampere and is proportional to the value of programming current. The ability to operate at very low currents makes these devices especially attractive for battery-powered applications. Additionally, the DC voltage requirements are generally quite flexible with

Sec. 11.2 Instrumentation Amplifiers 417

±1.2 to ±18 volts being a representative range. Here again, the programmable op amp is well suited for battery-powered applications. The DC supply current remains fairly constant with changes in supply voltage provided the programming current is held constant.

The open-loop voltage gain increases as the programming current increases. It is reasonable to expect as much as a 100:1 change in open-loop voltage gain as the programming current is varied over its operating range. This characteristic can be considered an advantage (e.g., programmable voltage gain) or as a disadvantage (e.g., unstable voltage gain) depending on the nature of the application.

Input bias current also increases as programming current increases. In this case, variations as great as 200:1 in input bias current are reasonable over the range of programming currents. The minimum input bias current is often in the fractional nanoampere range.

The slew rate of a programmable op amp increases as programming current increases. It can also be increased by using higher DC supply voltages. As the programming current is varied over its operating range, the slew rate can be expected to vary as much as 1500:1. The upper limit for slew rate is typically greater than the standard 741 op amp (i.e., greater than 0.5 V/μs).

The unity-gain frequency (or gain-bandwidth product) increases as the programming current increases. Ranges as great as 500:1 are reasonable changes to expect in unity-gain frequency as the programming current is varied over its operating range. The highest unity-gain frequency is typically higher than the standard 741 op amp rating (i.e., higher than 1.0 MHz).

The input noise voltage of a programmable op amp decreases as the programming current is increased. Ranges of as much as 200:1 in noise voltage are reasonable as the programming current is varied throughout its control range.

In general, the programmable amplifier can be used for most of the applications previously discussed for general-purpose op amps provided the appropriate specifications are adequate for a given application. The data sheets are interpreted in the same manner as other amplifiers with the exception that the effects of programming current are included. These effects may be shown by including multiple data sheets for different values of programming current and/or by presenting graphs which show the effects of programming current. These devices are often selected for low power applications or for applications which require a controllable parameter (typically voltage gain). Representative devices include the MC1776 and MC3476 manufactured by Motorola.

11.2 INSTRUMENTATION AMPLIFIERS

An instrumentation amplifier is essentially a high-gain differential amplifier that is internally compensated to minimize nonideal characteristics. In particular, the instrumentation amplifier has a very high common-mode rejection ratio. That is, signal voltages which appear on both input terminals are essentially ignored. The amplifier output only responds to the differential input signal. In general, the instrumentation amplifier is generally designed to achieve the following:

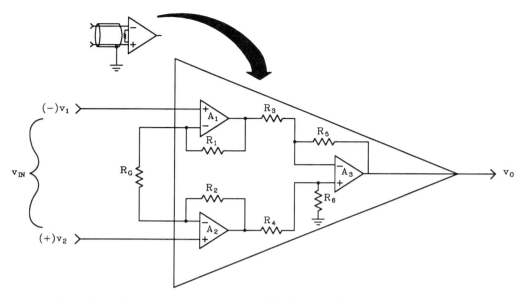

Figure 11.1 The basic instrumentation amplifier is essentially a subtraction circuit preceded by two buffer amplifiers.

1. offset voltages and drifts are minimized,
2. gain is stabilized,
3. nonlinearity is very low,
4. input impedance is very high,
5. output impedance is very low,
6. common-mode rejection is very high.

In Chap. 9, the subtractor circuit was presented. If we precede a subtractor circuit with two buffer amplifiers, we will have the basis for a fundamental instrumentation amplifier circuit. This configuration is shown in Fig. 11.1.

For purposes of the following discussion, let us use the values shown in Fig. 11.2.

Since amplifiers A_1 and A_2 are operated with a closed, negative-feedback loop, we can expect the voltages on the $(-)$ input terminals of the amplifiers to be equal to the voltages on the $(+)$ inputs of the respective amplifiers. This means that the voltage on the upper end of R_G will be equal to the voltage applied to the $(-)$ input of the overall instrumentation amplifier. In the present example, this voltage is $+2$ volts. Similarly, the voltage on the lower end of R_G will be the same as the voltage applied to the $(+)$ input of the overall instrumentation amplifier ($+2.1$ volts for this example). The voltage across R_G (v_G) is the difference between the two input voltages. That is

$$|v_G| = |v_1 - v_2|$$

Sec. 11.2 Instrumentation Amplifiers   419

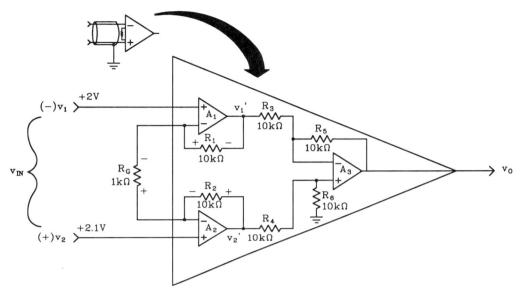

Figure 11.2 Basic instrumentation amplifier used for numerical analysis.

$$= |2 - 2.1|$$
$$= 0.1 \text{ V}$$

The polarity of the voltage drop depends on the relative polarities and magnitudes of the input voltages. For the present example, the lower end of R_G is the more positive since v_2 is more positive than v_1. The current through R_G can be computed with Ohm's Law as

$$i_G = \frac{v_G}{R_G}$$
$$= \frac{0.1}{1 \times 10^3}$$
$$= 100 \text{ }\mu\text{A}$$

Now, since none of this current can flow in or out of amplifiers A_1 and A_2 (ignoring the small bias currents), then i_G must also flow through the feedback resistors of amplifiers A_1 and A_2. Ohm's Law can be used to determine the resulting voltage drop across the feedback resistors.

$$v_{R_1} = v_{R_2} = i_G R_1$$
$$= v_{R_2} = 100 \times 10^{-6} \times 10 \times 10^3$$
$$= 1 \text{ V}$$

Since electron current is flowing downward through R_G, the polarity of R_1 and R_2 will be as shown in Fig. 11.2.

The voltage on the output of A_1 can be found by applying Kirchoff's Voltage Law. That is

$$v'_1 = v_1 - v_{R_1}$$
$$= +2 - 1$$
$$= +1 \text{ V}$$

Similarly, the voltage at the output of A_2 is computed as

$$v'_2 = v_2 + v_{R_2}$$
$$= +2.1 + 1$$
$$= +3.1 \text{ V}$$

The operation of the subtractor circuit (A_3) was discussed in Chap. 9 and will not be repeated here. The output, you will recall, is simply the difference between its inputs. In the present case, the output of A_3 will be

$$v_O = v'_2 - v'_1$$
$$= 3.1 - 1$$
$$= +2.1 \text{ V}$$

We can apply some basic algebraic manipulations to determine an important equation for voltage gain. We have already determined the following relationships (with polarities shown in Fig. 11.2)

$$v'_1 = v_1 - v_{R_1}$$
$$v'_2 = v_2 + v_{R_2}$$
$$i_G = \frac{v_2 - v_1}{R_G}$$
$$v_{R_1} = i_G R_1 = \frac{(v_2 - v_1)R_1}{R_G}$$
$$v_{R_2} = i_G R_2 = \frac{(v_2 - v_1)R_2}{R_G}$$
$$v_O = v'_2 - v'_1$$

Substituting and simplifying gives us the following results:

$$v_O = (v_2 + v_{R_2}) - (v_1 - v_{R_1})$$
$$= \left[v_2 + \frac{(v_2 - v_1)R_2}{R_G}\right] - \left[v_1 - \frac{(v_2 - v_1)R_1}{R_G}\right]$$

Sec. 11.3   Logarithmic Amplifiers 421

$$= \left[(v_2 - v_1) + \frac{(R_1 + R_2)(v_2 - v_1)}{R_G} \right]$$

Since resistors R_1 and R_2 are equal, we can replace the expression $R_1 + R_2$ with the expression 2R. Making this substitution and simplifying gives us the following results:

$$v_O = (v_2 - v_1)\left(1 + \frac{2R}{R_G}\right)$$

Since voltage gain is equal to the output voltage of an amplifier divided by its input voltage, and since the input voltage to our present circuit is $v_2 - v_1$, we can now obtain our final gain equation

$$\boxed{A_V = \frac{2R}{R_G} + 1} \qquad (11.1)$$

This shows us that the gain of the instrumentation amplifier is determined by the value of the external resistor R_G. In the case of the circuit in Fig. 11.2, the voltage gain is computed as

$$A_V = \frac{2R}{R_G} + 1$$

$$= \frac{2 \times 10 \times 10^3}{1 \times 10^3} + 1$$

$$= 21$$

This, of course, correlates to our earlier discovery that an input voltage of 0.1 volts (2.1 − 2) produced an output voltage of 2.1 volts.

Actual integrated instrument amplifiers may use either one or two external resistors to establish the voltage gain of the amplifier. Some devices have internal, precision resistors that can be jumpered into the circuit to obtain certain fixed gains (e.g., 10, 100, 1000). Additionally, they will generally have other inputs for such things as trimming offset voltage and modifying the frequency response (frequency compensation).

The instrumentation amplifier is an important building block based on op amps. An understanding of its general operation coupled with the data provided by the manufacturer will allow you to use this device effectively.

11.3 LOGARITHMIC AMPLIFIERS

Although it is certainly possible to construct discrete amplifier circuits based on op amps that produce outputs which are proportional to the logarithm or antilogarithm of the input voltage, the design is very critical if stable performance is to be expected. In many cases, it is more practical to utilize integrated log and antilog amplifiers that perform these functions in a stable and predictable manner.

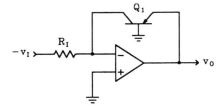

Figure 11.3 The basic logarithmic amplifier circuit relies on the nonlinear relationship between emitter current and base-emitter voltage in a bipolar transistor.

Whether the amplifiers are constructed from discrete components or purchased in an integrated form, the basic operation remains the same. In the case of the log amp, the output voltage is proportional to the logarithm of the input voltage. This relationship is obtained by utilizing the logarithmic relationship that exists between base-emitter voltage and emitter current in a bipolar transistor. Figure 11.3 shows a representative circuit that generates an output which is proportional to the logarithm of the input voltage.

The input current is computed in the same way as for a simple inverting amplifier (i.e., $i_I = v_I/R_I$). Since no substantial part of this current can flow in or out of the ($-$) input of the op amp, it all continues to become the collector current of Q_1. The base-emitter voltage will be controlled by the op amp to a value that allows the collector current to equal the input current.

The relationship between collector current and base-emitter voltage is given by the following equation:

$$i_C = I_{ES}\epsilon^{38.9 v_{BE}}$$

where I_{ES} is the saturation current of the emitter-base diode, v_{BE} is the base-to-emitter voltage and ϵ is the natural logarithm base (approximately 2.71828). Substituting v_I/R_I for i_C and v_O for v_{BE} gives us the following expression:

$$\frac{v_I}{R_I} = I_{ES}\epsilon^{38.9 v_O}$$

Transposing to solve for v_O gives us the expression for v_O in terms of v_I. That is

$$\boxed{v_O \approx 0.026 \ln\left(\frac{v_I}{I_{ES}R_I}\right)} \tag{11.2}$$

Once the circuit has been designed, the only variable is input voltage (v_I). The output voltage is clearly proportional to the logarithm of the input voltage. Figure 11.4 shows the actual response of the circuit shown in Fig. 11.3. The upper waveform is a linear voltage ramp which provides the input to the circuit. The lower waveform was taken at the output of the circuit. The logarithmic relationship is quite evident.

For more critical applications, two circuits similar to the one shown in Fig. 11.3 are connected via a subtractor circuit. One of the log amps is driven by the input signal. The input to the second log amp is connected to a reference voltage. If the transistors

Sec. 11.3 Logarithmic Amplifiers

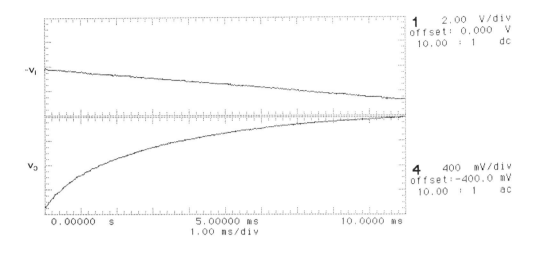

Figure 11.4 Oscilloscope display showing the actual behavior of the circuit shown in Fig. 11.3. (Test equipment courtesy of Hewlett-Packard Company)

are similar (e.g., a matched pair integrated on the same die), then the effects of saturation current (I_{ES}) will be eliminated. This is particularly important since this parameter varies directly with temperature.

There are numerous applications for logarithmic amplifiers. One of the most common applications is for signal compression. By passing an analog signal through a logarithmic amplifier, a very wide range of input signals can be accommodated without saturating the output. For example, an input swing of 1.0 millivolt to 10 volts might produce a corresponding output swing of 0 to 8 volts. While this may not sound too impressive at first glance, realize that the smaller signals will receive much greater gain than the larger signals. In this particular example, the relationship is 2 volts per decade. Thus, a 1- to 10-millivolt change on the input will cause a 0 to 2-volt change in the output. Similarly, a 0.1-volt to 1.0-volt input change will also cause a 2 (4 V to 6 V) volt change in the output. Since the smaller signals are amplified more, the signal-to-noise ratio can be improved. More specifically, the smaller signals become larger relative to the noise signals. When the composite signal is subsequently translated to its original form, the noise will also be reduced.

Another application for the logarithmic amplifier is to convert a linear transducer into a logarithmic response. Optical density of microfilm, for example, is measured as the logarithm of the light which passes through it. By using a light source to shine through the microfilm and onto a photodiode whose response is linear, we have an output voltage (or current) which varies linearly with optical density. Once this waveform is passed through a logarithmic amplifier, we will have the required logarithmic relationship that represents optical density.

11.4 ANTILOGARITHMIC AMPLIFIERS

An antilogarithmic amplifier provides an output that is exponentially related to the input voltage. If, for example, a linear ramp were passed through a logarithmic amplifier (as shown in Fig. 11.4), it would emerge as a logarithmic curve (similar in shape to the familiar RC time constant curve). If this logarithmic signal were then passed through an antilog amplifier, the output would again be a linear ramp. Figure 11.5 shows the basic antilog amplifier configuration.

The input current for this circuit can be estimated using the relationship between emitter-base voltage and collector current described earlier. That is

$$i_C = I_{ES}\epsilon^{38.9v_{BE}}$$

For proper operation of this particular circuit, the input voltage must be negative. Substituting v_I for v_{BE} gives us the following expression for input current:

$$i_I = I_{ES}\epsilon^{38.9v_I}$$

It is important to realize that the exponent in the equation should be positive even though the actual input voltage is negative.

Since no substantial part of this current flows in or out of the ($-$) input of the op amp, it must all continue through R_F. The voltage drop across R_F (and therefore, the output voltage) is determined with Ohm's Law.

$$v_O = v_{R_F} = R_F i_{R_F}$$
$$= R_F I_{ES}\epsilon^{38.9v_I}$$

This latter equation is repeated as Eq. (11.3) and clearly shows the exponential (antilogarithmic) relationship between input voltage and output voltage.

$$\boxed{v_O = R_F I_{ES}\epsilon^{38.9v_I}} \qquad (11.3)$$

The most obvious application for antilogarithmic amplifiers is to expand a signal that has undergone logarithmic compression. One particular application uses log/antilog circuits to reduce the number of bits needed to digitally represent an analog voltage. The signal is first compressed with a logarithmic amplifier. It is then converted to digital form with an analog-to-digital converter. After digital processing, the signal can be restored to proper analog form by utilizing a digital-to-analog converter followed by an antilog amplifier.

Although the average technician or engineer can easily construct one of these circuits

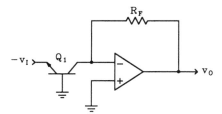

Figure 11.5 The basic antilogarithmic circuit configuration.

in the laboratory and confirm logarithmic operation, the design of a reliable circuit that is minimally affected by temperature and other nonideal conditions is anything but trivial. In most cases, the designer should consider employing a logarithmic device that is produced commercially. These devices contain closely matched components, track well with temperature, and are generally easy to implement. Additionally, the cost is very reasonable for many applications.

The model 755, 6-decade, high accuracy, wideband log/antilog amplifier manufactured by Analog Devices is an example of a commercially available logarithmic module. It requires no external components and can be configured to produce either a logarithmic or an antilogarithmic response. It comes as a module that is about 1.5 inches on a side and 0.4 inches thick.

11.5 MULTIPLIERS/DIVIDERS

Multiplier/divider circuits can be constructed from standard op amps and discrete components. However, the low cost and high performance of integrated circuit versions of these devices makes discrete designs a very unattractive alternative in most cases. In this section, we shall make frequent reference to the AD532 Integrated Circuit Multiplier manufactured by Analog Devices. (See Fig. 11.6.)

Regardless of the type of analog multiplier being considered, the device is essentially a variable gain amplifier. One of the multiplier inputs is amplified by the circuit and appears in the output. The second multiplier input is used to control the gain of the circuit. For example, if we assume that the inputs to the multiplier circuit shown in Fig. 11.6 are voltages called V_X and V_Y, and we further assume that the gain of the circuit (A_V) is established by V_X, then the output of the circuit can be expressed as

$$V_O = A_v \cdot V_Y$$

In this form, the device appears to be a simple linear amplifier whose output is determined by the input voltage times a voltage gain. The voltage gain, however, is not constant in the case of a multiplier circuit. More specifically, the voltage gain is determined by the voltage applied to one (V_X is the present example) of the inputs. The voltage gain, therefore, can be expressed as

$$A_V = k \cdot V_X$$

where k is a constant determined by the circuit configuration. Substitution of this latter equation into the preceding equation gives us a form that reveals the multiplier action of the multiplier circuit.

$$V_O = kV_XV_Y$$

Figure 11.6 A basic integrated multiplier schematic representation.

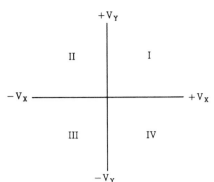

Figure 11.7 There are four possible modes or quadrants of operation for multiplier circuits.

Here we can see that the output voltage is clearly a result of multiplying the input voltage together with a circuit constant. The value of the constant *(k)* is typically 0.1 volt.

Since each of the input voltages may take on either of two polarities, this leads to four possible modes of operation. These four modes, or quadrants, are illustrated in Fig. 11.7.

If a particular multiplier circuit is designed to accept only one polarity of input voltage on each of its inputs, then its operation would be limited to a single quadrant and it would be called a one-quadrant multiplier. Similarly, if a given multiplier circuit requires a single polarity on one of its inputs but accepts both polarities on the other input, the device is called a two-quadrant multiplier. Finally, if a multiplier is designed to accept either polarity on both of its inputs, then the device is called a four-quadrant multiplier. The AD532 is a four-quadrant multiplier. It will accept voltages as large as ±10 volts on its inputs and produces output voltages as large as ±10 volts. The AD532 employs differential inputs and generates a single-ended output voltage described by the following expression:

$$V_{OUT} = \frac{(X_1 - X_2)(Y_1 - Y_2)}{10}$$

Figure 11.8 shows the schematic diagram of an AD532 circuit connected as a simple multiplier. The oscilloscope display in Fig. 11.9 indicates the actual behavior of the circuit.

In Fig. 11.9, waveforms 1 and 2 are the input signals. Waveform 4 is the output of the multiplier circuit. Again, its operation is easily understood by viewing it as a variable gain amplifier. One input (e.g., channel 1 in Fig. 11.9) controls the gain for the second input (e.g., channel 2 in Fig. 11.9).

The multiplier circuit can also be configured to perform several other mathematical operations including division, squaring, and square root functions. Figures 11.10, 11.11, and 11.12 show the connections required to utilize the AD532 to achieve these other functions.

Sec. 11.5 Multipliers/Dividers 427

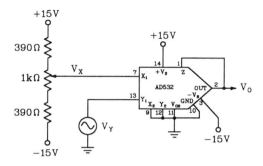

Figure 11.8 The AD532 connected as a simple multiplier circuit.

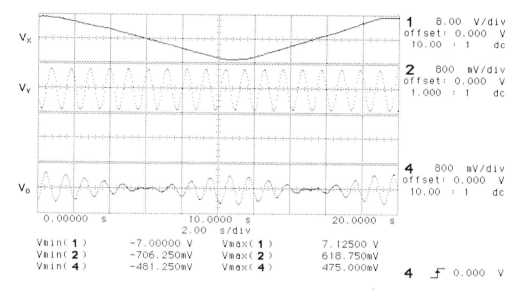

Figure 11.9 Oscilloscope display showing the actual behavior of the circuit shown in Fig. 11.8. (Test equipment courtesy of Hewlett-Packard Company)

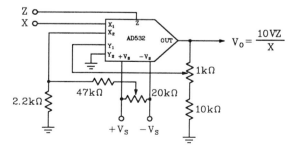

Figure 11.10 The AD532 connected as a divider circuit.

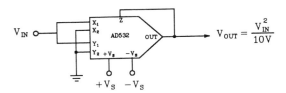

Figure 11.11 The AD532 connected to operate as a squaring circuit.

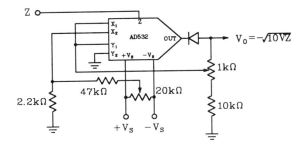

Figure 11.12 The AD532 connected to generate the square root function.

Still another interesting application of the basic multiplier circuit can be demonstrated by applying the same sinusoidal waveform to both multiplier inputs. In this case, the output will be a sinusoidal waveform with a frequency that is exactly double the input frequency. Thus the multiplier circuit can be used as a frequency doubler without resorting to the use of tuned tank circuits.

11.6 SINGLE-SUPPLY AMPLIFIERS

The 741, as well as other standard op amps, can be operated with a single-polarity power supply. We shall examine this type of operation in this section. Additionally, we will discuss the operation of two additional op amp types that are specifically designed to be powered from a single supply.

Figure 11.13 illustrates how to operate a standard 741 (or similar dual-supply op amp) from a single-polarity power source. Here the $-V_{CC}$ terminal is returned to ground and the $+V_{CC}$ terminal is connected to $+15$ volts. To establish midpoint bias, a voltage divider made up of two equal resistors (R_3 and R_4) provide a DC voltage of $1/2\ V_{CC}$ to the noninverting input terminal. Capacitor C_2 insures that the $(+)$ input is at AC ground potential and should have a low reactance at the lowest input frequency. Since under normal circuit conditions, the $(-)$ input will be at the same DC potential as the $(+)$ input ($+7.5$ volts in the present case), capacitor C_1 is included to provide DC isolation between the input source and the DC voltage on the $(-)$ input. Resistors R_1 and R_2 serve their usual function of establishing input impedance and voltage gain.

The oscilloscope displays in Fig. 11.14 further illustrate the operation of the circuit in Fig. 11.13. Waveform 1 is the signal at the input of the amplifier. Notice the DC voltage is about zero (60 μV). Waveform 4 is the signal at the output of the op amp.

Sec. 11.6 Single-Supply Amplifiers 429

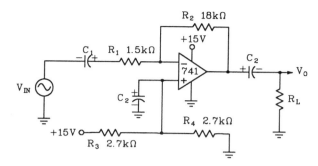

Figure 11.13 The standard 741 op amp can be operated from a single-polarity power source.

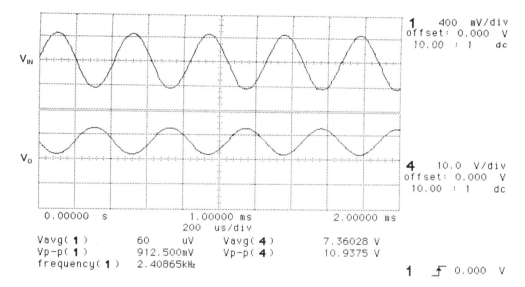

Figure 11.14 Actual performance of the circuit shown in Fig. 11.13.

Here we can see a DC offset of 7.36 volts or about 1/2 of V_{CC}. The voltage gain (12) and phase inversion are both evident from the oscilloscope display and correlate exactly with the performance of the normal dual-supply configuration. The $+V_{CC}$ pin is often connected to a higher voltage (e.g., +30V).

Some manufacturers provide op amps that are specifically designed to be operated from a single-polarity power supply. We shall now briefly examine the operation of two such devices (MC34071 and MC3401) both manufactured by Motorola.

Figure 11.15 shows an MC34071 connected as an AC-coupled inverting amplifier. In general, utilization of this device is similar to the basic 741 (e.g., $A_V = R_F/R_1$). Although the most conspicuous difference may be the fact that the MC34071 is specifically designed for single supply operation, there are a number of other features that clearly separate this device from the basic op amps. Consider the following highlights:

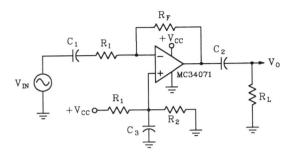

Figure 11.15 An AC coupled amplifier using the MC34071 op amp.

1. The output voltage can typically swing to within 1.0 volt of the $+V_{CC}$ and to within 0.3 volt of the $-V_{CC}$ level (ground in the case of single-supply operation).
2. The supply voltage can range between 3 and 44 volts.
3. Unity-gain frequency of about 4.5 megahertz.
4. Slew rates as high as 13 volts per microsecond.
5. Full output voltage swings at frequencies over 100 kilohertz.

Figure 11.16(a) shows an inverting amplifier configuration using an MC3401 op amp. Figure 11.16(b) shows a basic noninverting amplifier connection. This single-supply device is called a Norton operational amplifier or current differencing amplifier (CDA), and has an input structure that is considerably different than the other op amps discussed in earlier sections. Figure 11.17 shows a simplified schematic that helps clarify the behavior of the input circuitry on the Norton amplifier.

The op amp is generally biased by connecting a DC voltage (V_B) through a resistor (R_B) to the ($+$) input (see Fig. 11.16). This causes $I_I(+)$ to flow through diode D_1 in Fig. 11.17. The value of the $I_I(+)$ current can easily be estimated with Ohm's Law

$$I_I(+) = \frac{V_B - 0.6}{R_B}$$

as where V_B is the bias voltage source, R_B is the resistor in series with the ($+$) terminal, and 0.6 volt is the nominal voltage drop across the input diode (D_1 in Fig. 11.17). The range of normal operating currents for $I_I(+)$ is between 10 and 200 microamperes.

Transistor Q_1 in Fig. 11.17 has matched characteristics with D_1 and will thus develop a collector current that is very nearly identical to the value of $I_I(+)$. This type of circuit is called a current mirror and tends to cause a mirror current in the ($-$) input that is equal to the current in the ($+$) input. The base current of Q_2 is essentially negligible for this analysis. This tendency toward keeping similar currents in the two terminals is analogous to the tendency of a standard op amp to keep the voltage on the two input terminals at the same level. The actual "signal" seen by Q_2 is the difference between $I_I(+)$ and $I_I(-)$. Just as closed-loop operation in a standard op amp tends to reduce the differential input voltage to zero, closed-loop operation in a Norton amplifier tends to reduce the differential input current to zero. The output circuitry of the MC3401 behaves

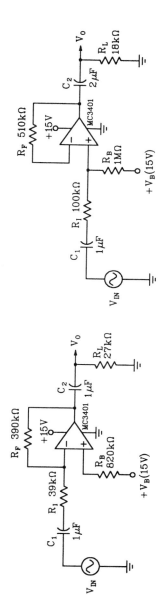

Figure 11.16 The MC3401 Norton operational amplifier is intended for single-supply operation. Circuit configurations for inverting (a) and noninverting (b) are shown.

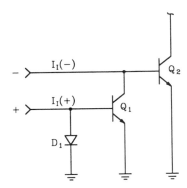

Figure 11.17 The simplified input circuitry for the Norton operational amplifier.

similar to a standard op amp although the output resistance is substantially higher. Some comparative characteristics are

1. The output voltage can typically swing to within 1.0 volt or less of $+V_{CC}$ and ground.
2. The supply voltage can range between 5 and 18 volts.
3. Unity-gain frequency of about 4 megahertz.
4. Slew rates of 0.5 volt per microsecond on the positive slope and 20 volts per microsecond on the negative slope.
5. Full output voltage swings up to about 8 kilohertz.

The DC output voltage (quiescent operating point) is established by the following manufacturer-provided equation:

$$V_{Odc} = \frac{(A_I)(V_B)(R_F)}{R_B} + \left(1 - \frac{R_F}{R_B}A_I\right)\phi$$

where A_I is the gain of the current mirror, V_B is the DC voltage source used to bias the (+) input, R_B is the bias resistor in the (+) input, R_F is the feedback resistor, and ϕ is base-emitter voltage drop of the input transistors. If the bias voltage is more than 5 volts, the second term of the equation may be discarded with only minimal error. Similarly, since the A_I factor is nominally equal to unity, we can simplify this expression to that given by Eq. (11.4).

$$\boxed{V_{Odc} = V_B\left(\frac{R_F}{R_B}\right)} \tag{11.4}$$

The DC output voltage for the circuit shown in Fig. 11.16(a), for example, can be estimated as follows:

$$V_{Odc} = V_B\left(\frac{R_F}{R_B}\right) = 15\left(\frac{390 \times 10^3}{820 \times 10^3}\right) = 7.13 \text{ V}$$

Sec. 11.6 Single-Supply Amplifiers

If the more exacting equation given previously were used, the output voltage would be computed as 7.44 volts.

In a similar manner, the DC output voltage for Fig. 11.16(b) can be estimated as follows:

$$V_{Odc} = V_B \left(\frac{R_F}{R_B} \right) = 15 \left(\frac{510 \times 10^3}{1 \times 10^6} \right) = 7.65 \text{ V}$$

The voltage gain for the inverting amplifier configuration, Fig. 11.16(a), is identical to the standard op amp circuit. That is

$$A_V = -\frac{R_F}{R_I}$$

The noninverting configuration, Fig. 11.16(b), is a little different. The manufacturer provides the following equation:

$$A_V = \frac{(R_F)(A_i)}{R_i + \frac{0.026}{I_i(+)}}$$

As long as the input resistor (R_I) is over 26 kilohms, you can get reasonable results from the simplified equation given in Eq. (11.5).

$$\boxed{A_V = \frac{R_F}{R_I}} \qquad (11.5)$$

It is very important to note that this is different than the standard noninverting amplifier circuit. If, for example, you wanted to build a voltage follower circuit (unity gain), then the feedback resistor (R_F) and the input resistor (R_I) would have to be equal. In the standard amplifier, on the other hand, unity gain is achieved in the noninverting configuration by reducing the feedback resistor to zero.

Since the input terminals, both (+) and (−), essentially have a PN junction to ground, the voltage at these points will not change substantially. Therefore, they may be considered as AC ground points for purposes of computing bandwidth and other AC values. For example, the lower cutoff frequency for the input coupling network is determined by the values of R_I and C_1. More specifically, the cutoff frequency for this portion of the circuit occurs when X_{C1} equals R_I. That is

$$f_C = \frac{1}{2 \pi C_1 R_I}$$

Although the circuit can be used as a DC or direct-coupled amplifier, it is frequently AC coupled as shown in Figs. 11.16(a) and 11.16(b) to avoid disturbing the DC bias on the input terminal and to eliminate the DC offset in the output (typically 1/2 of $+V_{CC}$).

11.7 MULTIPLE OP AMP PACKAGES

Although all of the circuits presented in this text have considered the op amp to be a single integrated circuit, many op amps are available that have more than one device in a single integrated package. With the single exception of $\pm V_{CC}$, these multiple op amps are independent. That is, they can be utilized in the same way as if they were packaged separately. The 747 op amp package, for example, is essentially two 741 op amps combined in a single package. Similarly, the MC3401 device presented in the previous section, actually provides four independent op amps in a single integrated package. The obvious advantages of selecting a multiple op amp package is to reduce the physical size of the circuit and to reduce cost (a package with four op amps is generally less expensive than four separate packages with similar op amps).

11.8 HYBRID OPERATIONAL AMPLIFIERS

Throughout the body of this text, we have generally confined our discussions and design examples to the basic 741 and the MC1741SC. These devices are certainly not the newest, fastest, or necessarily the best in all applications. They are, however, very common and the principles presented directly extend to other higher performance devices.

To provide the reader with a dramatic performance contrast, and to illustrate the performance characteristics that are available in op amps, we shall briefly examine the

TABLE 11.1 Comparison Between a General-Purpose Op Amp and a High Performance, Hybrid Op Amp

PARAMETER	GENERAL PURPOSE OP AMP	MSK 739
Input offset voltage	2 mV	± 25 μV
Input bias current	80 nA	± 75 pA
Input offset current	20 nA	25 pA
Output voltage swing	± 13 V	± 12 V
Full power bandwidth	6.12 kHz	30 MHz
Unity-gain frequency	1.0 MHz	210 MHz
Slew rate	0.5 V/μs	5500 V/μs
Output current	± 6.5 mA	± 120 mA
Power supply rejection ratio	90 dB	115 dB
Common mode rejection ratio	90 dB	110 dB

operational characteristics of a hybrid op amp manufactured by the M.S. Kennedy Corporation (MSK). Numerous devices are manufactured by MSK that offer outstanding performance. The MSK 739 Wideband Amplifier that is described was chosen as a representative device.

The MSK 739 is manufactured using hybrid (i.e., both thick-film and thin-film) technology. It is not appropriate for general-purpose applications any more than the standard 741 is appropriate for very high-performance applications. Nevertheless, the performance of the MSK 739 can be quickly appreciated by contrasting some of the key parameters with similar parameters of a general purpose amplifier. Table 11.1 provides comparative data.

The data in Table 11.1 is intended for comparison purposes only to illustrate some of the outstanding performance that is available in hybrid op amps. The characteristics cited for the general-purpose device are typical and do not necessarily represent a specific device.

REVIEW QUESTIONS

1. A programmable op amp is often selected for applications requiring low-power consumption. (True or False)
2. Circle the correct word to describe the relative value of each of the following parameters with reference to instrumentation amplifiers:
   a. input impedance   (High or low)
   b. output impedance   (High or low)
   c. common-mode rejection   (High or low)
3. When single-supply op amps are used in a multistage amplifier circuit, the individual stages must generally be AC coupled. Explain why.
4. A multiplier circuit that accepts either polarity of voltage on both of its inputs is called a _____ -quadrant multiplier.
5. What type of amplifier circuit is used to compress a wide dynamic range of input voltage amplitudes?

APPENDIX 1

Data sheets for Motorola MC1741 operational amplifier (Pages 2-170 through 2-174). Copyright of Motorola, Inc. Used by permission.

Appendixes

MOTOROLA SEMICONDUCTOR
TECHNICAL DATA

MC1741
MC1741C

OPERATIONAL AMPLIFIER

SILICON MONOLITHIC
INTEGRATED CIRCUIT

INTERNALLY COMPENSATED, HIGH PERFORMANCE OPERATIONAL AMPLIFIERS

. . . designed for use as a summing amplifier, integrator, or amplifier with operating characteristics as a function of the external feedback components.

- No Frequency Compensation Required
- Short-Circuit Protection
- Offset Voltage Null Capability
- Wide Common-Mode and Differential Voltage Ranges
- Low-Power Consumption
- No Latch Up

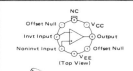
(Top View)

G SUFFIX
METAL PACKAGE
CASE 601

MAXIMUM RATINGS (T_A = +25°C unless otherwise noted)

Rating	Symbol	MC1741C	MC1741	Unit
Power Supply Voltage	V_{CC}	+18	+22	Vdc
	V_{EE}	−18	−22	Vdc
Input Differential Voltage	V_{ID}	±30		Volts
Input Common Mode Voltage (Note 1)	V_{ICM}	±15		Volts
Output Short Circuit Duration (Note 2)	t_S	Continuous		
Operating Ambient Temperature Range	T_A	0 to +70	−55 to +125	°C
Storage Temperature Range Metal and Ceramic Packages Plastic Packages	T_{stg}	−65 to +150 −55 to +125		°C

NOTES:
1. For supply voltages less than +15 V, the absolute maximum input voltage is equal to the supply voltage.
2. Supply voltage equal to or less than 15 V.

P1 SUFFIX
PLASTIC PACKAGE
CASE 626

U SUFFIX
CERAMIC PACKAGE
CASE 693

D SUFFIX
PLASTIC PACKAGE
CASE 751
(SO-8)

PIN CONNECTIONS

(Top View)

EQUIVALENT CIRCUIT SCHEMATIC

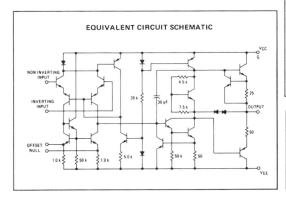

ORDERING INFORMATION

Device	Alternate	Temperature Range	Package
MC1741CD	—	0°C to +70°C	SO-8
MC1741CG	LM741CH, μA741HC		Metal Can
MC1741CP1	LM741CN, μA741TC		Plastic DIP
MC1741CU	—		Ceramic DIP
MC1741G	—	−55°C to +125°C	Metal Can
MC1741U	—		Ceramic DIP

MOTOROLA LINEAR/INTERFACE DEVICES

MC1741, MC1741C

ELECTRICAL CHARACTERISTICS (V_{CC} = +15 V, V_{EE} = −15 V, T_A = 25°C unless otherwise noted).

Characteristic	Symbol	MC1741 Min	MC1741 Typ	MC1741 Max	MC1741C Min	MC1741C Typ	MC1741C Max	Unit
Input Offset Voltage ($R_S \leq 10$ k)	V_{IO}	–	1.0	5.0	–	2.0	6.0	mV
Input Offset Current	I_{IO}	–	20	200	–	20	200	nA
Input Bias Current	I_{IB}	–	80	500	–	80	500	nA
Input Resistance	r_i	0.3	2.0	–	0.3	2.0	–	MΩ
Input Capacitance	C_i	–	1.4	–	–	1.4	–	pF
Offset Voltage Adjustment Range	V_{IOR}	–	±15	–	–	±15	–	mV
Common Mode Input Voltage Range	V_{ICR}	±12	±13	–	±12	±13	–	V
Large Signal Voltage Gain ($V_O = \pm 10$ V, $R_L \geq 2.0$ k)	A_v	50	200	–	20	200	–	V/mV
Output Resistance	r_o	–	75	–	–	75	–	Ω
Common Mode Rejection Ratio ($R_S \leq 10$ k)	CMRR	70	90	–	70	90	–	dB
Supply Voltage Rejection Ratio ($R_S \leq 10$ k)	PSRR	–	30	150	–	30	150	μV/V
Output Voltage Swing ($R_L \geq 10$ k) ($R_L \geq 2$ k)	V_O	±12 ±10	±14 ±13	– –	±12 ±10	±14 ±13	– –	V
Output Short-Circuit Current	I_{os}	–	20	–	–	20	–	mA
Supply Current	I_D	–	1.7	2.8	–	1.7	2.8	mA
Power Consumption	P_C	–	50	85	–	50	85	mW
Transient Response (Unity Gain – Non-Inverting) (V_I = 20 mV, $R_L \geq 2$ k, $C_L \leq 100$ pF) Rise Time (V_I = 20 mV, $R_L \geq 2$ k, $C_L \leq 100$ pF) Overshoot (V_I = 10 V, $R_L \geq 2$ k, $C_L \leq 100$ pF) Slew Rate	t_{TLH} os SR	– – –	0.3 15 0.5	– – –	– – –	0.3 15 0.5	– – –	μs % V/μs

ELECTRICAL CHARACTERISTICS (V_{CC} = +15 V, V_{EE} = −15 V, T_A = T_{low} to T_{high} unless otherwise noted).

Characteristic	Symbol	MC1741 Min	MC1741 Typ	MC1741 Max	MC1741C Min	MC1741C Typ	MC1741C Max	Unit
Input Offset Voltage ($R_S \leq 10$ kΩ)	V_{IO}	–	1.0	6.0	–	–	7.5	mV
Input Offset Current (T_A = 125°C) (T_A = −55°C) (T_A = 0°C to +70°C)	I_{IO}	– – –	7.0 85 –	200 500 –	– – –	– – –	– – 300	nA
Input Bias Current (T_A = 125°C) (T_A = −55°C) (T_A = 0°C to +70°C)	I_{IB}	– – –	30 300 –	500 1500 –	– – –	– – –	– – 800	nA
Common Mode Input Voltage Range	V_{ICR}	±12	±13	–	–	–	–	V
Common Mode Rejection Ratio ($R_S \leq 10$ k)	CMRR	70	90	–	–	–	–	dB
Supply Voltage Rejection Ratio ($R_S \leq 10$ k)	PSRR	–	30	150	–	–	–	μV/V
Output Voltage Swing ($R_L \geq 10$ k) ($R_L \geq 2$ k)	V_O	±12 ±10	±14 ±13	– –	– ±10	– ±13	– –	V
Large Signal Voltage Gain ($R_L \geq 2$ k, $V_{out} = \pm 10$ V)	A_v	25	–	–	15	–	–	V/mV
Supply Currents (T_A = 125°C) (T_A = −55°C)	I_D	– –	1.5 2.0	2.5 3.3	– –	– –	– –	mA
Power Consumption (T_A = +125°C) (T_A = −55°C)	P_C	– –	45 60	75 100	– –	– –	– –	mW

*T_{high} = 125°C for MC1741 and 70°C for MC1741C
T_{low} = −55°C for MC1741 and 0°C for MC1741C

MC1741, MC1741C

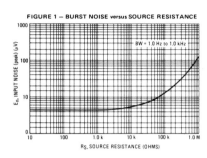

FIGURE 1 — BURST NOISE versus SOURCE RESISTANCE

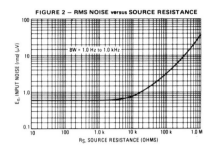

FIGURE 2 — RMS NOISE versus SOURCE RESISTANCE

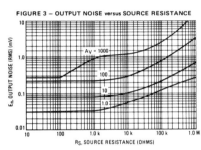

FIGURE 3 — OUTPUT NOISE versus SOURCE RESISTANCE

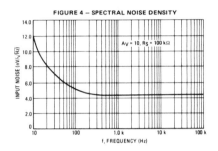

FIGURE 4 — SPECTRAL NOISE DENSITY

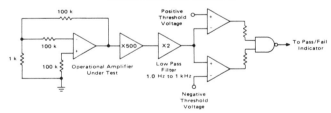

FIGURE 5 — BURST NOISE TEST CIRCUIT

Unlike conventional peak reading or RMS meters, this system was especially designed to provide the quick response time essential to burst (popcorn) noise testing.

The test time employed is 10 seconds and the 20 μV peak limit refers to the operational amplifier input thus eliminating errors in the closed-loop gain factor of the operational amplifier under test.

MC1741, MC1741C

TYPICAL CHARACTERISTICS
(V_{CC} = +15 Vdc, V_{EE} = -15 Vdc, T_A = +25°C unless otherwise noted)

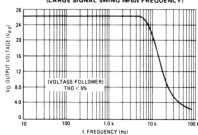

FIGURE 6 — POWER BANDWIDTH
(LARGE SIGNAL SWING versus FREQUENCY)

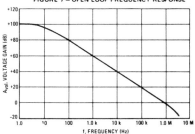

FIGURE 7 — OPEN LOOP FREQUENCY RESPONSE

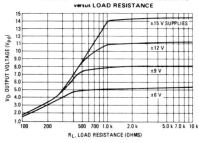

FIGURE 8 — POSITIVE OUTPUT VOLTAGE SWING versus LOAD RESISTANCE

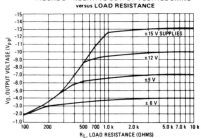

FIGURE 9 — NEGATIVE OUTPUT VOLTAGE SWING versus LOAD RESISTANCE

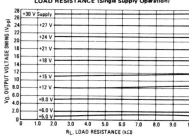

FIGURE 10 — OUTPUT VOLTAGE SWING versus LOAD RESISTANCE (Single Supply Operation)

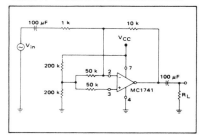

FIGURE 11 — SINGLE SUPPLY INVERTING AMPLIFIER

MC1741, MC1741C

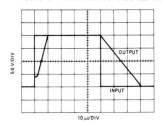

FIGURE 12 — NONINVERTING PULSE RESPONSE

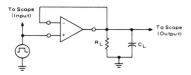

FIGURE 13 — TRANSIENT REPONSE TEST CIRCUIT

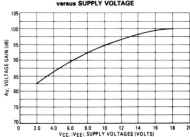

FIGURE 14 — OPEN LOOP VOLTAGE GAIN versus SUPPLY VOLTAGE

APPENDIX 2

Data sheets for Motorola MJE1103 transistor (Pages 3-474 through 3-475). Copyright of Motorola, Inc. Used by permission.

Appendixes 443

MJE1090 thru MJE1093 PNP (SILICON)
MJE2090 thru MJE2093
MJE1100 thru MJE1103 NPN
MJE2100 thru MJE2103

PLASTIC MEDIUM-POWER COMPLEMENTARY SILICON TRANSISTORS

Designed for use in driver and output stages in complementary audio amplifier applications.

- High DC Current Gain —
  h_{FE} = 750 (Min) @ I_C = 3.0 and 4.0 Adc
- True Three Lead Monolithic Construction — Emitter-Base Resistors to Prevent Leakage Multiplication are Built in.
- Available in Two Packages — Case 90 or Case 199

5.0 AMPERE DARLINGTON POWER TRANSISTORS COMPLEMENTARY SILICON

60-80 VOLTS
70 WATTS

MAXIMUM RATINGS

Rating	Symbol	MJE1090 MJE1091 MJE1100 MJE1101 MJE2090 MJE2091 MJE2100 MJE2101	MJE1092 MJE1093 MJE1102 MJE1103 MJE2092 MJE2093 MJE2102 MJE2103	Unit
Collector-Emitter Voltage	V_{CEO}	60	80	Vdc
Collector-Base Voltage	V_{CB}	60	80	Vdc
Emitter-Base Voltage	V_{EB}	5.0		Vdc
Collector Current	I_C	5.0		Adc
Base Current	I_B	0.1		Adc
Total Device Dissipation @ T_C = 25°C Derate above 25°C	P_D	70 0.56		Watts W/°C
Operating and Storage Junction Temperating Range	T_J, T_{stg}	−55 to +150		°C

THERMAL CHARACTERISTICS

Characteristic	Symbol	Max	Unit
Thermal Resistance, Junction to Case	θ_{JC}	1.8	°C/W

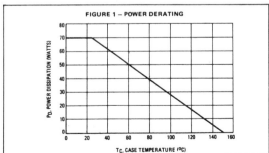

FIGURE 1 — POWER DERATING

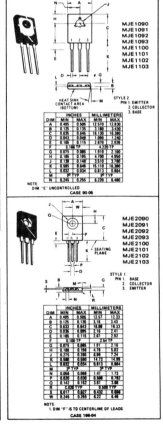

3-474

MJE1090 thru MJE1093 PNP/MJE1100 thru MJE1103 NPN (continued)
MJE2090 thru MJE2093 PNP/MJE2100 thru MJE2103 NPN

ELECTRICAL CHARACTERISTICS ($T_C = 25°C$ unless otherwise noted)

Characteristic		Symbol	Min	Max	Unit
OFF CHARACTERISTICS					
Collector-Emitter Breakdown Voltage[1]		BV_{CEO}			Vdc
($I_C = 100$ mAdc, $I_B = 0$)	MJE1090, MJE1091, MJE1100, MJE1101		60	–	
	MJE2090, MJE2091, MJE2100, MJE2101		60	–	
	MJE1092, MJE1093, MJE1102, MJE1103		80	–	
	MJE2092, MJE2093, MJE2102, MJE2103		80	–	
Collector Cutoff Current		I_{CEO}			µAdc
($V_{CE} = 30$ Vdc, $I_B = 0$)	MJE1090, MJE1091, MJE1100, MJE1101		–	500	
	MJE2090, MJE2091, MJE2100, MJE2101		–	500	
($V_{CE} = 40$ Vdc, $I_B = 0$)	MJE1092, MJE1093, MJE1102, MJE1103		–	500	
	MJE2092, MJE2093, MJE2102, MJE2103		–	500	
Collector Cutoff Current		I_{CBO}			mAdc
($V_{CB} = $ Rated BV_{CEO}, $I_E = 0$)			–	0.2	
($V_{CB} = $ Rated BV_{CEO}, $I_E = 0$, $T_C = 100°C$)			–	2.0	
Emitter Cutoff Current		I_{EBO}			mAdc
($V_{BE} = 5.0$ Vdc, $I_C = 0$)			–	2.0	
ON CHARACTERISTICS (1)					
DC Current Gain		h_{FE}			–
($I_C = 3.0$ Adc, $V_{CE} = 3.0$ Vdc)	MJE1090, MJE1092, MJE1100, MJE1102		750	–	
	MJE2090, MJE2092, MJE2100, MJE2102		750	–	
($I_C = 4.0$ Adc, $V_{CE} = 3.0$ Vdc)	MJE1091, MJE1093, MJE1101, MJE1103		750	–	
	MJE2091, MJE2093, MJE2101, MJE2103		750	–	
Collector-Emitter Saturation Voltage		$V_{CE(sat)}$			Vdc
($I_C = 3.0$ Adc, $I_B = 12$ mAdc)	MJE1090, MJE1092, MJE1100, MJE1102		–	2.5	
	MJE2090, MJE2092, MJE2100, MJE2102		–	2.5	
($I_C = 4.0$ Adc, $I_B = 16$ mAdc)	MJE1091, MJE1093, MJE1101, MJE1103		–	2.8	
	MJE2091, MJE2093, MJE2101, MJE2103		–	2.8	
Base-Emitter On Voltage		$V_{BE(on)}$			Vdc
($I_C = 3.0$ Adc, $V_{CE} = 3.0$ Vdc)	MJE1090, MJE1092, MJE1100, MJE1102		–	2.5	
	MJE2090, MJE2092, MJE2100, MJE2102		–	2.5	
($I_C = 4.0$ Adc, $V_{CE} = 3.0$ Vdc)	MJE1091, MJE1093, MJE1101, MJE1103		–	2.5	
	MJE2091, MJE2093, MJE2101, MJE2103		–	2.5	
DYNAMIC CHARACTERISTICS					
Small-Signal Current Gain		h_{fe}	1.0	–	–
($I_C = 3.0$ Adc, $V_{CE} = 3.0$ Vdc, $f = 1.0$ MHz)					

[1] Pulse Test: Pulse Width ≤ 300 µs, Duty Cycle ≤ 2.0%.

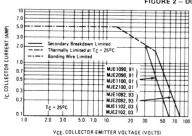

FIGURE 2 – DC SAFE OPERATING AREA

There are two limitations on the power handling ability of a transistor: junction temperature and secondary breakdown. Safe operating area curves indicate $I_C - V_{CE}$ limits of the transistor that must be observed for reliable operation; e.g., the transistor must not be subjected to greater dissipation than the curves indicate.

At high case temperatures, thermal limitations will reduce the power that can be handled to values less than the limitations imposed by secondary breakdown. (See AN-415)

FIGURE 3 – DARLINGTON CIRCUIT SCHEMATIC

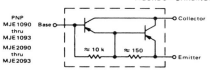
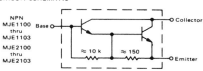

3-475

APPENDIX 3

Data sheets for Motorola 2N2222 transistor (Pages 2-255 through 2-261). Copyright of Motorola, Inc. Used by permission.

2N2218,A, 2N2219,A 2N2221,A (SILICON)
2N2222,A, 2N5581, 2N5582

NPN SILICON ANNULAR HERMETIC TRANSISTORS

... widely used "Industry Standard" transistors for applications as medium-speed switches and as amplifiers from audio to VHF frequencies.

- DC Current Gain Specified — 1.0 to 500 mAdc
- Low Collector-Emitter Saturation Voltage —
  $V_{CE(sat)}$ @ I_C = 500 mAdc
  = 1.6 Vdc (Max) — Non-A Suffix
  = 1.0 Vdc (Max) — A-Suffix
- High Current-Gain–Bandwidth Product —
  f_T = 250 MHz (Min) @ I_C = 20 mAdc — All Types Except
  = 300 MHz (Min) @ I_C = 20 mAdc — 2N2219A, 2N2222A, 2N5582
- Complements to PNP 2N2904,A thru 2N2907,A
- JAN/JANTX Available for all devices

NPN SILICON SWITCHING AND AMPLIFIER TRANSISTORS

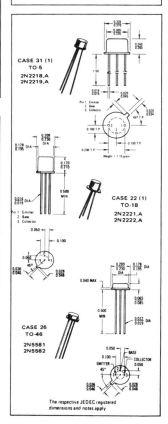

CASE 31 (1) TO-5
2N2218,A
2N2219,A

CASE 22 (1) TO-18
2N2221,A
2N2222,A

CASE 26 TO-46
2N5581
2N5582

The respective JEDEC registered dimensions and notes apply

SELECTION GUIDE

Device Type	Characteristic			Package
	BV_{CEO} I_C = 10 mAdc Volts	h_{FE} I_C = 150 mAdc Min/Max	h_{FE} I_C = 500 mAdc Min	
2N2218 2N2219	30	40/120 100/300	20 30	TO-5
2N2221 2N2222	30	40/120 100/300	20 30	TO-18
2N5581 2N5582	40	40/120 100/300	25 40	TO-46
2N2218A 2N2219A	40	40/120 100/300	25 40	TO-5
2N2221A 2N2222A	40	40/120 100/300	25 40	TO-18

*MAXIMUM RATINGS

Rating	Symbol	2N2218 2N2219 2N2221 2N2222	2N2218A 2N2219A 2N2221A 2N2222A	2N5581 2N5582	Unit
Collector-Emitter Voltage	V_{CEO}	30	40	40	Vdc
Collector-Base Voltage	V_{CB}	60	75	75	Vdc
Emitter-Base Voltage	V_{EB}	5.0	6.0	6.0	Vdc
Collector Current — Continuous	I_C	800	800	800**	mAdc
		2N2218,A 2N2219,A	2N2221,A 2N2222,A	2N5581 2N5582	
Total Device Dissipation @ T_A = 25°C Derate above 25°C	P_D	0.8 5.33	0.5 3.33	0.5 3.33	Watt mW/°C
Total Device Dissipation @ T_C = 25°C Derate above 25°C	P_D	3.0 20	1.8 12	2.0 11.43	Watts mW/°C
Operating and Storage Junction Temperature Range	T_J, T_{stg}	−65 to +200			°C

*Indicates JEDEC Registered Data.
**Motorola Guarantees this Data in Addition to JEDEC Registered Data.

2–255

Appendixes

2N2218,A, 2N2219,A, 2N2221,A, 2N2222,A, 2N5581, 2N5582 (continued)

*ELECTRICAL CHARACTERISTICS ($T_A = 25°C$ unless otherwise noted)

Characteristic		Symbol	Min	Max	Unit
OFF CHARACTERISTICS					
Collector-Emitter Breakdown Voltage ($I_C = 10$ mAdc, $I_B = 0$)	Non-A Suffix	BV_{CEO}	30	—	Vdc
	A-Suffix, 2N5581,2N5582		40	—	
Collector-Base Breakdown Voltage ($I_C = 10$ μAdc, $I_E = 0$)	Non-A Suffix	BV_{CBO}	60	—	Vdc
	A-Suffix, 2N5581,2N5582		75	—	
Emitter-Base Breakdown Voltage ($I_E = 10$ μAdc, $I_C = 0$)	Non-A Suffix	BV_{EBO}	5.0	—	Vdc
	A-Suffix, 2N5581,2N5582		6.0	—	
Collector Cutoff Current ($V_{CE} = 60$ Vdc, $V_{EB(off)} = 3.0$ Vdc)	A-Suffix, 2N5581,2N5582	I_{CEX}	—	10	nAdc
Collector Cutoff Current		I_{CBO}			μAdc
($V_{CB} = 50$ Vdc, $I_E = 0$)	Non-A Suffix		—	0.01	
($V_{CB} = 60$ Vdc, $I_E = 0$)	A-Suffix, 2N5581,2N5582		—	0.01	
($V_{CB} = 50$ Vdc, $I_E = 0$, $T_A = 150°C$)	Non-A Suffix		—	10	
($V_{CB} = 60$ Vdc, $I_E = 0$, $T_A = 150°C$)	A-Suffix, 2N5581,2N5582		—	10	
Emitter Cutoff Current ($V_{EB} = 3.0$ Vdc, $I_C = 0$)	A-Suffix, 2N5581,2N5582	I_{EBO}	—	10	nAdc
Base Cutoff Current ($V_{CE} = 60$ Vdc, $V_{EB(off)} = 3.0$ Vdc)	A-Suffix	I_{BL}	—	20	nAdc
ON CHARACTERISTICS					
DC Current Gain		h_{FE}			—
($I_C = 0.1$ mAdc, $V_{CE} = 10$ Vdc)	2N2218,A,2N2221,A,2N5581(1)		20	—	
	2N2219,A,2N2222,A,2N5582(1)		35	—	
($I_C = 1.0$ mAdc, $V_{CE} = 10$ Vdc)	2N2218,A,2N2221,A,2N5581		25	—	
	2N2219,A,2N2222,A,2N5582		50	—	
($I_C = 10$ mAdc, $V_{CE} = 10$ Vdc)	2N2218,A,2N2221,A,2N5581(1)		35	—	
	2N2219,A,2N2222,A,2N5582(1)		75	—	
($I_C = 10$ mAdc, $V_{CE} = 10$ Vdc, $T_A = -55°C$)	2N2218A,2N2221A,2N5581		15	—	
	2N2219A,2N2222A,2N5582		35	—	
($I_C = 150$ mAdc, $V_{CE} = 10$ Vdc)(1)	2N2218,A,2N2221,A,2N5581		40	120	
	2N2219,A,2N2222,A,2N5582		100	300	
($I_C = 150$ mAdc, $V_{CE} = 1.0$ Vdc)(1)	2N2218A,2N2221A,2N5581		20	—	
	2N2219A,2N2222A,2N5582		50	—	
($I_C = 500$ mAdc, $V_{CE} = 10$ Vdc)(1)	2N2218,2N2221		20	—	
	2N2219,2N2222		30	—	
	2N2218A,2N2221A,2N5581		25	—	
	2N2219A,2N2222A,2N5582		40	—	
Collector-Emitter Saturation Voltage(1)		$V_{CE(sat)}$			Vdc
($I_C = 150$ mAdc, $I_B = 15$ mAdc)	Non-A Suffix		—	0.4	
	A-Suffix, 2N5581,2N5582		—	0.3	
($I_C = 500$ mAdc, $I_B = 50$ mAdc)	Non-A Suffix		—	1.6	
	A-Suffix, 2N5581,2N5582		—	1.0	
Base-Emitter Saturation Voltage(1)		$V_{BE(sat)}$			Vdc
($I_C = 150$ mAdc, $I_B = 15$ mAdc)	Non-A Suffix		0.6	2.0	
	A-Suffix, 2N5581,2N5582		0.6	1.2	
($I_C = 500$ mAdc, $I_B = 50$ mAdc)	Non-A Suffix		—	2.6	
	A-Suffix, 2N5581,2N5582		—	2.0	

2N2218,A, 2N2219,A, 2N2221,A, 2N2222,A, 2N5581, 2N5582 (continued)

*ELECTRICAL CHARACTERISTICS (Continued)

Characteristic		Symbol	Min	Max	Unit
SMALL-SIGNAL CHARACTERISTICS					
Current-Gain–Bandwidth Product(2) (I_C = 20 mAdc, V_{CE} = 20 Vdc, f = 100 MHz)	All Types, Except 2N2219A, 2N2222A, 2N5582	f_T	250 300	— —	MHz
Output Capacitance(3) (V_{CB} = 10 Vdc, I_E = 0, f = 100 kHz)		C_{ob}	—	8.0	pF
Input Capacitance(3) (V_{EB} = 0.5 Vdc, I_C = 0, f = 100 kHz)	Non-A Suffix A-Suffix, 2N5581, 2N5582	C_{ib}	— —	30 25	pF
Input Impedance (I_C = 1.0 mAdc, V_{CE} = 10 Vdc, f = 1.0 kHz) (I_C = 10 mAdc, V_{CE} = 10 Vdc, f = 1.0 kHz)	2N2218A, 2N2221A, 2N5581 2N2219A, 2N2222A, 2N5582 2N2218A, 2N2221A, 2N5581 2N2219A, 2N2222A, 2N5582	h_{ie}	1.0 2.0 0.2 0.25	3.5 8.0 1.0 1.25	k ohms
Voltage Feedback Ratio (I_C = 1.0 mAdc, V_{CE} = 10 Vdc, f = 1.0 kHz) (I_C = 10 mAdc, V_{CE} = 10 Vdc, f = 1.0 kHz)	2N2218A, 2N2221A, 2N5581 2N2219A, 2N2222A, 2N5582 2N2218A, 2N2221A, 2N5581 2N2219A, 2N2222A, 2N5582	h_{re}	— — — —	5.0 8.0 2.5 4.0	$\times 10^{-4}$
Small-Signal Current Gain (I_C = 1.0 mAdc, V_{CE} = 10 Vdc, f = 1.0 kHz) (I_C = 10 mAdc, V_{CE} = 10 Vdc, f = 1.0 kHz)	2N2218A, 2N2221A, 2N5581 2N2219A, 2N2222A, 2N5582 2N2218A, 2N2221A, 2N5581 2N2219A, 2N2222A, 2N5582	h_{fe}	30 50 50 75	150 300 300 375	—
Output Admittance (I_C = 1.0 mAdc, V_{CE} = 10 Vdc, f = 1.0 kHz) (I_C = 10 mAdc, V_{CE} = 10 Vdc, f = 1.0 kHz)	2N2218A, 2N2221A, 2N5581 2N2219A, 2N2222A, 2N5582 2N2218A, 2N2221A, 2N5581 2N2219A, 2N2222A, 2N5582	h_{oe}	3.0 5.0 10 25	15 35 100 200	μmhos
Collector-Base Time Constant (I_E = 20 mAdc, V_{CB} = 20 Vdc, f = 31.8 MHz)	A-Suffix, 2N5581, 2N5582	$r_b'C_c$	—	150	ps
Noise Figure (I_C = 100 μAdc, V_{CE} = 10 Vdc, R_S = 1.0 k ohm, f = 1.0 kHz)	2N2219A, 2N2222A	NF	—	4.0	dB
SWITCHING CHARACTERISTICS (A-Suffix, 2N5581 and 2N5582)					
Delay Time	(V_{CC} = 30 Vdc, V_{BE}(off) = 0.5 Vdc, I_C = 150 mAdc, I_{B1} = 15 mAdc) (Figure 14)	t_d	—	10	ns
Rise Time		t_r	—	25	ns
Storage Time	(V_{CC} = 30 Vdc, I_C = 150 mAdc, I_{B1} = I_{B2} = 15 mAdc) (Figure 15)	t_s	—	225	ns
Fall Time		t_f	—	60	ns
Active Region Time Constant** (I_C = 150 mAdc, V_{CE} = 30 Vdc)		T_A	—	2.5	ns

*Indicates JEDEC Registered Data.
**Motorola Guarantees this Data in Addition to JEDEC Registered Data.
(1) Pulse Test: Pulse Width ≤ 300 μs, Duty Cycle ≤ 2.0%.
(2) f_T is defined as the frequency at which $|h_{fe}|$ extrapolates to unity.
(3) 2N5581 and 2N5582 are Listed C_{cb} and C_{eb} for these conditions and values.

2N2218,A, 2N2219,A, 2N2221,A, 2N2222,A, 2N5581, 2N5582 (continued)

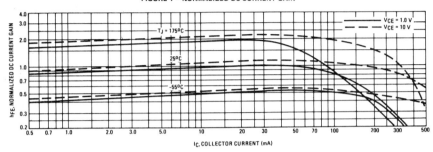

FIGURE 1 — NORMALIZED DC CURRENT GAIN

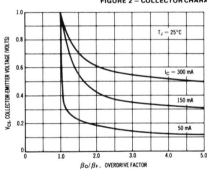

FIGURE 2 — COLLECTOR CHARACTERISTICS IN SATURATION REGION

This graph shows the effect of base current on collector current. β_o (current gain at the edge of saturation) is the current gain of the transistor at 1 volt, and β_s (forced gain) is the ratio of I_C/I_B in a circuit.

EXAMPLE: For type 2N2219, estimate a base current (I_B) to insure saturation at a temperature of 25°C and a collector current of 150 mA.

Observe that at $I_c = 150$ mA an overdrive factor of at least 2.5 is required to drive the transistor well into the saturation region. From Figure 1, it is seen that h_{FE} @ 1 volt is approximately 0.62 of h_{FE} @ 10 volts. Using the guaranteed minimum gain of 100 @ 150 mA and 10 V, $\beta_o = 62$ and substituting values in the overdrive equation, we find:

$$\frac{\beta_o}{\beta_s} = \frac{h_{FE} @ 1.0\,V}{I_C/I_B} \qquad 2.5 = \frac{62}{150/I_B} \qquad I_B \approx 6.0\ mA$$

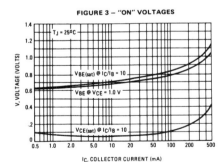

FIGURE 3 — "ON" VOLTAGES

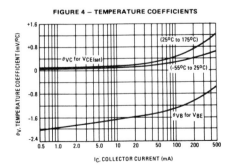

FIGURE 4 — TEMPERATURE COEFFICIENTS

2N2218,A, 2N2219,A, 2N2221,A, 2N2222,A, 2N5581, 2N5582 (continued)

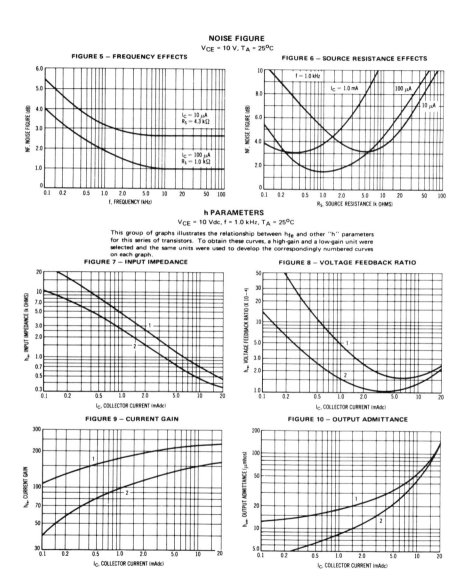

Appendixes

2N2218,A, 2N2219,A, 2N2221,A, 2N2222,A, 2N5581, 2N5582 (continued)

SWITCHING TIME CHARACTERISTICS

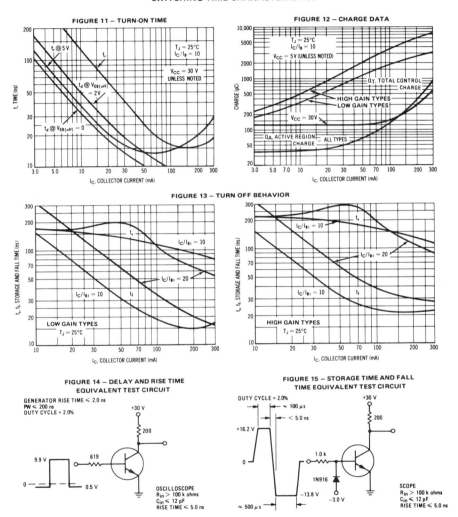

2N2218,A, 2N2219,A, 2N2221,A, 2N2222,A, 2N5581, 2N5582 (continued)

FIGURE 16 — CURRENT-GAIN–BANDWIDTH PRODUCT AND COLLECTOR-BASE TIME CONSTANT DATA

FIGURE 17 — CAPACITANCES

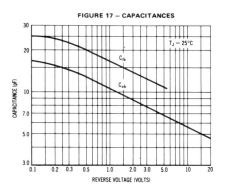

FIGURE 18 — ACTIVE-REGION SAFE OPERATING AREAS

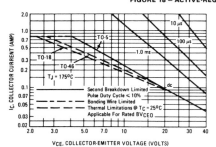

This graph shows the maximum $I_C \cdot V_{CE}$ limits of the device both from the standpoint of thermal dissipation (at 25°C case temperature), and secondary breakdown. For case temperatures other than 25°C, the thermal dissipation curve must be modified in accordance with the derating factor in the Maximum Ratings table.

To avoid possible device failure, the collector load line must not fall below the limits indicated by the applicable curve. Thus, for certain operating conditions the device is thermally limited, and for others it is limited by secondary breakdown.

For pulse applications, the maximum $I_C \cdot V_{CE}$ product indicated by the dc thermal limits can be exceeded. Pulse thermal limits may be calculated by using the transient thermal resistance curve of Figure 19.

FIGURE 19 — THERMAL RESPONSE

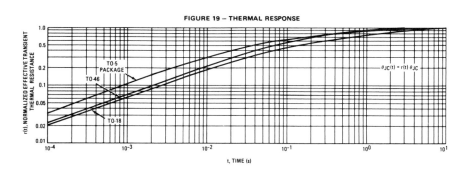

2N2223, A

For Specifications, See 2N2060 Data.

APPENDIX 4

Data sheets for Motorola MC1741SC op amp (Pages 2-175 through 2-180). Copyright of Motorola, Inc. Used by permission.

MOTOROLA SEMICONDUCTOR TECHNICAL DATA

MC1741S / MC1741SC

HIGH SLEW RATE, INTERNALLY COMPENSATED OPERATIONAL AMPLIFIER

The MC1741S/MC1741SC is functionally equivalent, pin compatible, and possesses the same ease of use as the popular MC1741 circuit, yet offers 20 times higher slew rate and power bandwidth. This device is ideally suited for D-to-A converters due to its fast settling time and high slew rate.

- High Slew Rate — 10 V/µs Guaranteed Minimum (for unity gain only)
- No Frequency Compensation Required
- Short-Circuit Protection
- Offset Voltage Null Capability
- Wide Common-Mode and Differential Voltage Ranges
- Low Power Consumption
- No Latch-Up

OPERATIONAL AMPLIFIER

SILICON MONOLITHIC INTEGRATED CIRCUIT

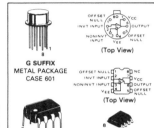

G SUFFIX
METAL PACKAGE
CASE 601

(Top View)

P1 SUFFIX
PLASTIC PACKAGE
CASE 626

D SUFFIX
PLASTIC PACKAGE
CASE 751
(SO-8)

ORDERING INFORMATION

Device	Temperature Range	Package
MC1741SG	−55°C to +125°C	Metal Can
MC1741SCD		SO-8
MC1741SCG	0°C to +70°C	Metal Can
MC1741SCP1		Plastic DIP

TYPICAL APPLICATION OF OUTPUT CURRENT TO VOLTAGE TRANSFORMATION FOR A D-TO-A CONVERTER

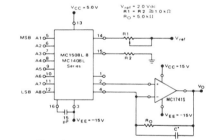

$V_{CC} = 5.0$ V
$V_{ref} = 2.0$ Vdc
$R1 = R2 \geq 10$ kΩ
$R_O = 5.0$ kΩ

Pins not shown are not connected.

Settling time to within 1/2 LSB (±19.5 mV) is approximately 4.0 µs from the time that all bits are switched.
*The value of C may be selected to minimize overshoot and ringing (C ≈ 150 pF).

Theoretical V_O

$$V_O = \frac{V_{ref}}{R1}(R_O)\left[\frac{A1}{2} + \frac{A2}{4} + \frac{A3}{8} + \frac{A4}{16} + \frac{A5}{32} + \frac{A6}{64} + \frac{A7}{128} + \frac{A8}{256}\right]$$

Adjust V_{ref}, R1 or R_O so that V_O with all digital inputs at high level is equal to 9.961 volts.

$$V_O = \frac{2 \text{ V}}{1 \text{ k}}(5\text{ k})\left[\frac{1}{2} + \frac{1}{4} + \frac{1}{8} + \frac{1}{16} + \frac{1}{32} + \frac{1}{64} + \frac{1}{128} + \frac{1}{256}\right] = 10 \text{ V}\left[\frac{255}{256}\right] = 9.961 \text{ V}$$

MC1741S LARGE-SIGNAL TRANSIENT RESPONSE

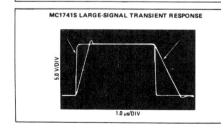

STANDARD MC1741 versus MC1741S RESPONSE COMPARISON

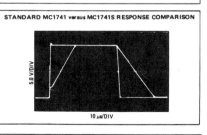

MOTOROLA LINEAR/INTERFACE DEVICES

Appendixes

MC1741S, MC1741SC

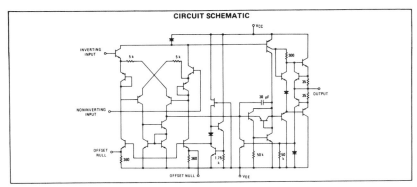

CIRCUIT SCHEMATIC

MAXIMUM RATINGS (T_A = +25°C unless otherwise noted.)

Rating	Symbol	Value MC1741SC	Value MC1741S	Unit
Power Supply Voltage	V_{CC}	+18	+22	Vdc
	V_{EE}	−18	−22	
Differential Input Signal Voltage	V_{ID}	±30		Volts
Common-Mode Input Voltage Swing (See Note 1)	V_{ICR}	±15		Volts
Output Short-Circuit Duration (See Note 2)	t_s	Continuous		
Power Dissipation (Package Limitation)	P_D			
Metal Package		680		mW
Derate above T_A = +25°C		4.6		mW/°C
Plastic Dual In-Line Package		625		mW
Derate above T_A = +25°C		5.0		mW/°C
Operating Ambient Temperature Range	T_A	0 to +75	−55 to +125	°C
Storage Temperature Range	T_{stg}			°C
Metal Package		−65 to +150		
Plastic Package		−55 to +125		

Note 1. For supply voltages less than ±15 Vdc, the absolute maximum input voltage is equal to the supply voltage.
Note 2. Supply voltage equal to or less than 15 Vdc.

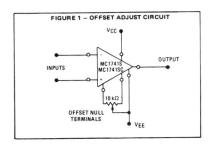

FIGURE 1 — OFFSET ADJUST CIRCUIT

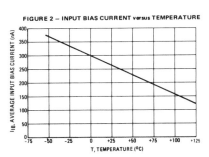

FIGURE 2 — INPUT BIAS CURRENT versus TEMPERATURE

MOTOROLA LINEAR/INTERFACE DEVICES

MC1741S, MC1741SC

ELECTRICAL CHARACTERISTICS (V_{CC} = +15 Vdc, V_{EE} = −15 Vdc, T_A = +25°C unless otherwise noted.)

Characteristic	Symbol	MC1741S Min	MC1741S Typ	MC1741S Max	MC1741SC Min	MC1741SC Typ	MC1741SC Max	Unit
Power Bandwidth (See Figure 3) A_v = 1, R_L = 2.0 kΩ, THD = 5%, V_O = 20 V(p-p)	BW_p	150	200	–	150	200	–	kHz
Large-Signal Transient Response Slew Rate (Figures 10 and 11)	SR							V/μs
V(−) to V(+)		10	20	–	10	20	–	
V(+) to V(−)		10	12	–	10	12	–	
Settling Time (Figures 10 and 11) (to within 0.1%)	t_{setlg}	–	3.0	–	–	3.0	–	μs
Small-Signal Transient Response (Gain = 1, E_{in} = 20 mV, see Figures 7 and 8)								
Rise Time	t_{TLH}	–	0.25	–	–	0.25	–	μs
Fall Time	t_{THL}	–	0.25	–	–	0.25	–	μs
Propagation Delay Time	t_{PLH}, t_{PHL}	–	0.25	–	–	0.25	–	μs
Overshoot	OS	–	20	–	–	20	–	%
Short-Circuit Output Currents	I_{OS}	±10	–	±35	±10	–	±35	mA
Open-Loop Voltage Gain (R_L = 2.0 kΩ) (See Figure 4)	A_{vol}							
V_O = ±10 V, T_A = +25°C		50,000	200,000	–	20,000	100,000	–	
V_O = ±10 V, T_A = T_{low}* to T_{high}*		25,000	–	–	15,000	–	–	
Output Impedance (f = 20 Hz)	z_o	–	75	–	–	75	–	Ω
Input Impedance (f = 20 Hz)	z_i	0.3	1.0	–	0.3	1.0	–	MΩ
Output Voltage Swing	V_O							V_{pk}
R_L = 10 kΩ, T_A = T_{low} to T_{high} (MC1741S only)		±12	±14	–	±12	±14	–	
R_L = 2.0 kΩ, T_A = +25°C		±10	±13	–	±10	±13	–	
R_L = 2.0 kΩ, T_A = T_{low} to T_{high}		±10	–	–	±10	–	–	
Input Common-Mode Voltage Range T_A = T_{low} to T_{high} (MC1741S)	V_{ICR}	±12	±13	–	±12	±13	–	V_{pk}
Common-Mode Rejection Ratio (f = 20 Hz) T_A = T_{low} to T_{high} (MC1741S)	CMRR	70	90	–	70	90	–	dB
Input Bias Current (See Figure 2)	I_{IB}							nA
T_A = +25°C and T_{high}		–	200	500	–	200	500	
T_A = T_{low}		–	500	1500	–	–	800	
Input Offset Current	$|I_{IO}|$							nA
T_A = +25°C and T_{high}		–	30	200	–	30	200	
T_A = T_{low}		–	–	500	–	–	300	
Input Offset Voltage (R_S ≤ 10 kΩ)	$|V_{IO}|$							mV
T_A = +25°C		–	1.0	5.0	–	2.0	6.0	
T_A = T_{low} to T_{high}		–	–	6.0	–	–	7.5	
DC Power Consumption (See Figure 9) (Power Supply = ±15 V, V_O = 0) T_A = T_{low} to T_{high}	P_C	–	50	85	–	50	85	mW
Positive Voltage Supply Sensitivity (V_{EE} constant) T_A = T_{low} to T_{high} on MC1741S	PSS+	–	2.0	100	–	2.0	150	μV/V
Negative Voltage Supply Sensitivity (V_{CC} constant)	PSS−	–	10	150	–	10	150	μV/V

*T_{low} = 0 for MC1741SC
     = −55 °C for MC1741S
T_{high} = +70°C for MC1741SC
     = +125 °C for MC1741S

Appendixes

MC1741S, MC1741SC

TYPICAL CHARACTERISTICS
(V_{CC} = +15 Vdc, V_{EE} = -15 Vdc, T_A = +25°C unless otherwise noted.)

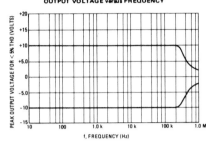

FIGURE 3 – POWER BANDWIDTH – NONDISTORTED OUTPUT VOLTAGE versus FREQUENCY

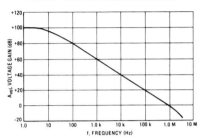

FIGURE 4 – OPEN-LOOP FREQUENCY RESPONSE

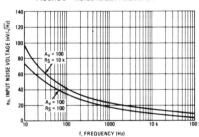

FIGURE 5 – NOISE versus FREQUENCY

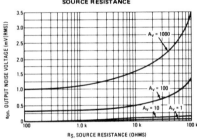

FIGURE 6 – OUTPUT NOISE versus SOURCE RESISTANCE

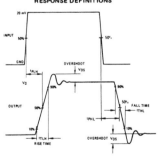

FIGURE 7 – SMALL-SIGNAL TRANSIENT RESPONSE DEFINITIONS

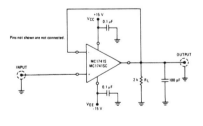

FIGURE 8 – SMALL-SIGNAL TRANSIENT RESPONSE TEST CIRCUIT

MOTOROLA LINEAR/INTERFACE DEVICES

MC1741S, MC1741SC

TYPICAL CHARACTERISTICS
(V_{CC} = +15 Vdc, V_{EE} = –15 Vdc, T_A = +25°C unless otherwise noted.)

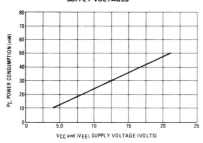

FIGURE 9 – POWER CONSUMPTION versus POWER SUPPLY VOLTAGES

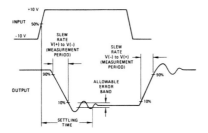

FIGURE 10 – LARGE-SIGNAL TRANSIENT WAVEFORMS

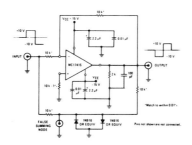

FIGURE 11 – SETTLING TIME AND SLEW RATE TEST CIRCUIT

SETTLING TIME

In order to properly utilize the high slew rate and fast settling time of an operational amplifier, a number of system considerations must be observed. Capacitance at the summing node and at the amplifier output must be minimal and circuit board layout should be consistent with common high-frequency considerations. Both power supply connections should be adequately bypassed as close as possible to the device pins. In bypassing, both low and high-frequency components should be considered to avoid the possibility of excessive ringing. In order to achieve optimum damping, the selection of a capacitor in parallel with the feedback resistor may be necessary. A value too small could result in excessive ringing while a value too large will degrade slew rate and settling time.

SETTLING TIME MEASUREMENT

In order to accurately measure the settling time of an operational amplifier, it is suggested that the "false" summing junction approach be taken as shown in Figure 11. This is necessary since it is difficult to determine when the waveform at the output of the operational amplifier settles to within 0.1% of it's final value. Because the output and input voltages are effectively subtracted from each other at the amplifier inverting input, this seems like an ideal node for the measurement. However, the probe capacitance at this critical node can greatly affect the accuracy of the actual measurement.

The solution to these problems is the creation of a second or "false" summing node. The addition of two diodes at this node clamps the error voltage to limit the voltage excursion to the oscilloscope. Because of the voltage divider effect, only one-half of the actual error appears at this node. For extremely critical measurements, the capacitance of the diodes and the oscilloscope, and the settling time of the oscilloscope must be considered. The expression

$$t_{setlg} = \sqrt{x^2 + y^2 + z^2}$$

can be used to determine the actual amplifier settling time, where
t_{setlg} = observed settling time
 x = amplifier settling time (to be determined)
 y = false summing junction settling time
 z = oscilloscope settling time

It should be remembered that to settle within ±0.1% requires 7RC time constants.

The ±0.1% factor was chosen for the MC1741S settling time as it is compatible with the ±1/2 LSB accuracy of the MC1508L8 digital-to-analog converter. This D-to-A converter features ±0.19% maximum error.

Appendixes

MC1741S, MC1741SC

FIGURE 12 — WAVEFORM AT FALSE SUMMING NODE

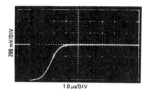

FIGURE 13 — EXPANDED WAVEFORM AT FALSE SUMMING NODE

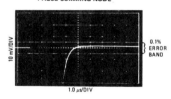

TYPICAL APPLICATION

FIGURE 14 — 12.5-WATT WIDEBAND POWER AMPLIFIER

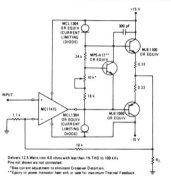

Delivers 12.5 Watts into 4.0 ohms with less than 1% THD to 100 kHz.
Pins not shown are not connected.
*Bias current adjustment to eliminate Crossover Distortion.
**Epoxy to power transistor heat sink or case for maximum Thermal Feedback.

APPENDIX 5

Data sheets for Motorola 1N5221 thru 1N5281 (Pages 1-21 through 1-26). Copyright of Motorola, Inc. Used by permission.

Appendixes

1N5221 (SILICON)
thru
1N5281 series

500 MILLIWATT SURMETIC▲ 20 SILICON ZENER DIODES
(SILICON OXIDE PASSIVATED)

...in answer to the Circuit Design and Component Engineers' many requests — A complete new series of Zener Diodes in the popular DO-204AA case with higher ratings, tighter limits, better operating characteristics and a full set of designers' curves that reflect the superior capabilities of silicon-oxide-passivated junctions. All this in an axial-lead, transfer-molded plastic package offering protection in all common environmental conditions.

- Proven Capability to MIL-S-19500 Specifications
- 10 Watt Surge Rating
- Weldable Leads
- Maximum Limits Guaranteed on Six Electrical Parameters

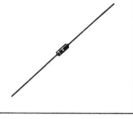

500 MILLIWATT ZENER REGULATOR DIODES

2.4 — 200 VOLTS

MAXIMUM RATINGS

Junction and Storage Temperature: —65 to +200°C

Lead Temperature not less than 1/16" from the case for 10 seconds: 230°C

DC Power Dissipation: 500 mW @ T_L = 75°C, Lead Length = 3/8"
(Derate 4.0 mW/°C above 75°C)

Surge Power: 10 Watts (Non-recurrent square wave @ PW = 8.3 ms, T_J = 55°C, Figure 16)

MECHANICAL CHARACTERISTICS

CASE: Void free, transfer molded, thermosetting plastic.

FINISH: All external surfaces are corrosion resistant. Leads are readily solderable and weldable.

POLARITY: Cathode indicated by color band. When operated in zener mode, cathode will be positive with respect to anode.

MOUNTING POSITION: Any.

WEIGHT: 0.18 gram (approximately).

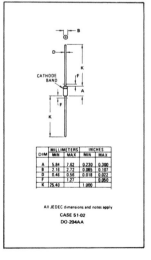

CASE 51-02
DO-204AA

FIGURE 1 — POWER-TEMPERATURE DERATING CURVE

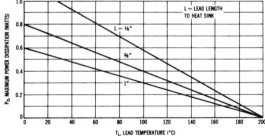

▲Trademark of Motorola Inc.

1-21

1N5221 thru 1N5281 series (continued)

ELECTRICAL CHARACTERISTICS ($T_A = 25°C$ unless otherwise noted). Based on dc measurements at thermal equilibrium; lead length = 3/8"; thermal resistance of heat sink = 30°C/W; $V_f = 1.1$ Max @ $I_f = 200$ mA for all types.

JEDEC Type No. (Note 1)	Nominal Zener Voltage V_Z @ I_{ZT} Volts (Note 2)	Test Current I_{ZT} mA	Max Zener Impedance A & B Suffix Only		Max Reverse Leakage Current					Max Zener Voltage Temp. Coeff. (A & B Suffix Only) θ_{VZ} (%/°C) (Note 3)
					A & B Suffix Only				Non-Suffix	
			Z_{ZT} @ I_{ZT} Ohms	Z_{ZK} @ $I_{ZK} = 0.25$ mA Ohms	I_R μA	@	V_R Volts		I_R @ V_R Used For Suffix A μA	
							A	B		
1N5221	2.4	20	30	1200	100		0.95	1.0	200	−0.085
1N5222	2.5	20	30	1250	100		0.95	1.0	200	−0.085
1N5223	2.7	20	30	1300	75		0.95	1.0	150	−0.080
1N5224	2.8	20	30	1400	75		0.95	1.0	150	−0.080
1N5225	3.0	20	29	1600	50		0.95	1.0	100	−0.075
1N5226	3.3	20	28	1600	25		0.95	1.0	100	−0.070
1N5227	3.6	20	24	1700	15		0.95	1.0	100	−0.065
1N5228	3.9	20	23	1900	10		0.95	1.0	75	−0.060
1N5229	4.3	20	22	2000	5.0		0.95	1.0	50	±0.055
1N5230	4.7	20	19	1900	5.0		1.9	2.0	50	±0.030
1N5231	5.1	20	17	1600	5.0		1.9	2.0	50	+0.030
1N5232	5.6	20	11	1600	5.0		2.9	3.0	50	+0.038
1N5233	6.0	20	7.0	1600	5.0		3.3	3.5	50	+0.038
1N5234	6.2	20	7.0	1000	5.0		3.8	4.0	50	+0.045
1N5235	6.8	20	5.0	750	3.0		4.8	5.0	30	+0.050
1N5236	7.5	20	6.0	500	3.0		5.7	6.0	30	+0.058
1N5237	8.2	20	8.0	500	3.0		6.2	6.5	30	+0.062
1N5238	8.7	20	8.0	600	3.0		6.2	6.5	30	+0.065
1N5239	9.1	20	10	600	3.0		6.7	7.0	30	+0.068
1N5240	10	20	17	600	3.0		7.6	8.0	30	+0.075
1N5241	11	20	22	600	2.0		8.0	8.4	30	+0.076
1N5242	12	20	30	600	1.0		8.7	9.1	10	+0.077
1N5243	13	9.5	13	600	0.5		9.4	9.9	10	+0.079
1N5244	14	9.0	15	600	0.1		9.5	10	10	+0.082
1N5245	15	8.5	16	600	0.1		10.5	11	10	+0.082
1N5246	16	7.8	17	600	0.1		11.4	12	10	+0.083
1N5247	17	7.4	19	600	0.1		12.4	13	10	+0.084
1N5248	18	7.0	21	600	0.1		13.3	14	10	+0.085
1N5249	19	6.6	23	600	0.1		13.3	14	10	+0.086
1N5250	20	6.2	25	600	0.1		14.3	15	10	+0.086
1N5251	22	5.6	29	600	0.1		16.2	17	10	+0.087
1N5252	24	5.2	33	600	0.1		17.1	18	10	+0.088
1N5253	25	5.0	35	600	0.1		18.1	19	10	+0.089
1N5254	27	4.6	41	600	0.1		20	21	10	+0.090
1N5255	28	4.5	44	600	0.1		20	21	10	+0.091
1N5256	30	4.2	49	600	0.1		22	23	10	+0.091
1N5257	33	3.8	58	700	0.1		24	25	10	+0.092
1N5258	36	3.4	70	700	0.1		26	27	10	+0.093
1N5259	39	3.2	80	800	0.1		29	30	10	+0.094
1N5260	43	3.0	93	900	0.1		31	33	10	+0.095
1N5261	47	2.7	105	1000	0.1		34	36	10	+0.095
1N5262	51	2.5	125	1100	0.1		37	39	10	+0.096
1N5263	56	2.2	150	1300	0.1		41	43	10	+0.096
1N5264	60	2.1	170	1400	0.1		44	46	10	+0.097
1N5265	62	2.0	185	1400	0.1		45	47	10	+0.097
1N5266	68	1.8	230	1600	0.1		49	52	10	+0.097
1N5267	75	1.7	270	1700	0.1		53	56	10	+0.098
1N5268	82	1.5	330	2000	0.1		59	62	10	+0.098
1N5269	87	1.4	370	2200	0.1		65	68	10	+0.099
1N5270	91	1.4	400	2300	0.1		66	69	10	+0.099
1N5271	100	1.3	500	2600	0.1		72	76	10	+0.110
1N5272	110	1.1	750	3000	0.1		80	84	10	+0.110
1N5273	120	1.0	900	4000	0.1		86	91	10	+0.110
1N5274	130	0.95	1100	4500	0.1		94	99	10	+0.110
1N5275	140	0.90	1300	4500	0.1		101	106	10	+0.110
1N5276	150	0.85	1500	5000	0.1		108	114	10	+0.110
1N5277	160	0.80	1700	5500	0.1		116	122	10	+0.110
1N5278	170	0.74	1900	5500	0.1		123	129	10	+0.110
1N5279	180	0.68	2200	6000	0.1		130	137	10	+0.110
1N5280	190	0.66	2400	6500	0.1		137	144	10	+0.110
1N5281	200	0.65	2500	7000	0.1		144	152	10	+0.110

NOTE 1 — TOLERANCE AND VOLTAGE DESIGNATION

Tolerance designation — The JEDEC type numbers shown indicate a tolerance of ±10% with guaranteed limits on only V_Z, I_R and V_f as shown in the above table. Units with guaranteed limits on all six parameters are indicated by suffix "A" for ±10% tolerance and suffix "B" for ±5.0% units.

Non-Standard voltage designation — To designate units with zener voltages other than those assigned JEDEC numbers, the type number should be used.

EXAMPLE:

```
0.5 M 90 Z S 5
```

- 0.5 Watt
- Manufacturer
- Nominal Voltage
- Zener Diodes
- Surmetic
- Tolerance (±%)

NOTE 2 — SPECIAL SELECTIONS AVAILABLE INCLUDE:

1 — Nominal zener voltages between those shown.

2 — Matched sets: (Standard Tolerances are ±5.0%, ±3.0%, ±2.0%, ±1.0%) depending on voltage per device.

  a. Two or more units for series connection with specified tolerance on total voltage. Series matched sets make zener voltages in excess of 200 volts possible as well as providing lower temperature coefficients, lower dynamic impedance and greater power handling ability.

  b. Two or more units matched to one another with any specified tolerance.

3 — Tight voltage tolerances: 1.0%, 2.0%, 3.0%.

NOTE 3 — TEMPERATURE COEFFICIENT (θ_{VZ})

Test conditions for temperature coefficient are as follows:
  a. $I_{ZT} = 7.5$ mA, $T_1 = 25°C$,
     $T_2 = 125°C$ (1N5221A, B thru 1N5242A, B.)
  b. I_{ZT} = Rated I_{ZT}, $T_1 = 25°C$,
     $T_2 = 125°C$ (1N5243A, B thru 1N5281A, B.)

Device to be temperature stabilized with current applied prior to reading breakdown voltage at the specified ambient temperature.

Appendixes

1N5221 thru 1N5281 series (continued)

TYPICAL REVERSE CHARACTERISTICS FOR SELECTED ZENER DIODES

Curves marked T_A were obtained from dc measurements at thermal equilibrium; lead length = 3/8"; thermal resistance of heat sink = 30°C/W. Curves marked T_J were obtained from pulse tests; mounting conditions are not a factor.

$V_{Z(Nominal)}$ = 3.3 Volts

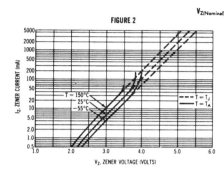
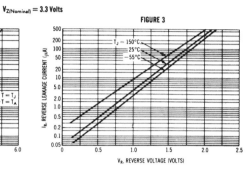

$V_{Z(Nominal)}$ = 5.1 Volts

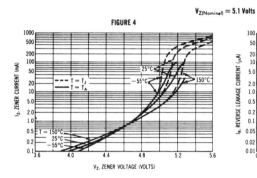
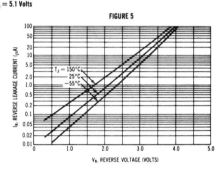

$V_{Z(Nominal)}$ = 27 Volts

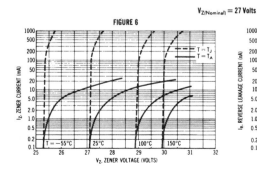
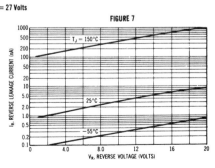

1N5221 thru 1N5281 series (continued)

TEMPERATURE COEFFICIENTS AND VOLTAGE REGULATION
(90% of the units are in the ranges indicated)

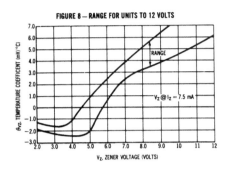

FIGURE 8 — RANGE FOR UNITS TO 12 VOLTS

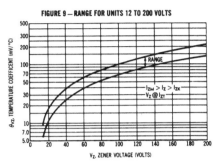

FIGURE 9 — RANGE FOR UNITS 12 TO 200 VOLTS

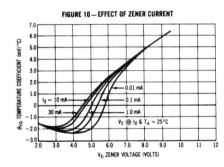

FIGURE 10 — EFFECT OF ZENER CURRENT

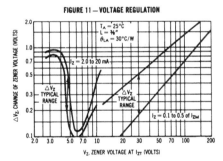

FIGURE 11 — VOLTAGE REGULATION

TYPICAL ZENER IMPEDANCE

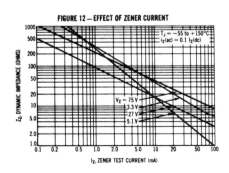

FIGURE 12 — EFFECT OF ZENER CURRENT

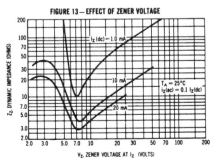

FIGURE 13 — EFFECT OF ZENER VOLTAGE

1-24

Appendixes

1N5221 thru 1N5281 Series (continued)

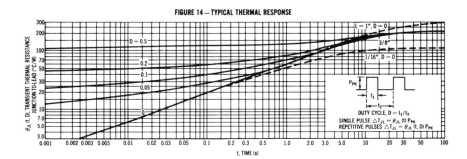

FIGURE 14 – TYPICAL THERMAL RESPONSE

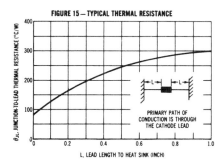

FIGURE 15 – TYPICAL THERMAL RESISTANCE

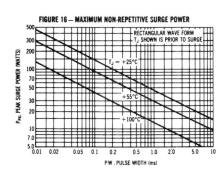

FIGURE 16 – MAXIMUM NON-REPETITIVE SURGE POWER

APPLICATION NOTE

Since the actual voltage available from a given zener diode is temperature dependent, it is necessary to determine junction temperature under any set of operating conditions, in order to calculate its value. The following procedure is recommended:

Lead Temperature, T_L, should be determined from:

$$T_L = \theta_{LA} P_D + T_A$$

θ_{LA} is the lead-to-ambient thermal resistance and P_D is the power dissipation. θ_{LA} is generally 30-40°C/W for the various clips and tie points in common use and for printed circuit board wiring.

Junction Temperature, T_J, may be found from:

$$T_J = T_L + \Delta T_{JL}$$

ΔT_{JL} is the increase in junction temperature above the lead temperature and may be found from Figure 14 for a train of power pulses or from Figure 15 for dc power.

For worst-case design, using expected limits of I_Z, limits of P_D and the extremes of $T_J (\Delta T_J)$ may be estimated. Changes in voltage, V_Z, can then be found from:

$$\Delta V = \theta_{VZ} \Delta T_J$$

θ_{VZ}, the zener voltage temperature coefficient, is found from Figures 8, 9, and 10.

Under high power-pulse operation, the zener voltage will vary with time and may also be affected significantly by the zener resistance. For best regulation, use short leads, especially to the cathode, and keep current excursions as low as possible.

Data of Figure 14 should not be used to compute surge capability. Surge limitations are given in Figure 16. They are lower than would be expected by considering only junction temperature, as current crowding effects cause temperatures to be extremely high in small spots resulting in device degradation should the limits of Figure 16 be exceeded.

1N5221 thru 1N5281 Series (continued)

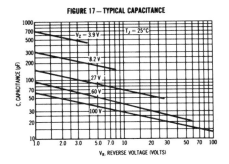

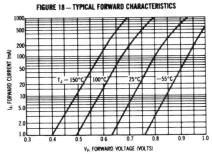

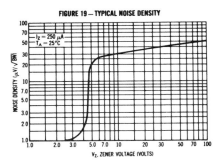

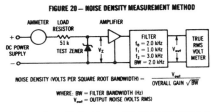

The input voltage and load resistance are high so that the zener diode is driven from a constant current source. The amplifier is low noise so that the amplifier noise is negligible compared to that of the test zener. The filter bandpass is known so that the noise density can be calculated from the formula shown. The data of Figure 19 and the formula can also be used to find noise for any system bandwidth.

APPENDIX 6

Table A6.1 lists the approximate electrical characteristics for the 1N914A diode. This device or its equivalent is available from multiple manufacturers.

TABLE A6.1 Approximate Ratings for a 1N914A Diode

PARAMETER	RATING
Peak inverse voltage	75 V
Average forward current	75 mA (below 25 °C) 10 mA (at 150 °C)
Reverse current	25 nA (at 25 °C) 50 μA (at 150 °C)
Power dissipation	250 mW
Capacitance	4 pF
Reverse recovery time	8 ns

APPENDIX 7

Data sheets for the Motorola MPF102 transistor (Pages 3-724). Copyright of Motorola, Inc. Used by permission.

MPF102 (SILICON)

PIN 1. DRAIN
2. SOURCE
3. GATE

CASE 29 (5)
(TO-92)

Drain and Source may be interchanged

Silicon N-channel junction field-effect transistor designed for VHF amplifier and mixer applications.

MAXIMUM RATINGS (T_A = 25°C unless otherwise noted)

Rating	Symbol	Value	Unit
Drain-Source Voltage	V_{DS}	25	Vdc
Drain-Gate Voltage	V_{DG}	25	Vdc
Gate-Source Voltage	V_{GS}	25	Vdc
Gate Current	I_G	10	mAdc
Total Device Dissipation @ T_A = 25°C Derate above 25°C	P_D [1]	310 2.82	mW mW/°C
Operating Junction Temperature	T_J [1]	125	°C
Storage Temperature Range	T_{stg}	-65 to +150	°C

ELECTRICAL CHARACTERISTICS (T_A = 25°C unless otherwise noted)

Characteristic	Symbol	Min	Max	Unit
OFF CHARACTERISTICS				
Gate-Source Breakdown Voltage (I_G = 10 μAdc, V_{DS} = 0)	BV_{GSS}	25	—	Vdc
Gate Reverse Current (V_{GS} = 15 Vdc, V_{DS} = 0) (V_{GS} = 15 Vdc, V_{DS} = 0, T_A = 100 C)	I_{GSS}	— —	2.0 2.0	nAdc μAdc
Gate-Source Cutoff Voltage (V_{DS} = 15 Vdc, I_D = 2.0 nAdc)	$V_{GS(off)}$	—	8.0	Vdc
Gate-Source Voltage (V_{DS} = 15 Vdc, I_D = 0.2 mAdc)	V_{GS}	0.5	7.5	Vdc
ON CHARACTERISTICS				
Zero-Gate-Voltage Drain Current [1] (V_{DS} = 15 Vdc, V_{GS} = 0 Vdc)	I_{DSS}	2.0	20	mAdc
DYNAMIC CHARACTERISTICS				
Forward Transfer Admittance [1] (V_{DS} = 15 Vdc, V_{GS} = 0, f = 1 kHz)	$|y_{fs}|$	2000	7500	μmhos
Input Capacitance (V_{DS} = 15 Vdc, V_{GS} = 0, f = 1 MHz)	C_{iss}	—	7.0	pF
Reverse Transfer Capacitance (V_{DS} = 15 Vdc, V_{GS} = 0, f = 1 MHz)	C_{rss}	—	3.0	pF
Forward Transfer Admittance (V_{DS} = 15 Vdc, V_{GS} = 0, f = 100 MHz)	$|y_{fs}|$	1600	—	μmhos
Input Conductance (V_{DS} = 15 Vdc, V_{GS} = 0, f = 100 MHz)	$Re(y_{is})$	—	800	μmhos
Output Conductance (V_{DS} = 15 Vdc, V_{GS} = 0, f = 100 MHz)	$Re(y_{os})$	—	200	μmhos

*Pulse Test: Pulse Width ≤ 630 ms; Duty Cycle ≤ 10%

[1] Continuous package improvements have enhanced these guaranteed Maximum Ratings as follows: P_D = 1.0 W @ T_C = 25°C. Derate above 25°C — 8.0 mW/°C, T_J = -65 to +150°C, θ_{JC} = 125° C/W.

APPENDIX 8

Data sheets for the Motorola 1N4728 through 1N4764 zener diodes (Pages 1-105 through 1-109). Copyright of Motorola, Inc. Used by permission.

1N4728 thru 1N4764 (SILICON)
1M110ZS10 thru 1M200ZS10

Designers Data Sheet

1.0 WATT SURMETIC 30 SILICON ZENER DIODES

... a complete series of 1.0 Watt Zener Diodes with limits and operating characteristics that reflect the superior capabilities of silicon-oxide-passivated junctions. All this in an axial-lead, transfer-molded plastic package offering protection in all common environmental conditions.

- To 80 Watts Surge Rating @ 1.0 ms
- Maximum Limits Guaranteed on Six Electrical Parameters
- Package No Larger Than the Conventional 400 mW Package

Designer's Data for "Worst Case" Conditions

The Designers Data sheets permit the design of most circuits entirely from the information presented. Limit curves — representing boundaries on device characteristics — are given to facilitate "worst case" design.

1.0 WATT ZENER REGULATOR DIODES
3.3–200 VOLTS

MAXIMUM RATINGS

Rating	Symbol	Value	Unit
*DC Power Dissipation @ T_A = 50°C	P_D	1.0	Watt
Derate above 50°C		6.67	mW/°C
DC Power Dissipation @ T_L = 75°C Lead Length = 3/8"	P_D	3.0	Watts
Derate above 75°C		24	mW/°C
*Operating and Storage Junction Temperature Range	T_J, T_{stg}	–65 to +200	°C

MECHANICAL CHARACTERISTICS

CASE: Void-free, transfer-molded, thermosetting plastic

FINISH: All external surfaces are corrosion resistant and leads are readily solderable and weldable

POLARITY: Cathode indicated by polarity band. When operated in zener mode, cathode will be positive with respect to anode

MOUNTING POSITION: Any

WEIGHT: 0.4 gram (approx)

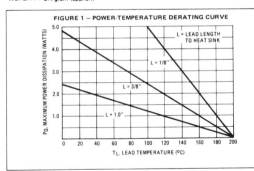

FIGURE 1 — POWER-TEMPERATURE DERATING CURVE

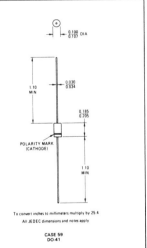

CASE 59
DO-41

*Indicates JEDEC Registered Data

Appendixes

1N4728 thru 1N4764 (continued)
1M110ZS10 thru 1M200ZS10

ELECTRICAL CHARACTERISTICS ($T_A = 25°C$ unless otherwise noted) *$V_F = 1.5$ V max, $I_F = 200$ mA for all types

JEDEC Type No. (Note 1)	Motorola Type No. (Note 2)	*Nominal Zener Voltage V_Z @ I_{ZT} Volts (Note 2 & 3)	*Test Current I_{ZT} mA	*Max Zener Impedance (Note 4) Z_{ZT} @ I_{ZT} Ohms	Z_{ZK} @ I_{ZK} Ohms	I_{ZK} mA	*Leakage Current I_R μA Max	V_R @ Volts	*Surge Current @ $T_A = 25°C$ i_r — mA (Note 5)
1N4728	1M3.3ZS10	3.3	76	10	400	1.0	100	1.0	1380
1N4729	1M3.6ZS10	3.6	69	10	400	1.0	100	1.0	1260
1N4730	1M3.9ZS10	3.9	64	9.0	400	1.0	50	1.0	1190
1N4731	1M4.3ZS10	4.3	58	9.0	400	1.0	10	1.0	1070
1N4732	1M4.7ZS10	4.7	53	8.0	500	1.0	10	1.0	970
1N4733	1M5.1ZS10	5.1	49	7.0	550	1.0	10	1.0	890
1N4734	1M5.6ZS10	5.6	45	5.0	600	1.0	10	2.0	810
1N4735	1M6.2ZS10	6.2	41	2.0	700	1.0	10	3.0	730
1N4736	1M6.8ZS10	6.8	37	3.5	700	1.0	10	4.0	660
1N4737	1M7.5ZS10	7.5	34	4.0	700	0.5	10	5.0	605
1N4738	1M8.2ZS10	8.2	31	4.5	700	0.5	10	6.0	550
1N4739	1M9.1ZS10	9.1	28	5.0	700	0.5	10	7.0	500
1N4740	1M10ZS25	10	25	7.0	700	0.25	10	7.6	454
1N4741	1M11ZS10	11	23	8.0	700	0.25	5.0	8.4	414
1N4742	1M12ZS10	12	21	9.0	700	0.25	5.0	9.1	380
1N4743	1M13ZS10	13	19	10	700	0.25	5.0	9.9	344
1N4744	1M15ZS10	15	17	14	700	0.25	5.0	11.4	304
1N4745	1M16ZS10	16	15.5	16	700	0.25	5.0	12.2	285
1N4746	1M18ZS10	18	14	20	750	0.25	5.0	13.7	250
1N4747	1M20ZS10	20	12.5	22	750	0.25	5.0	15.2	225
1N4748	1M22ZS10	22	11.5	23	750	0.25	5.0	16.7	205
1N4749	1M24ZS10	24	10.5	25	750	0.25	5.0	18.2	190
1N4750	1M27ZS10	27	9.5	35	750	0.25	5.0	20.6	170
1N4751	1M30ZS10	30	8.5	40	1000	0.25	5.0	22.8	150
1N4752	1M33ZS10	33	7.5	45	1000	0.25	5.0	25.1	135
1N4753	1M36ZS10	36	7.0	50	1000	0.25	5.0	27.4	125
1N4754	1M39ZS10	39	6.5	60	1000	0.25	5.0	29.7	115
1N4755	1M43ZS10	43	6.0	70	1500	0.25	5.0	32.7	110
1N4756	1M47ZS10	47	5.5	80	1500	0.25	5.0	35.8	95
1N4757	1M51ZS10	51	5.0	95	1500	0.25	5.0	38.8	90
1N4758	1M56ZS10	56	4.5	110	2000	0.25	5.0	42.6	80
1N4759	1M62ZS10	62	4.0	125	2000	0.25	5.0	47.1	70
1N4760	1M68ZS10	68	3.7	150	2000	0.25	5.0	51.7	65
1N4761	1M75ZS10	75	3.3	175	2000	0.25	5.0	56.0	60
1N4762	1M82ZS10	82	3.0	200	3000	0.25	5.0	62.2	55
1N4763	1M91ZS10	91	2.8	250	3000	0.25	5.0	69.2	50
1N4764	1M100ZS10	100	2.5	350	3000	0.25	5.0	76.0	45
—	1M110ZS10	110	2.3	450	4000	0.25	5.0	83.6	—
—	1M120ZS10	120	2.0	550	4500	0.25	5.0	91.2	—
—	1M130ZS10	130	1.9	700	5000	0.25	5.0	98.8	—
—	1M150ZS10	150	1.7	1000	6000	0.25	5.0	114.0	—
—	1M160ZS10	160	1.6	1100	6500	0.25	5.0	121.6	—
—	1M180ZS10	180	1.4	1200	7000	0.25	5.0	136.8	—
—	1M200ZS10	200	1.2	1500	8000	0.25	5.0	152.0	—

*Indicates JEDEC Registered Data

NOTE 1 — TOLERANCE AND TYPE NUMBER DESIGNATION

The JEDEC type numbers listed have a standard tolerance on the nominal zener voltage of ±10%. A standard tolerance of ±5% on individual units is also available and is indicated by suffixing "A" to the standard type number.

NOTE 2 — SPECIALS AVAILABLE INCLUDE:

(A) NOMINAL ZENER VOLTAGES BETWEEN THE VOLTAGES SHOWN AND TIGHTER VOLTAGE TOLERANCES: To designate units with zener voltages other than those assigned JEDEC numbers and/or tight voltage tolerances (±5%, ±3%, ±2%, ±1%), the Motorola type number should be used.

(B) MATCHED SETS: (Standard Tolerances are ±5.0%, ±3.0%, ±2.0%, ±1.0%).

Zener diodes can be obtained in sets consisting of two or more matched devices. The method for specifying such matched sets is similar to the one described in (A), except that two extra suffixes are added to the code number described.

These units are marked with code letters to identify the matched sets and, in addition, each unit in a set is marked with the same serial number, which is different for each set being ordered.

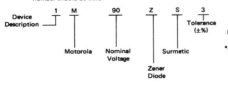

Example: 1M90ZS3

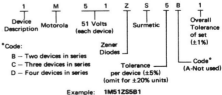

*Code:
B — Two devices in series
C — Three devices in series
D — Four devices in series

Tolerance per device (±5%) (omit for ±20% units)

Code* (A-Not used)

Example: 1M51ZS5B1

1-106

Appendixes

1N4728 thru 1N4764 (continued)
1M110ZS10 thru 1M200ZS10

(C) ZENER CLIPPERS: (Standard Tolerance ±10% and ±5%).
Special clipper diodes with opposing Zener junctions built into the device are available by using the following nomenclature:

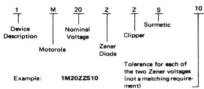

Example: 1M20ZZS10

NOTE 3 — ZENER VOLTAGE (V_Z) MEASUREMENT

Motorola guarantees the zener voltage when measured at 90 seconds while maintaining the lead temperature (T_L) at 30°C ± 1°C, 3/8" from the diode body.

NOTE 4 — ZENER IMPEDANCE (Z_Z) DERIVATION

The zener impedance is derived from the 60 cycle ac voltage, which results when an ac current having an rms value equal to 10% of the dc zener current (I_{ZT} or I_{ZK}) is superimposed on I_{ZT} or I_{ZK}.

NOTE 5 — SURGE CURRENT (i_r) NON-REPETITIVE

The rating listed in the electrical characteristics table is maximum peak, non-repetitive, reverse surge current of 1/2 square wave or equivalent sine wave pulse of 1/120 second duration superimposed on the test current, I_{ZT}, per JEDEC registration, however, actual device capability is as described in Figures 4 and 5.

APPLICATION NOTE

Since the actual voltage available from a given zener diode is temperature dependent, it is necessary to determine junction temperature under any set of operating conditions in order to calculate its value. The following procedure is recommended:

Lead Temperature, T_L, should be determined from:

$$T_L = \theta_{LA} P_D + T_A$$

θ_{LA} is the lead-to-ambient thermal resistance (°C/W) and P_D is the power dissipation. The value for θ_{LA} will vary and depends on the device mounting method. θ_{LA} is generally 30-40°C/W for the various clips and tie points in common use and for printed circuit board wiring.

The temperature of the lead can also be measured using a thermocouple placed on the lead as close as possible to the tie point. The thermal mass connected to the tie point is normally large enough so that it will not significantly respond to heat surges generated in the diode as a result of pulsed operation once steady-state conditions are achieved. Using the measured value of T_L, the junction temperature may be determined by:

$$T_J = T_L + \Delta T_{JL}$$

ΔT_{JL} is the increase in junction temperature above the lead temperature and may be found from Figure 2 for a train of power pulses (L = 3/8 inch) or from Figure 3 for dc power.

$$\Delta T_{JL} = \theta_{JL} P_D$$

For worst-case design, using expected limits of I_Z, limits of P_D and the extremes of $T_J (\Delta T_J)$ may be estimated. Changes in voltage, V_Z, can then be found from:

$$\Delta V = \theta_{VZ} \Delta T_J$$

θ_{VZ}, the zener voltage temperature coefficient, is found from Figures 6 and 7.

Under high power-pulse operation, the zener voltage will vary with time and may also be affected significantly by the zener resistance. For best regulation, keep current excursions as low as possible.

Data of Figure 2 should not be used to compute surge capability. Surge limitations are given in Figure 4. They are lower than would be expected by considering only junction temperature, as current crowding effects cause temperatures to be extremely high in small spots resulting in device degradation should the limits of Figure 4 be exceeded.

1N4728 thru 1N4764 (continued)
1M110ZS10 thru 1M200ZS10

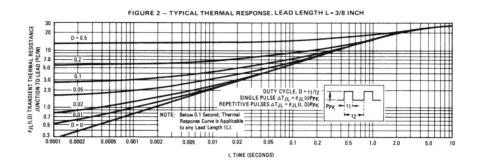

FIGURE 2 — TYPICAL THERMAL RESPONSE, LEAD LENGTH L = 3/8 INCH

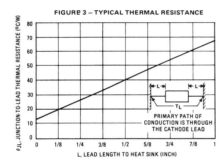

FIGURE 3 — TYPICAL THERMAL RESISTANCE

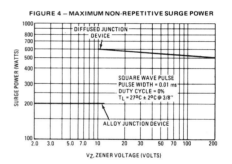

FIGURE 4 — MAXIMUM NON-REPETITIVE SURGE POWER

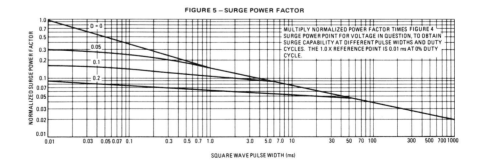

FIGURE 5 — SURGE POWER FACTOR

Appendixes

1N4728 thru 1N4764 (continued)
1M110ZS10 thru 1M200ZS10

TEMPERATURE COEFFICIENTS AND VOLTAGE REGULATION
(90% OF THE UNITS ARE IN THE RANGES INDICATED)

FIGURE 6 — TEMPERATURE COEFFICIENT-RANGE FOR UNITS TO 12 VOLTS

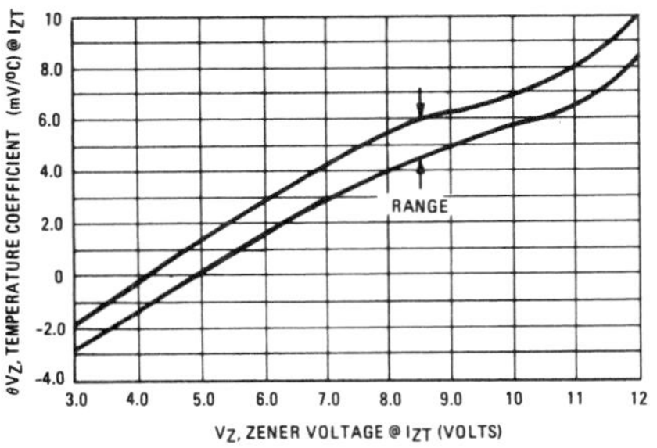

FIGURE 7 — TEMPERATURE COEFFICIENT-RANGE FOR UNITS 10 TO 200 VOLTS

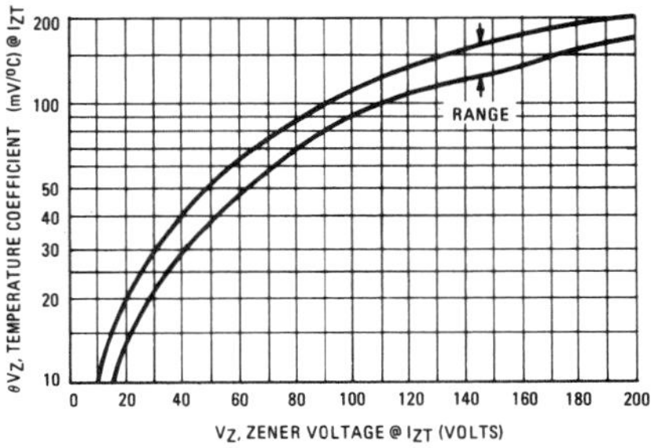

FIGURE 8 — VOLTAGE REGULATION

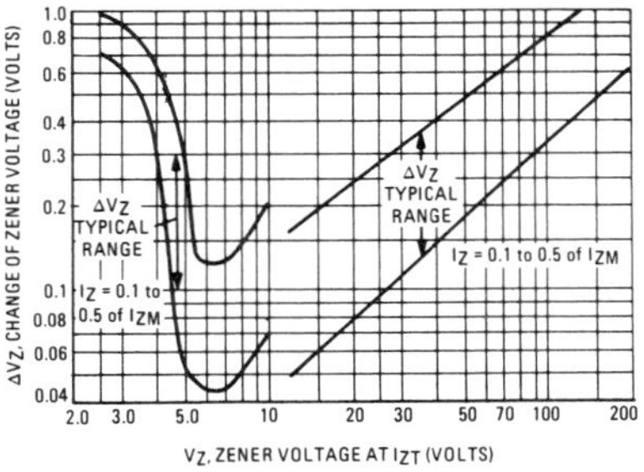

FIGURE 9 — MAXIMUM REVERSE LEAKAGE
(95% OF THE UNITS ARE BELOW THE VALUES SHOWN)

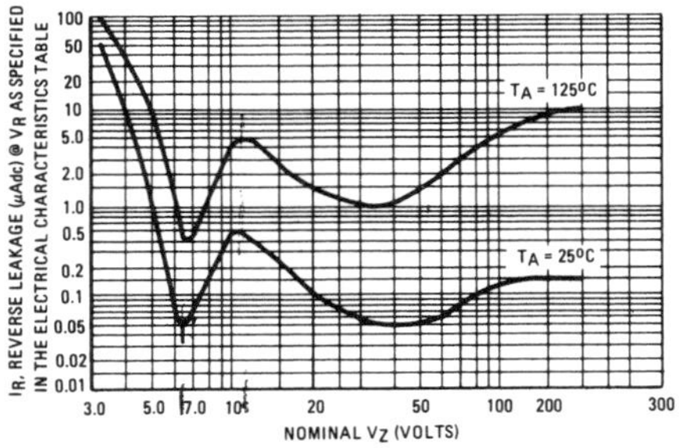

1–109

APPENDIX 9

Data sheets for the Motorola 2N3440 transistor (Pages 2-509 through 2-512). Copyright of Motorola, Inc. Used by permission.

2N3439 (SILICON)
2N3440

NPN SILICON HIGH VOLTAGE POWER TRANSISTORS

...designed for use in consumer and industrial line-operated applications. These devices are particularly suited for audio, video and differential amplifiers as well as high-voltage, low-current inverters, switching and series pass regulators.

- High DC Current Gain —
  h_{FE} = 40 - 160 @ I_C = 20 mAdc
- Current-Gain—Bandwidth Product —
  f_T = 15 MHz (Min) @ I_C = 10 mAdc
- Low Output Capacitance —
  C_{ob} = 10 pF (Max) @ f = 1.0 MHz

1 AMPERE POWER TRANSISTORS NPN SILICON

250-350 VOLTS
10 WATTS

* MAXIMUM RATINGS

Rating	Symbol	2N3439	2N3440	Unit
Collector-Emitter Voltage	V_{CEO}	350	250	Vdc
Collector-Base Voltage	V_{CB}	450	300	Vdc
Emitter-Base Voltage	V_{EB}	7.0		Vdc
Collector Current - Continuous	I_C	1.0		Adc
Base Current	I_B	0.5		Adc
Total Device Dissipation @ T_A = 25°C Derate above 25°C	P_D	1.0 5.7		Watts mW/°C
Total Device Dissipation @ T_C = 25°C Derate above 25°C	P_D	10 0.057		Watts W/°C
Operating and Storage Junction Temperature Range	T_J, T_{stg}	−65 to +200		°C

THERMAL CHARACTERISTICS

Characteristic	Symbol	Max	Unit
Thermal Resistance, Junction to Case	θ_{JC}	17.5	°C/W
Thermal Resistance, Junction to Ambient	θ_{JA}	175	°C/W

ELECTRICAL CHARACTERISTICS (T_C = 25°C unless otherwise noted)

Characteristic		Symbol	Min	Max	Unit
OFF CHARACTERISTICS					
*Collector-Emitter Sustaining Voltage (1) (I_C = 50 mAdc, I_B = 0)	2N3439 2N3440	V_{CEO}(sus)	350 250	— —	Vdc
Collector Cutoff Current (V_{CE} = 300 Vdc, I_B = 0) (V_{CE} = 200 Vdc, I_B = 0)	2N3439 2N3440	I_{CEO}	— —	20 50	μAdc
Collector Cutoff Current (V_{CE} = 450 Vdc, $V_{BE(off)}$ = 1.5 Vdc) (V_{CE} = 300 Vdc, $V_{BE(off)}$ = 1.5 Vdc)	2N3439 2N3440	I_{CEX}	— —	500 500	
*Collector Cutoff Current (V_{CB} = 360 Vdc, I_E = 0) (V_{CB} = 250 Vdc, I_E = 0)	2N3439 2N3440	I_{CBO}	— —	20 20	μAdc
*Emitter Cutoff Current (V_{BE} = 6.0 Vdc, I_C = 0)		I_{EBO}	—	20	μAdc
ON CHARACTERISTICS (1)					
DC Current Gain (I_C = 2.0 mAdc, V_{CE} = 10 Vdc) *(I_C = 20 mAdc, V_{CE} = 10 Vdc)	2N3439 Both Types	h_{FE}	30 40	— 160	—
*Collector-Emitter Saturation Voltage (I_C = 50 mAdc, I_B = 4.0 mAdc)		$V_{CE(sat)}$	—	0.5	Vdc
*Base-Emitter Saturation Voltage (I_C = 50 mAdc, I_B = 4.0 mAdc)		$V_{BE(sat)}$	—	1.3	Vdc
DYNAMIC CHARACTERISTICS					
Current-Gain — Bandwidth Product (I_C = 10 mAdc, V_{CE} = 10 Vdc)		f_T	15	—	MHz
Output Capacitance (V_{CB} = 10 Vdc, I_E = 0, f = 1.0 MHz)		C_{ob}	—	10	pF
Input Capacitance (V_{EB} = 5.0 Vdc, I_C = 0, f = 1.0 MHz)		C_{ib}	—	75	pF
Small-signal Current Gain (I_C = 5.0 mAdc, V_{CE} = 10 Vdc, f = 1.0 kHz)		h_{fe}	25	—	—
Real Part of Common Emitter Small-Signal Short-Circuit Input Impedance (V_{CE} = 10 Vdc, I_C = 5.0 mAdc, f = 1.0 MHz)		Re(h_{ie})	—	300	Ohms

*Indicates JEDEC Registered Data.
(1) Pulse Test: Pulse Width ≤ 300 μs, Duty Cycle ≤ 2.0%.

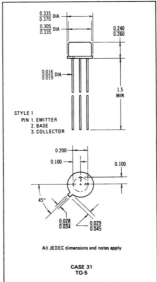

STYLE 1
PIN 1. EMITTER
2. BASE
3. COLLECTOR

All JEDEC dimensions and notes apply

CASE 31 TO-5

2—509

2N3439, 2N3440 (continued)

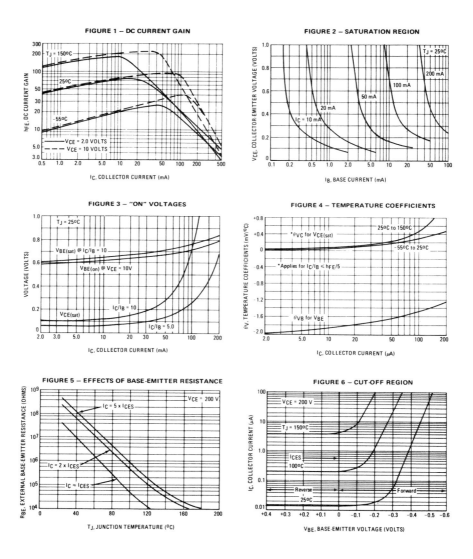

Appendixes

2N3439, 2N3440 (continued)

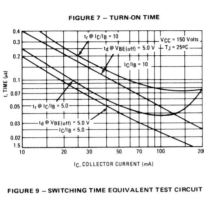

FIGURE 7 – TURN-ON TIME

FIGURE 8 – TURN-OFF TIME

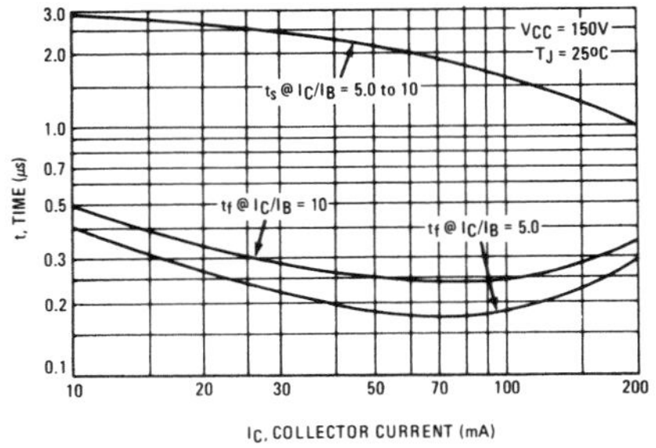

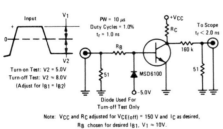

FIGURE 9 – SWITCHING TIME EQUIVALENT TEST CIRCUIT

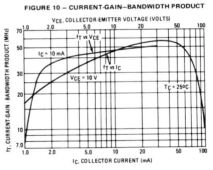

FIGURE 10 – CURRENT-GAIN-BANDWIDTH PRODUCT

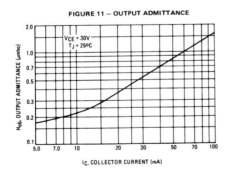

FIGURE 11 – OUTPUT ADMITTANCE

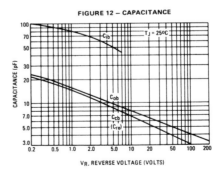

FIGURE 12 – CAPACITANCE

2–511

2N3439, 2N3440 (continued)

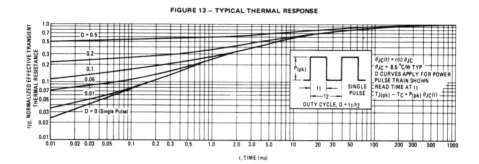

FIGURE 13 – TYPICAL THERMAL RESPONSE

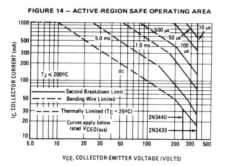

FIGURE 14 – ACTIVE-REGION SAFE OPERATING AREA

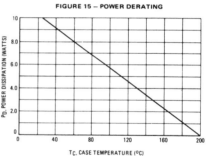

FIGURE 15 – POWER DERATING

There are two limitations on the power handling ability of a transistor; average junction temperature and second breakdown. Safe operating area curves indicate $I_C \cdot V_{CE}$ limits of the transistor that must be observed for reliable operation; i.e., the transistor must not be subjected to greater dissipation than the curves indicate.

The data of Figure 14 is based on $T_{J(pk)}$ = 200°C; T_C is variable depending on conditions. Second breakdown pulse limits are valid for duty cycles to 10% provided $T_{J(pk)}$ = 200°C. $T_{J(pk)}$ may be calculated from the data in Figure 13. At high case temperatures, thermal limitations will reduce the power that can be handled to values less than the limitations imposed by second breakdown. (See AN-415)

2N3444 (SILICON)

For Specifications, See 2N3252 Data.

APPENDIX 10

THERMAL CALCULATIONS FOR TRANSISTORS

The material contained in this section will enable you to compare the specifications of a given transistor with the requirements of a given application and to make a judgment regarding the use of heat sinks. You will be able to determine whether a heat sink is required, and if it is, be able to specify the particular heat sink needed.

When current flows through a semiconductor device, heat is generated. In the case of a forward-biased transistor, most of the heat is generated in the collector-base junction. The manufacturer specifies a maximum temperature (T_j) for the internal junction. Operation at higher temperatures will likely damage the transistor.

To keep the junction at an acceptable temperature, it is necessary for us to provide a path for the heat to escape the junction and reach the ambient air. As the heat travels from the junction to the air, it encounters opposition or *thermal resistance*. There is thermal resistance between the junction and the case (θ_{JC}), another resistance between the case and the heat sink (θ_{CS}) and an apparent resistance between the heat sink and the ambient air (θ_{SA}). The sum of these thermal resistances is the total thermal resistance between the junction and the ambient air (θ_{JA}). In equation form, we have

$$\boxed{\theta_{JA} = \theta_{JC} + \theta_{CS} + \theta_{SA}} \tag{A10.1}$$

In the case of a transistor with no heat sink, the total thermal resistance between the junction and the ambient air (θ_{JA}) is the sum of the junction-to-case thermal resistance (θ_{JC}) plus an effective resistance called the case-to-air thermal resistance (θ_{CA}). The primary mechanism for heat transfer between the junction and the case is through conduction. From the case to the air, on the other hand, the mechanism is primarily radiation and convection which appears as a much higher thermal resistance.

The manufacturer's data sheet will normally provide the value of junction-to-case thermal resistance (θ_{JC}). The overall junction-to-ambient thermal resistance (θ_{JA}), on the other hand is not always given in the data sheet. Table A10.1 can be used as an estimate of the θ_{JA} if the exact value is not available in the data sheet.

TABLE A10.1

TRANSISTOR CASE STYLE	THERMAL RESISTANCE (°C/W)
TO-3	32
TO-5	150
TO-18	300
TO-39	150
TO-66	62
TO-220	50

The thermal resistance between the heat sink and the air (θ_{SA}) is provided by the manufacturer of the heat sink. The value of thermal resistance between the transistor case and the heat sink (θ_{CS}) depends on how the transistor is mounted to the heat sink. That is, whether the transistor is screwed directly to the heat sink (metal-to-metal), whether heat conductive grease is used, whether an insulating wafer is used, and even how securely the mounting screws are tightened. Table A10.2 provides representative values for θ_{CS} under different mounting conditions for two of the most common power transistor packages.

TABLE A10.2

CASE STYLE	DRY METAL CONTACT	HEAT COMPOUND	INSULATING WAFER
TO-3	0.5 °C/W	0.12 °C/W	0.36 °C/W
TO-220	1.2 °C/W	1.0 °C/W	1.7 °C/W

The power dissipated in the collector-base junction can be estimated with Eq. (A10.2).

$$P_D = I_C V_{CE} \quad (A10.2)$$

where I_C and V_{CE} are DC or average values. The maximum power that can be dissipated in a junction without the use of an external heat sink can be determined with Eq. (A10.3).

$$P_D = \frac{T_j(\max) - T_A}{\theta_{JA}} \quad (A10.3)$$

where $T_j(\max)$ is the highest design temperature (not necessarily the highest temperature allowed by the manufacturer), T_A is the highest ambient air temperature to be encountered, and θ_{JA} is the junction-to-air thermal resistance described previously.

Appendixes

For a particular application, the required junction-to-air thermal resistance (whether a heat sink is used or not) can be determined by with Eq. (A10.4).

$$\theta_{JA}(\text{required}) = \frac{T_J(\text{max}) - T_A}{P_D} \quad \text{(A10.4)}$$

If it is found that the required value for θ_{JA} is less than the value of θ_{JC}, then a transistor with a higher power rating must be used. That is, even if we had a perfect heat sink, the junction temperature would still exceed the $T_J(\text{max})$ limit. If the required value for θ_{JA} is greater than the actual θ_{JA} (Table A10.1) for the particular package being considered, then no heat sink will be needed.

If a heat sink is needed for a particular application, then the required value of thermal resistance for the heat sink (θ_{SA}) is determined by Eq. (A10.5).

$$\theta_{SA} = \theta_{JA}(\text{required}) - \theta_{JC} - \theta_{CS} \quad \text{(A10.5)}$$

The transistor manufacturer does not always include values for θ_{JC} and/or θ_{JA} in the data sheet. However, the derating factor is usually available. The thermal resistance for either junction-to-case or junction-to-air is equal to the reciprocal of the respective derating factor. That is

$$\theta_{JC} = \frac{1}{\rho_C} \quad \text{(A10.6)}$$

where ρ_C is the derating factor for the case temperature. Similarly,

$$\theta_{JA} = \frac{1}{\rho_A} \quad \text{(A10.7)}$$

where ρ_A is the derating factor for ambient air.

Example 1

A particular transistor has a quiescent collector current of 1.2 amps and a quiescent collector-to-emitter voltage of 12 volts. The design goal allows a maximum junction temperature of 140 °C and a maximum ambient temperature of 65 °C. Determine whether this transistor can be used in this application. If it can, does it need a heat sink? If so, what is the required thermal resistance of the heat sink?

Solution The manufacturer's data sheet provides the following information:

1. θ_{JC} 1.8 °C/W
2. $T_J(\text{rating})$ −55 to +150 °C
3. Package   TO-220

The power actually dissipated in the junction can be estimated with Eq. (A10.2) as

$$P_D = I_C V_{CE} = 1.2 \times 12 = 14.4 \text{ W}$$

The required thermal resistance from junction-to-air is estimated with Eq. (A10.4).

$$\theta_{JA}(\text{required}) = \frac{T_j(\text{max}) - T_A}{P_D} = \frac{140 - 65}{14.4} = 5.21 \text{ °C/W}$$

Since the required value of θ_{JA} is greater than θ_{JC}, this transistor can be used for this application. Table A10.1 lists the value of θ_{JA} for a TO-220 package as 50 °C/W. Since the required value of θ_{JA} is less than the value of θ_{JA} for the package itself, a heat sink will be required. The heat sink will need a thermal resistance as determined by Eq. (A10.5).

$$\theta_{SA} = 5.21 - 1.8 - 1.7 = 1.71 \text{ °C/W}$$

Example 2

A certain transistor has the following thermal characteristics listed on the manufacturer's data sheet:

1. θ_{JC} 3.125 °C/W
2. T_j(maximum)   150 °C
3. Package   TO-3

What is the most power that can be dissipated at room temperature (25 °C) without requiring a heat sink?

Solution The maximum power without a heat sink can be estimated with Eq. (A10.3) as

$$P_D = \frac{T_j(\text{max}) - T_A}{\theta_{JA}} = \frac{150 - 25}{32} = 3.91 \text{ W}$$

Example 3

A particular design requires an NPN transistor to conduct 75 milliamperes with a collector-to-emitter voltage of 15 volts. The highest ambient temperature is expected to be 70 °C, and it is a goal to keep the junction at 125 °C or below. Can a 2N2222A transistor be used in this application? If so, is a heat sink required? If a heat sink is required, what is the required thermal resistance (θ_{SA})?

Solution The manufacturer's data sheet for a 2N2222A provides the following information:

1. T_j (maximum)   200 °C
2. Package   TO-18
3. I_C (maximum)   0.8 amps
4. V_{CE} (maximum)   40 volts
5. Derating factor above $T_C = 25$ °C   12 mW/°C
6. Derating factor above $T_A = 25$ °C   3.33 mW/°C

The required junction-to-air resistance can be computed with Eq. (A10.4) as

$$\theta_{JA}(\text{required}) = \frac{T_J(\text{max}) - T_A}{P_D}$$

$$= \frac{125 - 70}{75 \times 10^{-3} \times 15}$$

$$= 48.9 \text{ °C/W}$$

The junction-to-case thermal resistance can be computed with Eq. (A10.6) as

$$\theta_{JC} = \frac{1}{\rho_C} = \frac{1}{12 \times 10^{-3}} = 83.3 \text{ °C/W}$$

Since the required θ_{JA} is less than the θ_{JC} of the transistor, we cannot use this device; it would overheat even with an ideal heatsink.

APPENDIX 11

INTERPRETATION OF OSCILLOSCOPE DISPLAYS

There are numerous examples throughout the text which include an oscilloscope display to reveal the operational characteristics of the actual circuit. All of these figures are actual plotted outputs from a digitizing oscilloscope. Although much of the display output is similar to more familiar analog oscilloscopes, Fig. A11.1 and the following descriptions will assist the reader who is unfamiliar with this type of equipment.

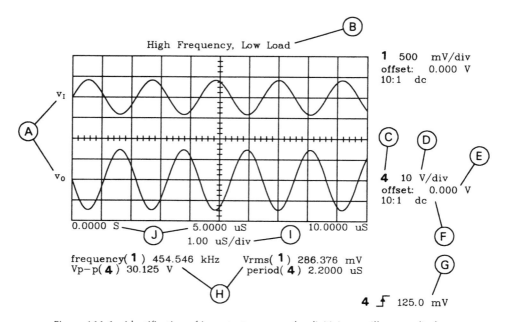

Figure A11.1 Identification of important areas on the digitizing oscilloscope displays.

486

Each of the following headings describes the purpose of the corresponding item in Fig. A11.1.

Item A. This is a reference to the schematic diagram. These annotations indicate where on the circuit each waveform was taken.

Item B. This annotation provides a brief description of the circuit conditions at the time of the measurements. Descriptions include such things as relative frequency, relative line voltage, or load conditions.

Item C. The oscilloscope is a four-channel device. That is, it is capable of displaying four different waveforms at the same time. The larger, bold numbers indicate which channel is associated with the subsequent parameter list.

Item D. This displays the vertical sensitivity of the indicated channel. It has the same meaning as the Volts/cm or Volts/Division parameter on an analog oscilloscope.

Item E. The offset specification indicates the amount of DC offset that has been set into the indicated channel. A positive offset has the effect of moving the waveform upward. A negative offset shifts the waveform downward. In either case, the amount of shift is established by the Volts/Div setting (Item D).

Item F. This ratio indicates the type of probe being used (e.g., 1:1 or 10:1). The attenuation effects of the probe are automatically accounted for in the display, so no mental arithmetic is required to obtain the correct answer. Additionally, the type of coupling (AC or DC) is displayed.

Item G. This portion of the display describes the trigger conditions. First, the larger, bold number indicates which channel served as the trigger source. Second, the arrow, pointing up or down on the rising or falling edge, respectively, indicates the slope of the trigger (i.e., positive or negative). Finally, the voltage value is the setting for the trigger level.

Item H. The lower portion of each display will vary depending on what characteristics of the displayed waveform are of interest. Each of the listed parameters will have a larger, bold number to indicate the channel reference. The remaining descriptive label and value are self-explanatory. Some example parameters include frequency, +width (positive pulse width), −width (negative pulse width), period, Vrms, Vavg, Vp-p, Vmin, Vmax, and so on.

Item I. This is the Time/Division setting for the horizontal sweep of the oscilloscope. It is interpreted in the same way as the Time/cm or Time/Division setting on an analog oscilloscope.

Item J. These labels simply indicate the relative time at various points across the screen (i.e., left, center, and right).

A few figures in the text (e.g., Fig. 3.16) have some additional dashed lines superimposed on the oscilloscope display. These lines are used for measuring the time (horizontal) and voltage (vertical) between two points on the waveform. When these lines are visible, there are normally corresponding time and/or voltage parameters listed in the lower portion of the display (i.e., in the area indicated by Item H in Fig. A11.1). These additional parameters are each associated with a particular channel and include such measurements as those cited

Vmarker1. This indicates the voltage level represented by one of the horizontal dashed lines.

Vmarker2. This is the voltage level of the second horizontal dashed line.

delta V. This is the difference in voltage between the levels of Vmarker2 and Vmarker1.

start marker. This is the time position (as indicated by Item J in Fig. A11.1) of one of the vertical dashed lines.

stop marker. This is the time position of the second vertical dashed line.

delta t. This is the difference in time between the setting of the stop marker and the start marker as described.

INDEX

Absolute value circuit, 381-87
  design, 385-87
  equivalent circuits, 383-84
  numerical analysis, 382-85
  operation, 382
  schematic, 382
AC coupled amplifier, *See* amplifier
  important nonideal parameters, 413
  single supply, 428-33
Active filter:
  bandpass, 233-41
    bandwidth, 236
    design, 236-41
    equivalent circuits, 234
    numerical analysis, 235-36
    operation, 233-34
    Q, 235
    resonant frequency, 235-36
    schematic, 234
    voltage gain, 236
  band reject, 241-51
    center frequency, 243-44
    design, 244-51
    input impedance, 244
    numerical analysis, 243-44
    operation, 241-43
  Butterworth, 219,226
  fundamentals, 217-18
  high-pass, 226-33
    cutoff frequency, 228
    design, 229-33
    equivalent circuits, 227
    input impedance, 228
    numerical analysis, 228
    operation, 226-28
    Q, 228
    schematic, 227
  low-pass, 219-26

    cutoff frequency, 221
    design, 222-26
    equivalent circuits, 220
    input impedance, 221-22
    numerical analysis, 221-22
    operation, 220
    Q, 221
    schematic, 219
  peaking, 219
  stability, 218,233
Adder, 370-75
  design, 373-75
  numerical analysis, 371-73
  operation, 370
  schematic, 371
Adjustable gain, 82
Amplifier:
  AC coupled, 97-113
    bandwidth, 105-9
    design, 109-13
    input impedance, 104-5
    numerical analysis, 99-109
    operation, 97-99
    schematic, 98
    single supply, 428-33
    voltage gain, 99-103
  antilogarithmic, 424-25
  averaging, 380-81
  current, 113-20
    design, 118-20
    gain, 114-15
    input current, 116
    input resistance, 117-18
    load current, 115-16
    maximum load resistance, 117
    numerical analysis, 114-18
    operation, 113-14
    output resistance, 118

    schematic, 114
  differential, 2
    double-ended, 2
    in instrument amplifier, 417
    in subtractor, 375-80
    phase relationships, 3
    single-ended, 2
  frequency response, 38-39
  hi-current, 120-31
    bias voltage, 123
    design, 125-31
    effective load resistance, 122
    input current, 125
    input voltage swing, 123-24
    numerical analysis, 121-25
    operation, 120-21
    output current, 122
    output voltage swing, 123
    peak load current, 122
    peak load voltage, 122
    schematic, 121
  instrumentation, 349,417-21
  inverting:
    bandwidth, 52-53
    basic rules, 42
    design, 54-59
    input current, 44-45
    input impedance, 44
    input voltage swing, 46-47
    minimum load resistance, 50-51
    numerical analysis, 42-54
    operation, 40-42
    output current, 49-50
    output impedance, 47-49
    output voltage swing, 45
    power supply rejection ratio, 53-54
    schematic, 40
    slew rate limiting frequency, 45-46

489

Index

Amplifier (cont.)
  inverting (cont.)
    voltage gain, 43-44
  inverting summing, 80-94
    as D/A converter, 354
    bandwidth, 88-89
    design, 90-94
    input current, 84
    input impedance, 84
    input voltage swing, 85-86
    numerical analysis, 83-90
    operation, 80-82
    output current, 88
    output impedance, 86-88
    output voltage swing, 84-85
    schematic, 82
    slew rate limiting frequency, 89-90
    voltage gain, 83-84
  logarithmic, 421-23
  noninverting:
    bandwidth, 67-68
    design, 69-73
    effects of slew rate limiting, 68
    input current, 64
    input impedance, 63-64
    input voltage swing, 65-66
    numerical analysis, 60-69
    operation, 59-60
    output current, 67
    output impedance, 66-67
    output voltage swing, 64-65
    power supply rejection ratio, 68-69
    schematic, 60
    slew rate limiting frequency, 65
    voltage gain, 61-63
  noninverting summing:
    adder, 370-75
    numerical analysis, 95-97
    operation, 95
    schematic, 95
  single supply, 428-33
  stability, 409
  voltage follower:
    bandwidth, 77-78
    design, 78-80
    input current, 75-76
    input impedance, 75
    input voltage swing, 76
    numerical analysis, 74-78
    operation, 74
    output current, 77
    output impedance, 76-77
    output voltage swing, 76
    power supply rejection ratio, 78
    schematic, 74
    slew rate limiting frequency, 76
    voltage gain, 75
  wideband power, 459
Analog:
  contrasted with digital, 346
  multiplexer, 350-52
Analog-to-digital conversion, 358-69
  accuracy, 349
  conversion rate, 349
  conversion time, 349
  defined, 346
  dual-slope, 362-67
    schematic, 363
  flash, 358-60
  fundamentals, 347-52
  gain error, 348
  linearity, 348
  monotonicity, 348
  offset error, 349
  parallel, 358-60
    hybrid, 360
    schematic, 359
  quantization, 347
  quantization uncertainty, 348
  scaling error, 348
  successive approximation, 367-69
    schematic, 367
  tracking, 360-62
    schematic, 361
Antilogarithmic amplifier, 424-25
  schematic, 424
Averaging amplifier, 380-81

Bandpass filter, 233-41, See also active filter
  definition, 217
Band reject filter, 241-51, See also active filter
  definition, 217
Bandstop filter, See active filter, band reject
Bandwidth, 16,233,402-3
  definition, 22
  general equation, 39
  power for MC1741, 440
  power for MC1741SC, 457
Break frequency, 224,406
Burst regulator, 284
Butterworth filter, 219,226

Capacitive reactance, 100
Center frequency, 233
Circuit construction:
  component placement, 27
  grounding, 31-32
  methods, 27
  performance example, 32-34
  power supply decoupling, 28-30
  power supply distribution, 28
  routing of leads, 27-28
Clamper, 314-23, See also ideal clamper
  design, 321-23
  numerical analysis, 316-21
  operation, 315-16
  schematic, 315-16
Clipper, 306-14, See also ideal biased clipper
  design, 309-14
  numerical analysis, 308-9
  operation, 306-7
  schematic, 307
Closed loop:
  definition, 20
  gain, 20,402-3
  output resistance, 47-49
CMRR, 21
Common-mode rejection ratio, 21

Comparator, See voltage comparator
Compensation:
  bias current, 42
  frequency, See frequency compensation
Constant current limiting, 285-87
Constant current source:
  charging a capacitor, 209,332-33
  in current amplifier, See amplifier
  inside op amp, 2-3,394-95
Conversion:
  analog-to-digital, 346-52,358-69
  current-to-voltage, 454
  decibel form of CMRR, 21
  decibel form of current gain, 38
  decibel form of power gain, 38
  decibel form of voltage gain, 38
  digital-to-analog, 346-47,352-58
  linear ramp to DC, 338
  peak to RMS, 45
  peak-to-peak to peak, 46
  peak-to-peak to RMS, 124
  RMS to peak, 71
  square wave to triangle wave, 332
  triangle wave to sinewave, 332
Corner frequency, 224,406
Coupling:
  AC, 97-98,413
  DC, 413
Crosstalk, 352
Crowbar circuit, 289
Current:
  effects of increasing load, 49-50
  feedback, 49-50
  load, 49-50
  op amp output, 49-50
  short circuit, 49
Current amplifier, See amplifier
Current boost, 121,257
Current limiting, 285-89
Current mirror, 430,432
Current-to-voltage converter, 454
Cutoff frequency, 39,219

Data sheets:
  diode:
    1N914A, 467
  op amp:
    MC1741, 437-41
    MC1741SC, 454-59
  transistor family:
    MJE1103, 443-44
    MPF102, 469
    2N2222, 446-52
    2N3440, 477-80
  zener diodes:
    1N4728-1N4764, 471-75
    1N5221-1N5281, 461-66
DC coupled amplifier:
  important nonideal parameters, 413
DC offset, 4,174-76,333,399-400,455
DC power supply, See voltage regulator
DC regulation, See voltage regulator
Decibel, 38
Decoupling, 28-30
  circuit, 30

Index

equivalent circuit, 30
power-entry, 30
Derating factor, 483
Design:
  AC coupled amplifier, 109-13
    compensation resistor, 110
    design results, 112-13
    feedback resistor, 109-10
    input coupling capacitor, 111
    input resistor, 109
    op amp selection, 110
    output coupling capacitor, 111
  adder, 373-75
    design results, 375
    feedback resistor, 374
    input resistor, 373-74
    op amp selection, 374-75
  bandpass filter:
    capacitors, 237
    compensation resistor, 238
    design results, 239-41
    op amp selection, 238-39
    resistors, 237-38
  band reject filter, 244-51
    compensation resistor, 246-47
    design results, 247-51
    filter components, 244-46
    op amp selection, 247
  current amplifier, 118-20
    current gain, 118-19
    current resistors, 119
    design results, 120
  differentiator, 340-44
    bypass capacitor, 342
    compensation resistor, 341
    design results, 342-44
    feedback resistor, 340-41
    input capacitor, 341
    input resistor, 341
  dual half-wave rectifier, 300-306
    compensation resistor, 302
    design results, 303-6
    feedback resistors, 301
    input resistor, 301
    op amp selection, 302-3
    rectifier diodes, 301-2
  hi-current amplifier, 125-31
    bias potentiometer, 129
    bias voltage, 128
    compensation resistor, 129-30
    current-boost transistor, 125-26
    design results, 130-31
    feedback resistor, 129
    input coupling capacitor, 130
    input resistor, 129
    op amp output voltage, 127-28
    op amp selection, 129-30
    output voltage swing, 128
    voltage gain, 128
  high-pass filter, 229-33
    capacitors, 229-30
    compensation resistor, 230
    design results, 231-33
    op amp selection, 230-31
    resistors, 229
  ideal biased clipper, 309-14

  clipper diode, 311
    design results, 312-14
    filter capacitor, 311-12
    op amp selection, 310-11
    voltage divider, 310
  ideal clamper, 321-23
  integrator, 334-38
    capacitor, 335
    compensation resistor, 335-36
    design results, 336-38
    feedback resistor, 335
    input resistor, 334-35
    op amp selection, 334
  inverting amplifier, 54-59
    compensation resistor, 57-58
    design results, 58-59
    feedback resistor, 55
    input resistor, 55
    op amp selection, 55-57
    supply voltage, 56-57
  inverting summing amplifier, 90-94
    compensation resistor, 92
    design results, 94
    feedback resistor, 91
    input resistors, 90-92
    op amp selection, 93
    power supply voltages, 92-93
    worst-case input, 90
  low-pass filter, 222-26
    capacitors, 222
    design results, 224-26
    op amp selection, 223-24
    resistors, 222-23
  noninverting amplifier, 69-73
    compensation resistor, 72
    design results, 72-73
    feedback resistor, 70
    input resistor, 69-70
    minimum supply voltages, 70-71
    op amp selection, 70-71
  peak detector, 326-31
    current limiting resistors, 329
    design results, 329-31
    detector diode, 327
    filter components, 327-28
    op amp selection, 326-27
  series voltage regulator, 264-70
    design results, 267-70
    feedback network, 266-67
    pass transistor, 264-65
    voltage gain, 265-66
  shunt voltage regulator, 274-80
    current limiting resistor, 276
    design results, 277-80
    error amp gain, 275
    feedback network, 275-76
    transistor selection, 276-77
  sign-changing circuit, 389-91
    design results, 390-91
    op amp selection, 389-90
    resistor, 389
  subtractor, 378-80
    design results, 379-80
    op amp selection, 378-79
    resistors, 378
  triangle wave oscillator, 210-14

    design results, 213-14
    integrator, 211-12
    op amp selection, 212-13
    voltage comparator, 210-11
  variable duty oscillator, 202-8
    design results, 207-8
    isolation diodes, 206
    op amp selection, 206-7
    output zener regulator, 202
    reference zener regulator, 203-4
    timing components, 204-6
  voltage comparator with hysteresis, 156-60
    compensation resistor, 157-58
    design results, 159-60
    feedback resistor, 157
    op amp selection, 156-57
    reference voltage, 158-59
  voltage comparator with output limiting, 169-74
    compensation resistor, 171
    design results, 173-74
    feedback resistor, 169-70
    hysteresis voltage, 169
    output zener diodes, 170
    output zener resistors, 170-71
    reference current limiting resistor, 172
    reference zener, 172
  voltage controlled oscillator, 191-97
    design results, 194-97
    integrator, 193-94
    op amp selection, 194
    zener current limiting resistor, 192-93
    zener diodes, 191-92
  voltage follower, 78-82
    compensation resistor, 80
    design results, 79-82
    op amp selection, 79
    power supply voltages, 79-80
  voltage reference, 257-60
    current-boost transistor, 258-59
    current resistor, 258
    design results, 259-60
    regulator diode, 258
  Wein bridge oscillator, 181-84
    amplitude control, 183-84
    design results, 183-84
    feedback resistor, 181-82
    frequency determining components, 181
    op amp selection, 183
    rectifier and filter, 182-83
  window voltage comparator, 162-65
    design results, 164-65
    isolation diodes, 164
    op amp selection, 162
    pull-up resistor, 163
    zener diodes, 163
    zener resistors, 163
  zero-crossing detector, 140-44
    compensation resistor, 142
    design results, 143-44
    input resistor, 142
    op amp selection, 140-41

Design *(cont.)*
  zero-crossing detector with hysteresis, 149-52
    compensation resistor, 151
    design results, 151-52
    feedback resistor, 150
    op amp selection, 149-50

Differential amplifier, *See* amplifier
Differential input voltage:
  affected by frequency, 63
  ideal value, 42
Differentiator, 338-44
  design, 340-44
  numerical analysis, 340
  operation, 338-40
  schematic, 339
  stability, 339
Digital:
  contrasted with analog, 346
  states, 346
Digital-to-analog conversion, 352-58
  accuracy, 352
  defined, 346
  gain error, 353
  glitches, 353
  monotonicity, 353
  offset error, 353
  propagation time, 352-53
  R2R ladder, 356-58
    schematic, 356
  scaling error, 353
  settling time, 353
  slew rate, 353
  weighted, 354-56
    schematic, 354
Diode, *See also* ideal diode
  data sheet:
    1N914A, 467
  in ideal rectifier, 297-305
  silicon versus ideal, 295-97
Divider, *See also* multipliers/dividers
  schematic, 427
Divider circuit, 427
Drift, 400-401
Dual-slope A/D conversion, 362-67
Duty cycle:
  in switching regulator, 281,284
  oscillator with variable, 198-208

Efficiency:
  in power supplies, 283
Electrostatic discharge, 35
Emissions:
  from power supplies, 283-84
  regulatory agencies, 283-84
Equivalent input noise, 404
Error amplifier, 261
ESD, 35
External frequency compensation, 411

Fast attack, 324
Feedback:
  definition, 39
  frequency selective, 40,177-78
  in series voltage regulator, 261

in shunt voltage regulator, 271-72
in switching voltage regulator, 281-82
positive, 144-45, 177
types of, 39-40
Feedforward frequency compensation, 411-12
Filter, *See* active filter
Flash A/D conversion, 358-60
Foldback current limiting, 287-89
  schematic, 288
Frequency compensation, 405-14
  external, 411
  feedforward, 411-12
  internal, 410
  principles, 406-10
  rule for stability, 409
  safety margins, 414
Frequency doubler, 428
Frequency response, 38-39,402-3, *See also* bandwidth
  AC coupled amplifier, 105-9
  inverting amplifier, 52-53
  MC1741, 440
  MC1741SC, 457
  noninverting amplifier, 67-68
  steepness of slope, 218-19
  voltage follower, 77-78
Full power bandwidth, *See* Bandwidth

Gain:
  decibel form, 38
  definition, 37
  fractional, 38
  voltage, 37
Gain-bandwidth product:
  definition, 22
Gain margin, 414
Gain-source-resistance product, 90
Ground:
  virtual, 42
Grounding considerations:
  analog versus digital, 31-32
  ground bounce, 31-32
  ground noise, 31-32
  ground plane, 31
  quiet ground, 31-32

Half-power frequency, 39,219
Half-power point, 39,219
Harmonics:
  in switching regulators, 283-84
Heat sink:
  selection, 481-85
Hi-current amplifier, *See* amplifier
High-pass filter, 226-33, *See also* active filter
  definition, 217
Holding current, 289
Hold-up time, 291
Hybrid op-amps, 434-35
Hysteresis, 145,166

Ideal based clipper, 306-14
  clipping levels, 308
  design, 309-14
  fundamentals, 306

highest frequency, 308-9
input impedance, 309
maximum input, 309
numerical analysis, 308-9
operation, 306-7
output impedance, 309
schematic, 307
Ideal clamper, 314-23
  adjustment range, 316-17
  design, 321-23
  frequency range, 317-19
  fundamentals, 314-15
  input impedance, 319
  maximum input voltage, 317
  numerical analysis, 316-21
  operation, 315-16
  output impedance, 319-21
  schematic, 316
Ideal diode:
  concept, 295-97
  in biased clamper, 315-16
  in biased clipper, 306-7
  in peak detector, 323-25
Ideal rectifier circuit, 297-306
  design, 300-306
  dual half-wave:
    schematic, 298
  half-wave:
    schematic, 297
  highest frequency, 300
  maximum input, 299-300
  maximum output, 298-99
  numerical analysis, 298-300
  operation, 297-98
  voltage gain, 299
Impedance:
  input:
    ideal, 16-17
    nonideal, 23-24,401,412
  output:
    effects of, 48-49
    ideal, 17
    nonideal, 24,401-2,412-13
  series RC circuit, 100
  transformation, 74
Inductance:
  power supply wires, 28-30
Input bias current, 176,394-98
  minimizing effects, 397-98
  model, 395
  versus temperature, 455
Input offset current, 398-99
Input offset voltage, 175-76,399-400
  model, 400
  nulling, 4,175-76,400,455
Input resistance:
  AC, 412
  DC, 401
Instrumentation amplifier, 349,417-21
  equivalent circuit, 418
Integration, 332
Integrator, 331-38
  as low-pass filter, 332
  definition, 331-32
  design, 334-38
  frequency of operation, 333-34

Index

numerical analysis, 333-34
operation, 332-33
used in VCO, 185
Internal frequency compensation, 410
Inverting amplifier, *See* amplifier
Inverting summing amplifier, *See* amplifier
   used in VCO, 186

Kirchoff's Current Law, 7-8
Kirchoff's Voltage Law, 8-9

Level detector, *See* voltage comparator
Limit detector, *See* window voltage comparator
Linear ramp, 185,332-33,363-65
Logarithmic amplifiers, 421-23
   schematic, 422
Lower cutoff frequency, 39
Lower threshold voltage, 144
Low-pass filter, 22,219-26
   definition, 217, *See also* active filter

Multiplexer, 350-52
   crosstalk, 352
Multiplier:
   modes of operation, 426
   scale factor, 425
   schematic, 427
   symbol, 425
Multipliers/dividers, 425-28
   as frequency doubler, 428
   divider circuit, 427
   multiplier circuit, 427
   square root circuit, 428
   squaring circuit, 428

Negative threshold voltage, 144
Noise, 404-5
   effect on differentiator, 339
   equivalent input noise, 404
   graphs for MC1741, 439
   graphs for MC1741SC, 457
   rejection of power supply noise, 53-54
   sources in power supplies, 25-26
Noise immunity:
   dual slope A/D, 367
   increased with hysteresis, 145
Noninverting amplifier, *See* amplifier
Noninverting summing amplifier, *See* amplifier
Norton's Theorem, 12
Notch filter:
   definition, 218,241-51, *See also* band reject filter
Nulling:
   output offset voltage, 4,175-76,400,455

Ohm's Law, 5-7
Op amp:
   applications survey, 4-5
   bias current, 394-98
   corner frequency, 223-24,406
   data sheets:
      MC1741, 437-41

      MC1741SC, 454-59
   equivalent input noise, 404
   frequency response, 16,22,39,402-3
   history, 1
   hybrid, 434-35
   ideal parameters:
      bandwidth, 16
      common mode voltage gain, 15-16
      differential voltage gain, 14-15
      input impedance, 16-17
      noise generation, 18
      output impedance, 17
      slew rate, 16
      table of, 18
      temperature effects, 18
   impedance:
      input, 16-17,23-24,401,412
      output, 17,24,48-49,401-2,412-13
   inputs:
      inverting, 3,14
      noninverting, 3,14
   internal circuitry, 3,394-95,430-32,437,455
   multiple-device packages, 434
   nonideal AC characteristics, 402-13
   nonideal DC characteristics, 394-402
   nonideal parameters:
      bandwidth, 22,402-3
      common-mode voltage gain, 20-22
      differential voltage gain, 20,402-3
      drift, 400-401
      input bias current, 394-98
      input impedance, 23-24,401,412
      input offset current, 398-99
      input offset voltage, 399-400
      noise generation, 24-25,404-5
      output impedance, 24,401-2,412
      slew rate, 22-23,403-4
      temperature effects, 24
   offset current, 398-99
   offset null, 4,175-76,400,455
   offset voltage, 175-76,399-400
   output voltage:
      versus frequency, 440
      versus load resistance, 440
   packages, 434,437,454
   power supply requirements, 25-26
   programmable, 416-17
   stability, 409
   symbol, 14
Open-loop:
   output resistance, 24,47
   voltage gain, 20,402-3
      MC1741, 440
      MC1741SC, 457
      versus supply voltage, 441
Oscillator:
   frequency selective, 177-78
   fundamentals, 177-78
   loading, 215-16
   nonideal considerations, 215-16
   triangle wave, 209-14
      design, 210-14
      frequency, 210
      numerical analysis, 209-10
      operation, 209

      schematic, 209
   variable duty cycle, 198-208
      design, 202-8
      frequency, 200-201
      numerical analysis, 199-202
      operation, 198-99
      percent duty, 202
      schematic, 198
   voltage-controlled, 185-97
      design, 191-97
      frequency, 191
      numerical analysis, 186-91
      operation, 185-86
      schematic, 185
   Wein bridge, 178-84
      design, 181-84
      frequency, 179
      numerical analysis, 179-80
      operation, 178-79
      schematic, 179
Oscilloscope:
   interpretation, 486-88
Output impedance:
   effects of, 48-49,401-2,412-13
Output resistance:
   AC, 412-13
   DC, 401-2
Over-current protection, 285-89
   constant current limiting, 285-87
      schematic, 286
   foldback current limiting, 287-89
      schematic, 288
   load interruption, 285
      schematic, 285
Over-voltage protection, 289
   schematic, 290

Parallel A/D conversion, 358-60
Peak detector, 323-31
   design, 326-31
   low-frequency limit, 325
   numerical analysis, 325-26
   operation, 324-25
   response time, 325-26
   schematic, 324
Phase margin, 414
Phase shift, 405-8
Positive threshold voltage, 144
Power dissipation:
   heat sink requirements, 481-85
   switching versus linear regulators, 283
Power-fail sensing, 289-91
   schematic, 291
Power supply noise, 25-26
Power supply rejection ratio:
   definition, 26
   inverting amplifier, 53-54
   noninverting amplifier, 68-69
Preamplifier, 349
Programmable op amp, 416-17
PSRR:
   definition, 26
   inverting amplifier, 53-54
   noninverting amplifier, 68-69
Pulse width modulator, 281-82

Q:
 definition, 219
 for narrow band filters, 233
 for wide band filters, 233
 value for minimum peaking, 221
Quadrants, multiplier, 426

R2R ladder D/A converter, 356-58
Ramp generator:
 in A/D converter, 363-65
 integrator circuit, 332-33
 in VCO, 185
Regulator:
 current, *See* amplifier, current
 voltage, *See* voltage regulator
Rejection:
 common-mode, 21
 power supply rejection ratio, 26,53-54,68-69
Relative magnitude rule, 414
Resolution, 347,352
Resonant frequency, 233,235-36,242-44
Response time:
 in peak detector, 325-28
 switching versus linear regulators, 284
Ripple voltage:
 in peak detector, 324,327-28
Rolloff, 63,410

Sample-and-hold, 349-50
 acquisition time, 350
 aperture time, 350
 droop rate, 350
 hold command, 350
 sampling rate, 350
 track command, 349-50
SAR, 367
Saturation voltage, 26
Sawtooth wave generator, 213-14
Scale factor, multiplier, 425
SCR, 289
Series voltage regulator, 261-70
 design, 264-70
 numerical analysis, 263-64
 operation, 261-63
 schematic, 261
Shunt voltage regulator, 271-80
 design, 274-80
 numerical analysis, 272-74
 operation, 271-72
 schematic, 271
Sign-changing circuit, 388-91
 design, 389-91
 operation, 388
 schematic, 388
Silicon-controlled rectifier, 289
Single-supply amplifiers, 428-33
 schematic, 429-31,440
Slew rate, 403-4
 definition, 16
 digital-to-analog conversion, 353
 effect on clamper, 319-20
 effect on integrator, 333-34
 effects on comparator, 138-39
 ideal, 16
 limiting frequency, 23

 nonideal, 22-23,403-4
Square root circuit, 428
Squaring circuit, 428
Stability:
 active filter, 218,233
 amplifiers, 409
 differentiator, 339
 gain margin, 414
 phase margin, 414
Subtractor:
 design, 378-80
 numerical analysis, 376-78
 operation, 376
 schematic, 376
Successive approximation A/D, 367-69
Summing amplifier, *See* amplifier
Superposition Theorem, 12-14
Switching voltage regulator, 280-84
 classes, 284
 principles, 280-82
 response time, 284
 schematic, 282
 versus linear, 283-84

Temperature:
 drift, 400-401
 effects, 18,24
 effects on bias current, 455
Thermal resistance:
 definition, 481
 equation, 481
 relationship to derating factors, 483
 required for heat sink, 483
 table of, 482
Thévenin's Theorem, 10-12
Trace inductance, 28-30
Tracking A/D converter, 360-62
Transient response:
 MC1741, 441
 MC1741SC, 458
 voltage regulator, 284
Transient suppressor, 289
Transistor:
 as a logarithmic element, 422-24
 as a switch, 363
 as a variable resistor, 178-80, 254
 current boost, 121,257
 current limiting, 285-88
 current mirror, 430,432
 data sheets:
  MJE1103, 443-44
  MPF102, 469
  2N2222, 446-52
  2N3440, 477-80
 pass, 261
 switching regulator, 281
 thermal calculations, 481-85
Triangle wave generator, *See* oscillator
Troubleshooting:
 active filters, 251
 amplifier circuits, 133-34
 arithmetic circuits, 391-92
 basic tips, 18,19,26
 oscillator circuits, 214-15
 power supply circuits, 291-93
 procedure, 132-33

 observation, 132
 resistance measurements, 133
 signal injection/tracing, 133
 voltage measurements, 133
signal processing circuits, 344
voltage comparators, 174-75
Twin T filter, 241-43, *See* active filter, band reject

Unity-gain frequency:
 definition, 22
Upper cutoff frequency, 39,106
Upper threshold voltage, 144

VCO, 185-97, *See* voltage controlled oscillator
VFC, 185-97, *See* voltage controlled oscillator
Virtual ground:
 definition, 42
Voltage:
 between two points, 8
 comparator:
  fundamentals, 136
 comparator with hysteresis, 152-60,291
  schematic, 153
 differential input:
  affected by frequency, 63
  ideal value, 42
 high-frequency drops, 28-30
 multiple sources, 12
 negative threshold, 144
 offset, 175-76,399-400
 positive threshold, 144
 reference, 256-60
 regulation:
  fundamentals, 253-60
  purpose, 253-54
 regulator:
  zener, 158-59
 saturation, 45
  definition, 26
 unregulated source, 253
Voltage comparator:
 fundamentals, 136-37
 nonideal considerations, 175-76
 used in A/D, 361,363,367
 used in VCO, 185-86
Voltage comparator with hysteresis, 152-60,291
 design, 156-60
 hysteresis, 154-55
 input resistance, 153-54
 lower threshold voltage, 154
 maximum frequency, 155-56
 numerical analysis, 153-56
 operation, 152-53
 schematic, 153
 upper threshold voltage, 154
Voltage comparator with output limiting, 165-74
 design, 169-74
 hysteresis, 168
 lower threshold voltage, 167-68
 numerical analysis, 166-69

Index

operation, 165-66
output voltage, 169
schematic, 166
upper threshold voltage, 166-67
zener currents, 168-69
Voltage controlled oscillator, 185-97
 design, 191-97
 frequency, 191
 numerical analysis, 186-91
 operation, 185-86
 schematic, 185
Voltage divider equation, 48
Voltage follower amplifier, *See* amplifier
Voltage gain:
 closed-loop, 20,402-3
 common-mode, 15-16,20,417
 differential, 14-15,20
 general, 37-38
 inverting amplifier, 41,43
 large signal, 20
 noninverting amplifier, 60-63
 open-loop, 20,402-3
  MC1741, 440
  MC1741SC, 457
  versus supply voltage (MC1741), 441
Voltage reference:
 design, 257-60
 operation, 256-57
Voltage regulation:
 fundamentals, 253-60

line, 255
load, 256
Voltage regulator:
 references, 256-60
 series, 254,261-70
  design, 264-70
  numerical analysis, 263-64
  operation, 261-63
  output current, 263-64
  output voltage, 263
  schematic, 261
 shunt, 254-55,271-80
  design, 274-80
  numerical analysis, 272-74
  operation, 271-72
  output current, 273
  output voltage, 272-73
  schematic, 271
 switching, 255,280-84
  classes, 284
  fundamentals, 280-81
  switching versus linear, 283-84
  types, 254
Voltage-to-frequency converter, 185-97, *See* voltage controlled oscillator

Weighted D/A converter, 354-56
Wein bridge oscillator, 178-84, *See also* oscillator
Window voltage comparator, 160-65
 design, 162-65

 numerical analysis, 160-62
 operation, 160
 output voltage, 161-62
 reference voltages, 161
 schematic, 161

Zener diode:
 data sheets:
  1N4728-1N4764, 471-75
  1N5221-1N5281, 461-66
Zener diode tester, 120
Zero-crossing detector, 137-44
 design, 140-44
 input current, 139
 input impedance, 138-39
 numerical analysis, 138-40
 operation, 137-38
 output voltage, 140
 schematic, 138
Zero-crossing detector with hysteresis, 144-52
 design, 149-52
 hysteresis, 147
 input resistance, 146
 lower threshold voltage, 147
 maximum frequency, 147-48
 noise immunity, 145
 numerical analysis, 146-49
 operation, 144-45
 schematic, 144
 upper threshold voltage, 146